Introduction to Pointcloudmetry

Introduction to Pointcloudmetry

Point Clouds from Laser Scanning and Photogrammetry

Mathias Lemmens
Geomatics Consultant, Teacher and Researcher
former Director MSC Geomatics, Delft University of Technology, and
Editor-in-Chief GIM International

WHITTLES PUBLISHING

Whittles Publishing Ltd.,
Dunbeath,
Caithness, KW6 6EG,
Scotland, UK

www.whittlespublishing.com

mlemmens@geotexs.nl

ISBN 978-184995-479-2

Printed and bound by CPI Group (UK) Ltd, Croydon, CR0 4YY

Contents

Preface

Since the year 2000 point clouds have evolved into a major geomatics data source and the importance of point clouds for many geo-related applications is steadily growing. As a special branch of informatics, geomatics focuses on the acquisition, processing and use of geodata. Geomatics is a technology with a nearly infinite number of applications.

The main geodata acquisition technologies for detailed 3D mapping of sites, including the generation of point clouds, are based on imaging devices (photogrammetry) and Lidar sensors (laser scanning). These technologies, which routinely acquire point clouds of many millions or even billions of points during one and the same survey, have matured rapidly since the mid-1990s and have turned to belong to the key geodata sources for 3D city modelling, inspection of roads, railways and dunes, and a multitude of other applications, as exposed in Chapter 13.

The acquisition, processing, and analysis of point clouds are key activities of geomatics – my own professional background. But also many other (engineering) disciplines and geo-sciences exploit point clouds. The large diversity of backgrounds affects the terminology, which is often diverse and hence confusing. The rapidly growing importance of point clouds is not yet reflected in a substantial number of textbooks. Existing books often are a collection of edited conference papers and accordingly highly technical and written in an academic tongue directed towards peer experts. This affects coherency in terminology and elucidation of concepts. These observations have been my main motivation to write a monograph on point clouds. The sweeping efforts resulted in a comprehensive book consisting of 13 chapters with a consistent and uniform set of terms and concepts underpinned by coherent educational principles, based on my over 35 years of academic teaching experience as well as my background as Editor-in-Chief of *GIM International,* the geomatics journal read worldwide. The chapters may be studied independently from each other, enabling teachers of university courses to select dedicated chapters best meeting the study goals of their courses. This is enabled through recapturing essential basics in each chapter. Those who study the entire book from cover to cover may accordingly encounter some repetition in the diverse chapters, not only in text but also in the presentation of formulas.

In order to tackle the same problem, e.g. interpolation and ground filtering, many methods have been developed over time. I have made major efforts to identify and elucidate the common grounds underlying the design principles of these methods as well as the ways in which they differ and their applicability for different point cloud types.

For many readers, English will not be their mother tongue. To ease the study of this book, I have strived to use simple wordings, without violating exactness, and to avoid complex academic formulations.

Technology as it exists today has been enabled by the hard work of many researchers and developers. Often a long and winding road had to be travelled before practically feasible methods were found, while solutions to many problems are still in the realm of inventions yet to be made. Therefore, I attached much value to placing the present technologies in a historical context and to creating awareness that exploiting point clouds is not just a matter of following a recipe or pushing buttons.

The robust seed of this book was planted in 1988 when I published a paper on automated matching of overlapping images for the creation of point clouds, entitled: 'A survey of stereo image matching techniques.' This paper brought together photogrammetry and computer vision for the first time and inspired many researchers through the years until the present day. From 1996 onwards I conducted fundamental and exploratory research on the use of Lidar technology for the creation of digital elevation models, which are extensively used by many GIS analysts, practitioners and researchers. This book helps to spread the essential knowledge about the core, idiosyncrasies and unparalleled potential of point cloud technology, a technology for which I have coined the term pointcloudmetry. This subbranch of geomatics can be defined as: *Pointcloudmetry encompasses the technologies for obtaining accurate and detailed information about earth-related objects, including the bare earth surface, by acquiring and processing point clouds.*

I have written this book primarily for an audience of geography and geomatics students, GIS analysts and practitioners using geodata in their daily work, as well as geo-scientists. Others, who have no background in geoinformation technology but occasionally deal with this topic, will discover many cues in this book to enhance their knowledge on point clouds. Software engineers dealing with point clouds may benefit equally well from the book's contents. University teachers can use this book as core course material for undergraduate and graduate students. Based on the study goals and level of pre-knowledge, they can develop their own assignments. This monograph will also help staff to get acquainted with point cloud producing sensors or point cloud processing software through courses from manufacturers. It will enable geodata surveyors to better understand the deficiencies in their survey results and to improve their design. One thing is for sure: no matter how advanced a device may be, if it is not in the hands of a skilled person, the output will be worth little.

Initially this book was planned for release in autumn 2020. However, due to the measures necessary to cope with the Covid-19 pandemic, which affected the entire world, the book is being published with a delay of over two years. Many thanks to all who pre-ordered the book and apologies for the adjournment.

A word of caution. Mathematics lies at the root of any technology, and pointcloudmetry is no exception. Readers are assumed to have basic knowledge of mathematics and statistics. The concepts presented are neither oversimplified nor unnecessarily complex. Furthermore, I have eased knowledge absorption by including over 200 insightful colour illustrations, graphs and other elucidating portrayals clarifying concepts and formulas.

The structure and contents of this book are presented in Section 1.7. This book has succeeded in its intent when it helps you to become a knowledgeable and skilled *pointcloudmetrist.* I would like to end this preface with a quote from Einstein: 'Science can flourish only in an atmosphere of free speech.'

Mathias Lemmens
Delft, October 2022

List of Abbreviations

2D	two-dimensional
3D	three-dimensional
AGL	above ground level
AI	artificial intelligence
ALS	airborne laser scanning
BIM	Building Information Model
CAD	computer aided design
CCD	charge-coupled device
CNN	convolution neural network
CPU	central processing unit
DBMS	database management system
DCM	digital canopy model
DEM	digital elevation model
DGNSS	differential GNSS
DGS	digital ground surface
DIM	dense image matching
DLM	digital landscape model
DoG	difference-of-Gaussians
DPW	digital photogrammetric workstation
DSM	digital surface model
DTM	digital terrain model
EDM	electronic distance measurement
EM	electromagnetic
FAST	features from accelerated segment test
FBM	feature-based image matching
FoV	field of view
FWD	full waveform digitisation
GCP	ground control point
GIS	Geographic Information System
GNSS	Global Navigation Satellite System
GPS	Global Positioning System
GPU	graphical processing unit

GSD	ground sample distance
GTM	geometric transformation model
Hz	Hertz (measure of frequency)
ICP	iterative closest point
ICT	information and communication technology
IDW	inverse distance weighted interpolation
IMU	inertial measurement unit
INS	inertial navigation system
InSAR	interferometric synthetic aperture Radar
Lidar	Light Detection and Ranging
LoD	level of detail
LoG	Laplacian of Gaussian
LSA	least squares adjustment
LSM	least squares image matching
MCC	multi-scale curvature classification
MEM(S)	microelectromechanical system
MLS	mobile laser scanner
MMS	mobile mapping system
MPiA	multiple pulses in air
NaN	natural nearest interpolation
NCC	normalised cross correlation
NDVI	Normalised Difference Vegetation Index
NIR	near infrared band
NPS	nominal point spacing
NSI	normalised shape indicator
OK	ordinary Kriging
PD	point density
PMF	progressive morphological filtering
POS	positioning and orientation system
Radar	radio detection and ranging
RANSAC	random sample consensus algorithm
RGB	red / green / blue (natural colour)
RMSE	root mean square error
SAR	synthetic aperture radar
SE	systematic error
SfM	structure from motion
SGM	semi-global matching
SIFT	Scale Invariant Feature Transform
SLAM	simultaneous localisation and mapping
SNR	signal-to-noise ratio
SPL	single-photon Lidar
SURF	speeded up robust features
TIN	triangulated irregular network

TIR	thermal infrared band
TLS	terrestrial laser scanning
ToF	time-of-flight
UAS	unmanned airborne system
UTM	Universal Transverse Mercator
WGS84	World Geodetic System 1984
WLS	weighted linear least squares

Part I
Embrace Light

Setting Light on the Landscape

CHAPTER 1

Laser scanning and automated image matching of overlapping photogrammetric images have matured rapidly since the mid-1990s. These technologies routinely acquire point clouds of many millions or even billions of points during the same survey. *Photogrammetric* means that great care has been taken to ensure that the images have high geometric fidelity, allowing high-quality geo-information to be obtained. This is done, among other things, by calibrating the cameras and using ground control points for georeferencing purposes. The term commonly used to denote laser scanning technology is Lidar – an acronym of Light Detection and Ranging; some prefer to read Lidar as Laser Imaging Detection and Ranging. Cameras and Lidar sensors can be mounted on a wide variety of platforms or carriers, including vehicles such as cars and boats, aeroplanes and copters. Platforms operating outdoors, such as manned aircraft and cars, are usually equipped with a Global Navigation Satellite System (GNSS) positioning device and an inertial navigation system (INS) to accurately determine the six position and orientation parameters of the sensors (three position parameters and three angles defining the orientation of the sensors in a three-dimensional space). A well-known example of GNSS is Global Positioning System (GPS). To improve the reliability of georeferencing, additional sensors are often used such as wheel counters, compasses and barometers. The use of ground control points for georeferencing further enhances the geometric fidelity of the data.

This book is about the acquisition and processing of point clouds (1) captured by Lidar devices and (2) derived from overlapping photogrammetric images through image-matching techniques. We have baptised this specialised branch of the geomatics discipline *Pointcloudmetry*, a term best understood by recognising the analogy with photogrammetry. Pointcloudmetry can be defined as follows:

> **Pointcloudmetry** encompasses the technologies for obtaining accurate and detailed information about earth-related objects, including the bare earth surface, by acquiring and processing point clouds.

Usually, the feasibility and effectiveness of Pointcloudmetry technologies have been thoroughly substantiated by scientific research. The importance of having a thorough knowledge of the principles of the underlying technology when using point clouds is difficult to underestimate. These technologies have made tremendous progress since the turn of the millennium. A major gain of these developments is that the technologies provide tremendous opportunities for tackling the many challenges our planet and its human population are steadily and increasingly confronted with. Major challenges include (1) population growth coupled with continued population shifts from rural areas to cities and (2) climate change. The most threatening effects of climate change are sea-level rise and lengthy heavy rainfall putting valleys, rivers, lowlands and deltas at increased risk of flooding and destruction by mudslides. In addition, point clouds are becoming increasingly important geodata sets for producing digital elevation models (DEMs) and realistic 3D models of urban areas, which form the elementary geodata sources for the creation of so-called digital twins. In the next section, we give an example of how point clouds, combined with simultaneously collected images, can help to diminish the frustration of motorists trying to grab a parking space in urban conglomerates, and thus to reduce air pollution and carbon emission.

1.1 App Based on Point Clouds

Motorists who regularly drive around busy megacities can waste a week or more per year hunting for a parking space. The quest to park one's car burns fuel unnecessarily pollutes the air with particulate matter and increases greenhouse gas emissions. To relieve the seekers of parking spaces while helping to preserve the environment, a smart parking app has been developed in the London Area (Lemmens, 2019). The app shows all on-street and off-street parking areas, restrictions and operating hours (Figure 1.1). While on the move, motorists can check their smartphones, make informed decisions about where to look for a parking lot and find the nearest and cheapest options. At the very core of such an app lies an accurate, detailed and up-to-date map, with detailed additional information on parking spaces, such as restrictions and opening hours. The

Figure 1.1 Parking app, called AppyParking, installed on a smartphone, developed by London-based company AppyWay

data used to create the map consists of street-level Lidar point clouds and imagery. The laser scanners and cameras – rigidly integrated into the same casing – were mounted on a car (Figure 1.2). Figure 1.3 shows examples of the collected point clouds. Using specialised software, the information on signs along the parking zones was digitised from the images, mainly by using machine-learning techniques and, if necessary, complemented by manual interpretation. The outlines of each parking zone were mapped as polygons. Wherever outlines were occluded – that means visibility obstructed by cars or other objects in the line of sight – the visible lines were extended and intersected with each other to obtain the coordinates of invisible corner points of parking zones. During the 100-day duration of the mobile mapping survey, the mobile mapping car drove 6,500 km and captured one million laser points per second. Only a few London drivers will be aware of the immense efforts it took to create the parking app, which conveniently lies in the palm of one's hand.

Figure 1.2 All sensors of the Pegasus:Two Ultimate are rigidly integrated into the same casing (right): the rack can be mounted on a variety of vehicles including cars (left)

Figure 1.3 Mobile laser scan of Admiralty Arch on The Mall in London, coloured from simultaneously captured images (left); laser point cloud of Tower Bridge in London, intensity coloured (top right) and colourised point cloud (bottom right)

The above is just one of the many examples demonstrating that the demand for detailed maps of major cities and many other places on our planet is rapidly increasing. For the creation of these maps, point clouds and photogrammetric images are often preferred as primary data sources. The products derived from point clouds include 3D city models, road cross-sections, triangular irregular networks (TIN), 3D virtual landscapes, DEMs and many more (Figure 1.4). The products can be used for a myriad of applications including asset management, flood prevention, drainage projects, road safety improvements, highway maintenance, 5G telecoms and environmental analysis. This means pointcloudmetry is primarily a technology with nearly an infinite number of applications. A word of warning but also of appreciation is appropriate here. It is annoying for some geomatics specialists that their efforts are so invisible and their products are usually hidden under many layers of user interfaces. However, others are proud that they have the deep knowledge and skills to ease the life of so many people and help solve severe perils. They are also flattered that there is an eager demand for their unique knowledge and skills.

The point clouds, we consider within this book, are a specific type of geodata. Earlier we have seen the example of how mobile mapping point clouds can help to tackle car parking problems in major cities. Point clouds are also beneficially used in forest inventories. For example, the height of individual trees, the density of the canopy, biomass and

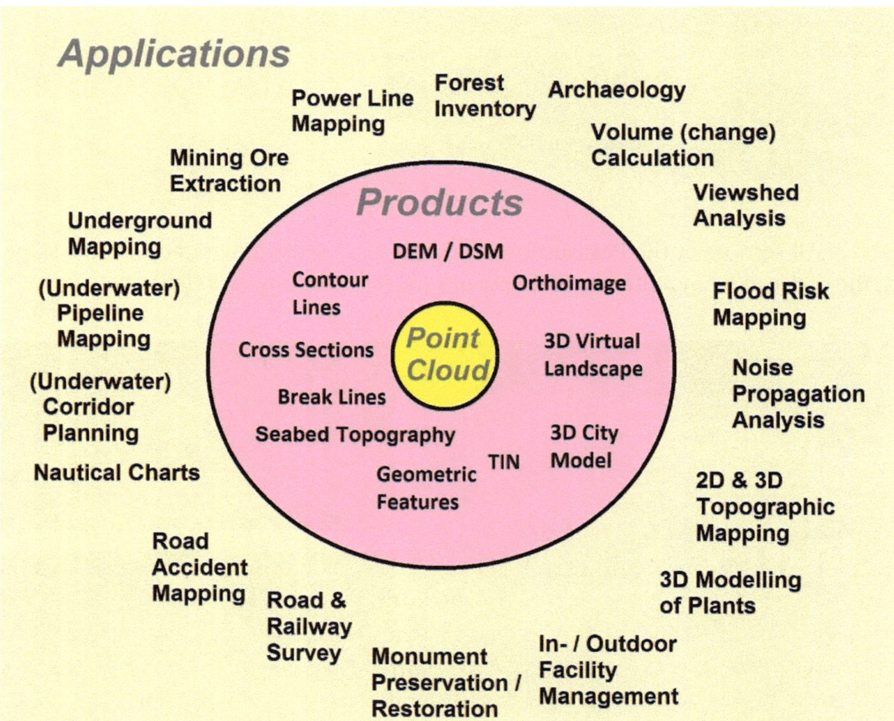

Figure 1.4 The products derived from point clouds are well suited as the information source for a myriad of applications; the listed products and applications are examples and non-limitative (source: Lemmens, 2014)

the stand volume can be calculated. In road construction, volumes and masses can be accurately calculated from point clouds to support earthworks. By comparing time series of point clouds, ore extraction of an open-pit mine can be determined as well as erosion and accumulation of sand dunes to support flood risk management. Chapter 13 of this book provides many examples of a large diversity of applications of point clouds, captured from a variety of platforms.

POPULATION AND POINT CLOUDS

Each year, the world's population expands by more than the equivalent of the total number of inhabitants in Australia and Canada combined. Less than 250 years ago – say, around the time of the founding of the United States, the rise of Napoleon and the first signs of deterioration of the Chinese Qing Empire – just one billion people were living on this planet. By the end of the year 2022 that number reached eight billion. This represents an annual population growth rate of over 1% and a doubling of the population every 70 years – which is less than a lifetime for many people. Remember this when you complain about overcrowded cities! The Industrial Revolution brought the world machinery to plough, sow and harvest fields – which freed peasants from hard labour on farms, but also transformed smallholdings into industrial operations and signalled the end of the idyllic pastoral scenes immortalised in 19th-century paintings. Since then, those peasants' descendants have continued to move around in search of work, contributing to the rapid growth of urban agglomerations. The resulting – and ongoing – societal developments have continuously increased the need for highly detailed, accurate and timely geodata. The exploration of point clouds in conjunction with geographical information systems (GISs) and software, which models earth-related physical and societal phenomena, can help solve major problems that modern society is confronted with, mainly as a result of rapid population growth.

1.2 Point Cloud Characterisation

A point cloud is a set of data points represented in a preferred coordinate system. Usually, this will be a geodetic reference system. Such a system may cover an entire country, a conglomerate of countries, a continent or the entire world, such as the World Geodetic System 1984 (WGS84). The position in space is determined by three coordinates: X, Y and Z, which are only meaningful if the coordinate system is known and well defined. They provide the *where* component of each point. Knowing where something is located without knowing *what* is the meaning of this something is of little value. In many standardised mapping processes, the what component is predefined as a set of class labels. A well-known example is the topographic map, which represents buildings, roads, railways, forests, meadows and arable land, amongst others. Buildings may be further subdivided into dwellings, offices, schools, factories, greenhouses and so on. It is important to understand that the what component depends on the tasks for which the mapping has

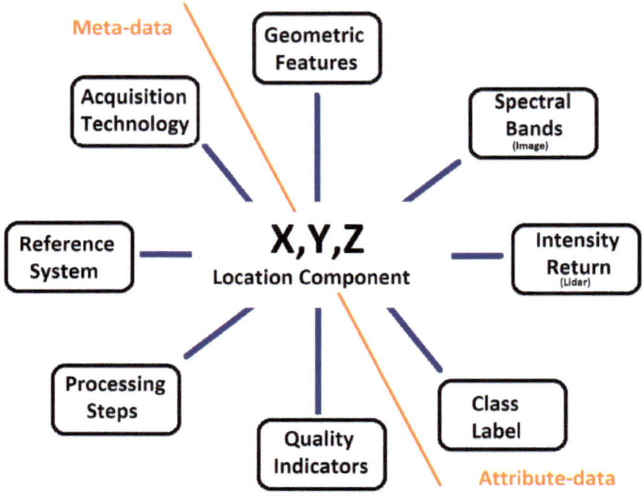

Figure 1.5 Diagram representing the nucleus of a point cloud consisting of three location coordinates, measured and derived attributes, and meta-data

been defined. The attributes which are collected together with the *X, Y, Z* positions may consist of the reflection value of a laser beam (intensity return), the red, green and blue (RGB) values in a natural colour image, or other spectral bands (Figure 1.5). They may be assigned as attributes to the (*X, Y, Z*) coordinates and provide clues about what the point may represent. Is it a point on a building, street, tree or bare ground underneath the tree? Analysing the shape properties of a neighbourhood of points in a point cloud together with the reflectance data may provide a further clue about the what component. The extraction of features from point clouds is the topic of Chapter 11. The assignment of the what component to a point or set of neighbouring points is called classification. The classes form the thematic information of the scene. The result is a description of the scene in both geometric and thematic terms. This classification process is sometimes called semantic enrichment of the point cloud, while thematic information is also called semantic information. The automated classification of point clouds is still far from a solved problem and a topic extensively investigated by researchers all over the world. Classification of point clouds is outside the scope of this book, but we will touch on this topic in Sections 1.5, 6.4 and 11.7.2.

In the characterisation of point clouds, it is convenient to distinguish 2.5D point clouds and 3D point clouds. This distinction makes sense since GIS users will frequently encounter DEM. A DEM describes the surface of the earth, either bare ground or with objects such as buildings, bridges and other construction on top of it, as a sampled curved surface in 3D space. To each planar coordinate (*X, Y*) one height value is assigned. In the 2.5D description, the height value is often considered as an attribute similar to the reflection value of the laser return or natural colour (RGB) values of an image. It may be described as a digital representation of the earth's topographic relief or simply as an elevation map consisting of *X, Y* positions to which height values – either indicated by *H* or *Z* – are assigned. In contrast to 2.5D point clouds, 3D point clouds aim at representing the full 3D space through three coordinates. This means that when looking at a planar position (*X*, *Y*), multiple *Z*-coordinates may be assigned to the same *X*, *Y* coordinates. DEMs are a major and crucial source of data for many GIS applications for solving earth-related problems. In order to acknowledge the enormous importance of DEMs in many applications, this book will pay major attention to 2.5D point clouds in three chapters brought together in Part III (Chapters 8–10).

1.3 Georeferencing

In the preceding considerations, we have seen that the core of the nucleus of a point cloud is the *X*, *Y*, *Z* coordinates of the individual points. This means that the points are rivetted in a reference system by three numbers which have metric dimension. Usually, this dimension will be the metre or a multiplicative derivative of the metre such as kilometre or centimetre. In the imperial system, the dimensions may be yard, mile or inch. The three numbers do not only have a suffix but also relate to a particular coordinate or reference system. And there are many coordinate systems! This section provides the basics of coordinate systems. The many coordinate systems can be grouped into two fundamental categories: (1) those based on a geometrical view of the earth and (2) those based on a geophysical view. The geometrical view has resulted in ellipsoidal coordinate systems and map projections of these. The geophysical view, in which the earth's gravity field plays a major role, has resulted in a representation in the form of the geoid. We end this section with some warning words on the pitfalls one may encounter when using coordinates just as universal numbers, which means ignoring the coordinate system they belong to.

1.3.1 Ellipsoid

The simplest model of the shape of the earth is a sphere. Proposed by the Greek mathematician Pythagoras (582–500 BC), the sphere model of the earth became generally accepted in antiquity. The dimension of the sphere could be estimated by the founder of Geodesy Eratosthenes of Alexandria (276–195 BC) by carrying out arc measurements. A better approximation of the shape of the earth is the ellipsoid. The ellipsoid is a geometric figure obtained by rotating an ellipse about its minor axis. The main parameters are the semi-major axis, the semi-minor axis (which approximately coincides with the rotation axis of the earth), the origin and the orientation in space (Seeber, 1993). A great variety of ellipsoids have been developed since the early 19th century. Each ellipsoid fits best the shape and size of a particular country, a conglomerate of adjacent countries, a continent and so on. For example, the ellipsoid used for mapping Britain, the Airy 1830 ellipsoid (the semi-major axis 6,377,563.396 m; the semi-minor axis 6,356,256.909 m), is designed to best fit Britain and is not so useful in Africa, Australia or other parts of the world far away from Britain.

A reference system on an ellipsoid is introduced through parallels (φ), which are curves parallel to the equator or, in other words, the curves of constant latitude and meridians (λ), which are curves running from the North Pole to the South pole and back (Figure 1.6). Meridians and parallels are orthogonal to each other. The height of a point

P above or beneath the ellipsoid is indicated by *h*. The triplet (φ, λ, h) is called geodetic or geographical coordinates. The orthogonal projection of *P* on the ellipsoid cuts the ellipsoid at point P'. *P* and P' have the same latitude and longitude coordinates. The extension of the line created by the orthogonal projection of *P* on the ellipsoid cuts the equatorial plane at an angle φ. Important geodetic reference ellipsoids are the Bessel ellipsoid, mainly used in Europe, and the Hayford ellipsoid, initially developed for use by the United States.

Since the introduction of global navigation satellite systems, today we have four major systems: GPS, Galileo, Glonass and Beidou. In the early 1980s, the WGS84 was widely used by surveyors. The origin of the WGS84 system is in the earth's centre of mass, the semi-major axis is 6,378.137 km and its semi-minor axis is 6,356.752 km. Often the WGS84 coordinates are converted into coordinates of a national or local geodetic reference system in order to make them compatible with older measurements. The WGS84 ellipsoid is a so-called best-fitting ellipsoid – that is, the ellipsoid which best fits the shape and the size of the earth as a whole.

1.3.2 Map Projections

A special type of transformation is map projections in which the triplet (φ, λ, h) ellipsoid coordinates are projected on a 2D space called a map. One of the most used projections worldwide is the Universal Transverse Mercator (UTM), named after the Flemish cartographer and geographer Gerard Mercator Rupelmundanus (1512–1594), who introduced this projection method in 1569 for ship navigation purposes – any course of constant bearing can be drawn as a straight line on the map. UTM is often represented as an *X*, *Y* coordinate system, which means the height is neglected. For national use, countries may employ their own map projection. For example, the Netherlands exploits an *X*, *Y* coordinate system based on the projection of the Bessel ellipsoid on a plane with its origin in the middle of the country (Amersfoort). The third coordinate (height) is based on the geoid (see Section 1.3.4).

1.3.3 X, Y, Z Coordinates

Using the earth's centre of mass as origin, a three-dimensional Cartesian coordinate system can be defined. The three orthogonal axes are defined as follows: The *Z*-axis is defined by the origin located at the earth's centre of mass and the pole or, in other words, the *Z*-axis coincides with the rotation axis of the earth. The *X*, *Y* plane lies in the equatorial plane of the earth, and by rendering a line from the origin to the point where the Greenwich meridian intersects the equator, the *X*-axis is defined. The *Y*-axis is drawn perpendicular to the *X*-axis to span up a 3D right-handed rectangular Cartesian system (Figure 1.6). GNSS receivers collect 3D coordinates in this system before they are transformed to WGS84. The advantage of this Cartesian coordinate system is that it provides a universal foundation for incongruent coordinate systems, simple equations, worldwide standardisation and convenient accuracy calculation; the technology is in place and all equations and procedures are available as open-source software (Burkholder, 2008). However, such a geocentric coordinate system is not convenient in practice because the *X*, *Y*, *Z* coordinates give no clue about the difference in elevation between two points.

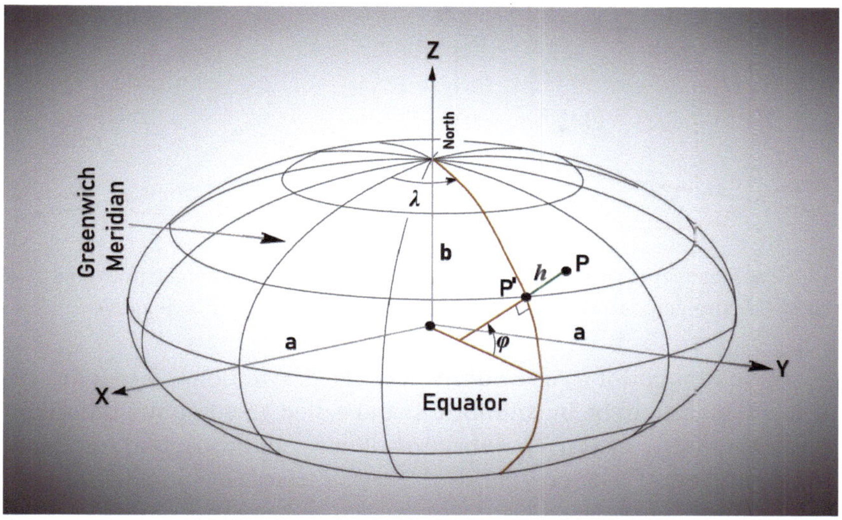

Figure 1.6 Geocentric coordinates (X, Y, Z) versus geodetic coordinates (φ, λ, h); geodetic latitude (φ) is defined by the angle between the normal of the point to the ellipsoid and the plane of the equator, and it does not refer to the origin

Furthermore, the X, Y, Z coordinates refer to the earth's centre of mass and are consequently in large numbers and thus impractical to handle. Therefore, geodetic coordinates (φ, λ, h) are preferred.

1.3.4 Geoid

However, also the use of geodetic coordinates has practical drawbacks for the computation of the differences in elevations between a series of points in a number of applications, particularly in the field of engineering. Many people live close to seas and oceans. That is where the majority of the megalopolises are located. That is where the land is fertile. Engineers all over the world work on getting rid of superfluous water and protecting lowlands from flooding. Proper water drainage and flood prevention require accurate elevation data. Dike constructors, dune maintainers, drainage engineers and other water managers and engineers should be certain that water flows in the direction as computed from the elevations in their datasets – that is, from locations with higher elevation values to locations with lower elevation levels. But what do we mean by "higher" and "lower," and how do we know that one position lies higher than the other position? Do the terms high and low refer to the height above the ellipsoid, which is an artificial geometric construct? The answer is a definite *No.* Geodetic coordinates are well suited for determining the distance between two positions or calculating the area of a polygon. However, the flow of water cannot be determined from the heights above an artificial mathematical body because it does only approximate the actual shape of the earth and has no geophysical meaning. The flow of water is determined by gravitational forces. Consequently, to be practically useful, elevation values should refer to the gravity field of the earth. This means the height values should refer to an equipotential surface. Such

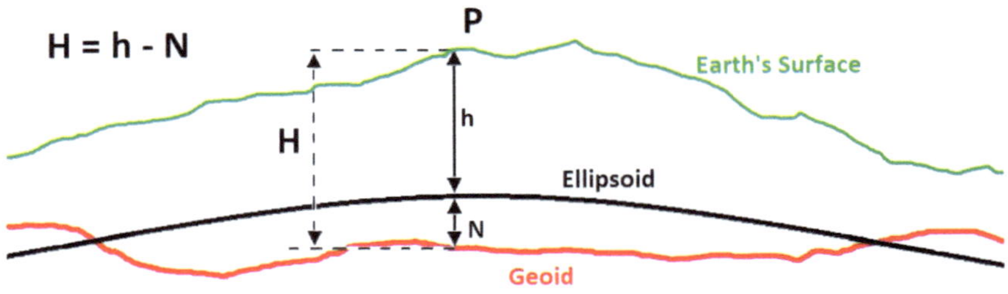

Figure 1.7 Graphical representation of the relationship between height above ellipsoid, orthometric height (*H*) and geoid undulation *N*; note that at location *P*, *N* is negative

a surface can be thought of as a surface where the gravitational forces are everywhere the same; it is then said to be in equilibrium and called an equipotential surface. Water on an equipotential surface is in equilibrium – it does not flow, it is at rest. Since water is free to flow in the presence of gravitational forces, surfaces of which the shape is almost entirely determined by the earth's gravity are sea and ocean surfaces. When no other forces than gravitational forces would act on water bodies, the water level would form an equipotential surface. Water above this equipotential surface will flow to this surface. As a result, the mean sea level is ideally suited as the height reference system for water management purposes.

The mean sea level or an approximation of this gravitational equipotential surface is called *geoid*. This equipotential surface continues underneath land bodies. Therefore, for being practically usable for water engineering works, heights above the ellipsoid should be transformed to heights above the geoid. This can only be done when the location of the geoid with respect to the reference ellipsoid is known. The determination of the position of the geoid with respect to a reference ellipsoid is one of the major tasks of global geodesy. In the pre-satellite era, this task was carried out by accurately measuring gravity at a large number of points all around the world both on land and at sea (Heiskanen and Vening-Meinesz, 1958; Heiskanen and Moritz 1967), a time-consuming, cumbersome and expensive job. With the launching of satellites, the task has become easier. The orbit of satellites is determined by the gravity field and can be inferred from tracking the satellites from ground stations (Sansó and Rummel, 1989). As is the case for ellipsoids, different countries use different geoids. Since gravity changes over time due to earthquakes, volcanic eruptions and other tectonic activities, the location of the geoid with respect to the reference ellipsoid has to be adjusted regularly. Once the location of the geoid is known with respect to the reference ellipsoid, any height above the geoid (*H*), called orthometric height, can be computed from the ellipsoidal height (*h*) by a simple subtraction operation (Figure 1.7). Expressing the height of the geoid with respect to the reference ellipsoid as *N*, the orthometric height (*H*) becomes $H = h - N$, where *N* is positive when the geoid is above the ellipsoid and negative when below (Moritz, 1990).

1.3.5 Coordinate Transformations

Frequently it will be necessary to transfer coordinates from one coordinate system to another coordinate system. This is necessary when combining two data sets, for example a point cloud given in a national geodetic reference system and geodata available in UTM projection. Transformations are also necessary when converting image coordinates into an earth-fixed coordinate system or when creating one point cloud from a diversity of laser scans acquired in different coordinate systems, which depend on the location of the scanner. In Chapters 4 and 6, we will regularly encounter geometric transformation models, including 2D similarity transformation; affine transformation (also called first-order polynomial transformation); 3D similarity transformation (also called seven-parameter transformation and Helmert transformation); and second-order polynomial transformation. The location component of geoinformation may be rivetted in a multitude of local, national and global coordinate systems. Textbooks in geodesy develop the equations for transforming one coordinate system into another (e.g. Bomford, 1980). Most GIS software contains functionality to carry out the transformations by computer. Yielding high-quality outcomes in earth-related analysis tasks requires sufficient understanding of the coordinate systems in use today and their mutual transformations. A clear compendium directed towards practitioners, analysts and GIS users, reviewing and outlining coordinate systems, is provided by Janssen (2009). A few examples of what may happen when the user of geoinformation is not sufficiently coordinate-aware are given in the next subsection.

1.3.6 Coordinate Awareness

For those not trained in geography or geomatics, the abundance of coordinate systems may be so dizzying that it is often a source of confusion and mistakes. To illustrate this point, let's give a few examples. In heavily urbanised areas, highways built in the 1950s or earlier consist of two lanes per travel direction. In today's countries where there is, on average, one car per adult fit for driving available, the road standards of the fifties result in daily congestion, especially during rush hours or after the occurrence of a traffic accident. The construction of additional lanes requires the purchase of pieces of land alongside the highway. The coordinate system of the maps used by the highway authorities often might differ from the system used by the cadastre to outline property boundaries. What may happen when the planners don't take account of these differences and don't convert the two different coordinate systems to the same system? The imperfect match between highway design and property boundaries causes pieces of land to be taken without providing the lawful owners with adequate financial compensation. Lawsuits emerge and the construction budget is grossly exceeded. Another less harmful but nonetheless inconvenient example may occur when you rent a car to enjoy the countryside of a sparsely populated area praised for its arcadian appearance in tourist guides. You type the address of the lodge you want to spend the night. The navigation system converts the address into coordinates of the underlying map, which uses a UTM projection. The GNSS device continuously measures your position in WGS84. In the middle of nowhere, the friendly computer voice says: "Destination reached." You can see a farmstead in the distance. The farmer explains you are 10 km off-track. Dangerous situations can occur when an airport in Europe uses the ED50 (European Datum 1950) geodetic reference system for following the tracks of aeroplanes by Radar control. The pilot of an aeroplane from an Asian airline has forgotten to switch the position button and navigates on GNSS using WGS84. The shift between Radar track and aeroplane bearing may induce wrong manoeuvres and create precarious situations. The lesson is: whenever using maps or other geodata, *be coordinate aware...*

1.4 Sources of Point Clouds

The devices used to acquire point clouds include levelling; GNSS devices; terrestrial laser scanning (TLS); mobile mapping systems, airborne laser scanning; airborne or spaceborne Radar; terrestrial, aerial or spaceborne photogrammetry; or multibeam sonar. Another source is contour lines digitised from maps. This book focuses on point clouds acquired by Lidar devices and photogrammetric cameras. Both devices can be mounted on a variety of platforms. We limit our considerations to these devices because automatic matching of overlapping images and laser scanning are the main geomatics technologies for capturing dense point clouds of earth-related scenes. It is important to consider the origin of the point cloud: Has it been measured by using Lidar devices or by photogrammetric means? Laser scanners emit electromagnetic beams themselves – they are said to be active systems and can directly create point clouds without time-consuming intermediate steps. In contrast to laser scanning, the creation of point clouds from overlapping images requires an intermediate step called image matching, by non-photogrammetrists also called *structure from motion* (SfM). Added to this, laser beams can partially pass through foliage and reflect from dune surfaces, which avoids data gaps. So the capability of laser scanning for rapid acquisition of dense point clouds from aircraft and ground-based platforms, either static or mobile, worked very much in favour of laser scanning. Around the year 2000, it seemed that laser scanning would surpass and scrap photogrammetry as a major source of point clouds. With the growing interest in 3D city models and the advancement of new technologies, the feasibility of imagery for the creation of point clouds has been hoisted into the zenith again. Since the turn of the millennium, charge-coupled device (CCD) arrays have replaced film in the image plane. The resulting digital images combined with an ever-increasing computer power enabled the astonishing acceleration of the processing steps from image capture to the production of point clouds. The emergence of oblique digital cameras enabled efficient and accurate combining of straight vertical views with calibrated oblique imagery. This facilitates the capture of highly overlapping images for extracting point clouds as well as textures for the generation of 3D city models. A major breakthrough was the invention of dense image matching (DIM). The stereo image-matching algorithms introduced in the early 1980s were not able to create point clouds with high point densities. This limitation disappeared with the introduction of DIM, one of the technological innovations contributing to the automation of the creation of high-density point clouds from overlapping images. DIM is based on semi-global matching, the basics and potentials of which have been developed and demonstrated by Hirschmüller (2008). The topic is addressed in Chapter 4 of this book. As a result, DIM enables the generation of DEMs, which may have densities similar to the pixel size in ground units, called ground sample distance (GSD), of the imagery from which they are derived. For example, images with a GSD of 5 cm may deliver a density of up to 400 points per square metre. The rise of DIM has been propelled by the advent of programmable graphical processing units (GPU) and conversion of the DIM concept into algorithms which together allow amazingly fast image matching. Moreover, the costs are low since today's computer power is cheap, and the same is true for digital cameras which provide high-quality imagery, while large overlaps do not add to costs. The pull of using imagery has increased further since open-source packages have become available from the field of computer vision.

The capturing of point clouds of buildings has gained increased attention from architects, constructions and facility managers, especially because these geomatic data sets allow these users to populate their Building Information Models (BIM) with reliable 3D models. To demonstrate the potential and high quality of point clouds derived from images, we touch upon a project carried out by students from the Delft University of Technology (TU Delft), the Netherlands (Arapakis et al., 2018). Working towards a master's degree in Geomatics at the Faculty of Architecture and the Built Environment, they carried out a group project to create building models of the TU Delft campus using overlapping images captured by a camera mounted on an unmanned airborne system (UAS). They designed and wrote feature-based image matching software themselves. Feature-based matching traces distinctive points in the images and assigns descriptors to these points. The descriptors are functions of the grey value or colour of the pixels in a window around the distinctive point. The corresponding points in overlaps are then sought based on similarities between these descriptors. When little or no texture is present, no reliable matches can be found. Providing that the overlaps were sufficiently high, the software the students developed was able to create point clouds even when the façades consisted of a variety of materials with complex textures. The point clouds of the buildings have been compared with the point cloud of the same buildings generated with commercial software in which DIM was implemented (Figure 1.8). Of course, the point clouds generated by the latter are much denser and less noisy than point clouds generated by the students' software. The project lasted just two months; the students started from scratch and went through an impressive learning curve.

Since the year 2000, laser scanning has also made admirable progress. The acquisition of over one million points per second is now a matter of routine. Sometimes, laser scanning and photogrammetry are portrayed as competitors. It is better to consider them as complementors.

Figure 1.8 Point clouds of one of the TU Delft campus buildings captured by UAS photogrammetry and processed by feature-based matching software (left) and commercial software incorporating semi-global matching for obtaining dense point clouds (right)

1.5 Semantic Enrichment

The processing software to extract meaningful information from the geodata is greatly supported by the achievements of the computer vision research community over the last four decades. However, since the emergence of computers, it seems to have been a rule of thumb that the amount of data acquired by sensors is ten times as much as the processing capacity of computers. So, the production of images and point clouds using a wide variety of geodata acquisition technologies is just one side of the coin; on the flip side is the extraction of meaningful information which is useful for a specific geo-related task at hand. Information extraction is a skilled, labour-intensive and tedious task which is mainly executed by specialists. To diminish manual involvement, many researchers are working on developing automated mapping methods. Although an in-depth benching of assigning labels to points by automated means is beyond the scope of this book, we will briefly touch on the topic of automatically assigning scene-related classes to individual points in a point cloud. Examples of scene-related classes are road, building façade, tree, utility pole and traffic sign. They describe the *what* component, as discussed earlier, within the specific context of a (3D) mapping task. Classification or semantic enrichment, as the class-assignment process is also called, converts data into meaningful information with respect to a certain task at hand. The following discussion is not meant to treat classification methods but particularly aims at making some qualifying comments on unrealistic and exaggerated expectations some may have on the achievable automation level.

The drive for automation has resulted in the revival of a research area which was first presented some 50 years ago as the Holy Grail for automatic object recognition from images, but eventually died a silent death along its winding road. It was called artificial intelligence (AI). Just two of the catchy AI-related terms that have found their way into today's vocabulary are "machine learning" and "deep learning." The sheer number of papers written in recent years on the application of AI in the processing of images and Lidar point clouds shows that many believe deep learning based on convolution neural networks (CNNs) is the definitive solution for automating mapping. Other popular methods are random forest classification and assigning class labels using the method of support vector machines. However, a CNN is not a magic box of tricks, but rather software built on a sequence of 2D differential filters, such as the Laplace operator, and 2D integrating filters – these digital filters are further outlined in Chapter 11 – to construct a hierarchy of image pyramids by aggregating small neighbourhoods, for example, windows of 3 × 3 or 5 × 5 pixels, in a process called pooling. This makes the approach sensitive to noise and texture. Many claim that this approach gives promising results, although the researchers also admit that there are several challenges to overcome due to the complexity of earth-related scenes, such as the presence of shadows and occlusions. In other words, there are still mountainous problems to be resolved. A generally accepted research method is to collect a huge amount of prototypes, of which 80% are used as training samples and 20% for validation. Indeed, deep learning requires massive training samples. To circumnavigate the burden associated with the quest for obtaining sufficient training samples, Lemmens (2018) proposed a knowledge-based approach, which was implemented for classifying mobile laser scanning point clouds of urban scenes by Zheng et al. (2019). The fundamental premise of the knowledge-based approach is that urban scenes are constructed according to man-made rules and these rules can be stored in a knowledge base and fruitfully exploited as features in the classification process possibly supported by additional geodata (Figure 1.9). One of these rules is that different objects, which extend in the vertical direction, are characterised by differing heights. For example, a building façade varies in height which may start at 7 m, or higher, depending on the urban area, while the height of a traffic sign mounted on a pole from ground level upwards usually does not exceed 3 m. The research of Zheng et al. (2019) demonstrated that the knowledge-based approach is feasible.

The results of automated classification are sensitive to noise, occlusion, clutter and decreasing point densities with increasing distances of the objects to the Lidar device. One of the major challenges is occlusion: Cars and other irrelevant objects in line of sight make parts of building façades and other relevant objects invisible. Many algorithms are developed to get rid of occlusion. They are implicitly based on assuming that the occluded areas show similar characteristics to their visible vicinity. Tobler's law – "everything is related to everything else, but near things are more related than distant things" – is valid in many cases. However, one may not rely on its ubiquitous validity. Using algorithms to invent data and presenting it as original data is sheeting and a surveyor sin of the first kind. The search for workable automated methods for the semantic enrichment of point clouds will remain a fierce area of research in the coming years. In order to demonstrate some of the issues one may encounter when attempting to automatically classify mobile laser scanning point clouds, we touch briefly upon this topic in Section 6.4. The knowledge-based approach is also briefly outlined.

We would like to end this section on the semantic enrichment of point clouds with an assertion, which may be disappointing for some. They may even classify it as too negative. Nevertheless, here we go. The thought that automated classification of point clouds – today you would better say, as stated earlier, semantic enrichment – could be at all difficult does not get through to those who have never tried it.

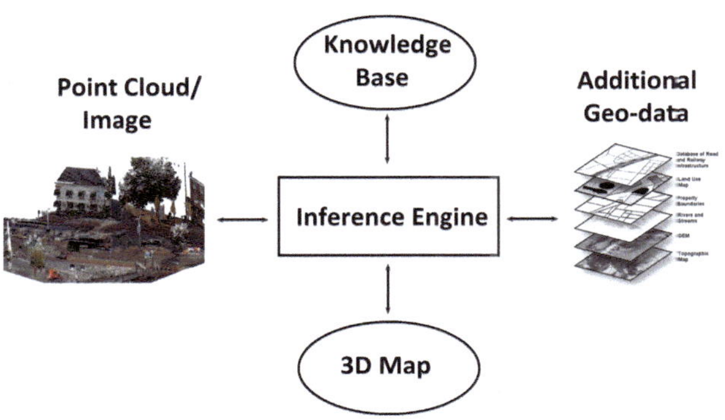

Figure 1.9 Schematic overview of a knowledge-based system for 3D mapping from point clouds

1.6 Point Clouds – Where Are They Heading?

The key trend in the evolution of pointcloudmetry can be summarised as the miniaturisation of components of devices accompanied by a continuous and seemingly neverending increase in computer power. Cameras and Lidar sensors suited for capturing high-quality data are becoming smaller and lighter, propped up by advanced processing software which facilitates the use of calibrated metric cameras and heavy Lidar sensors for precision solutions. Today's positioning and orientation systems (POS) based on GNSS and INS can be held in the palm of one's hand. The miniaturisation of rotors, electric engines and batteries, in combination with carbon-fibre frames, has enabled the construction of lightweight platforms. For example, today's lightweight UASs can carry camera(s) and Lidar (sensors) abreast for the simultaneous capture of images and Lidar point clouds without compromising air stability. Concurrent capturing of Lidar point clouds and photogrammetric images has proven to be beneficial for 3D mapping of built-up areas, for which application the acquisition of point clouds is a key intermediate step.

The ongoing miniaturisation of carriers and sensors in conjunction with simultaneous localisation and mapping (SLAM) algorithms for positioning and orientation purposes has also made it possible to easily manoeuvre lightweight and agile trolley-based, backpack, handheld mobile mapping systems or copters through indoor spaces, where GNSS signals are blocked. Equipped with cameras and/or laser scanners, they can collect high-density point clouds. The high level of detail and accuracy of the data helps facility managers to inspect their property. It also supports the creation of 3D cadastres, which are aimed at recording the ownership of volumetric parts of buildings and other constructions. On the flip side of technological advances are societal developments. Authorities and citizens alike are convinced that wasting fuel and other resources, as well as the emission of harmful substances, should be minimised through reuse, refurbishment and/or the use of alternatives in pursuit of the *circular economy*. The main consequence is that sites where humans are active, including agricultural lands and mines, need to be mapped and monitored in ever-greater detail. Within today's industrial agriculture, for example, the collection of point clouds and other geodata supports regular inspections to avoid waste of fertilisers, fuel, seeds and water. An agile platform such as a UAS is well suited for capturing such geodata on a regular basis. When it comes to indoor mapping, UAS and trolley-based, backpack and handheld mobile mapping systems complement rather than compete with one another. For example, if used indoors a UAS could collide with objects or people, causing damage and possibly injuries, making it useless in crowded indoor environments. In such a setting, mobile mapping is a perfect solution. In addition, the two platforms have different perspectives (i.e. view angles): sideways-looking versus image capture from above.

A major bottleneck, which may delay the introduction of point clouds in several applications, is that many professionals seem reluctant to replace tried-and-tested technology with a novelty that has a non-proven outcome – even though it may be convincingly cheaper and demonstrably more efficient.

1.7 Structure of the Book

This book is subdivided into four parts (Figure 1.10). Part I, entitled "Embrace Light" and consisting of three chapters, focuses on preliminaries and fundamentals. The present chapter introduced the characterisation of point clouds, the basics of georeferencing, and discussed the jeopardy of injudicious application of sophisticated technology. The elementary source of the generation of point clouds is light. Therefore, it is important to have basic insight into the characteristics of electromagnetic energy, which is the subject of Chapter 2. In Chapter 3, we further zoom in on a man-made type of electromagnetic energy: laser light. Part II, entitled "Acquisition of Point Clouds," focuses on the sensors and technologies to acquire point clouds from overlapping images as well as from laser scanners. Chapter 4 considers the derivation of point clouds from overlapping images. The depiction of the basics of laser scanning is distributed over two chapters. Chapter 5 considers the technologies of laser scanners mounted on aircraft, called airborne Lidar. Chapter 6 focuses on laser scanners mounted on ground-based platforms, such as cars exposed in Section 1.1 of this chapter. Extracting meaningful information from camera data and laser data starts with a good survey. Under the heading "Survey Peculiarities," Chapter 7 discusses issues on bringing sensors to work. Part III, called "2.5D Mapping from Point Clouds," focuses on the generation of DEMs in three chapters. Chapter 8 aims at providing a basic insight into how terrain surfaces are represented in numbers so that calculations can be carried out using a GIS or other software. Chapter 9 addresses the topic of computing height values of unmeasured points on the terrain surface from measured points, a process called *Interpolation*. Chapter 10 considers the binary classification problem of separating bare ground points from building, vegetation and other above-mentioned ground points. This classification problem is usually indicated by the term *Ground Filtering*. The last part of this book, Part IV focuses on "3D Mapping from Point Clouds." A basic step in the 3D mapping process is feature extraction. This topic is addressed in Chapter 11. The theoretical concepts of pointcloudmetry can only be deployed in practice by virtue of the existence of dedicated point cloud processing software. Chapter 12 briefly explores

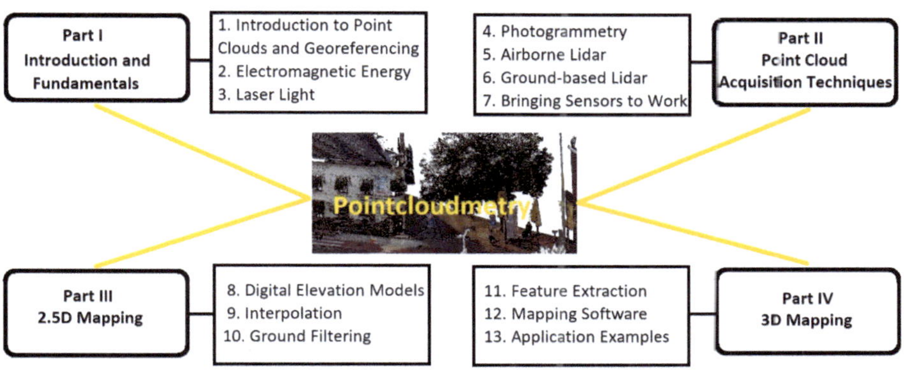

Figure 1.10 Graphical representation of the structure of this book on pointcloudmetry

the characteristics and properties of software developed for processes geo-related point clouds. The proof of the pudding is in eating it. Therefore, we present a great variety of applications of point clouds in the last chapter, Chapter 13, under the heading "Pilot Studies and Applications."

1.8 Concluding Remarks

The main geodata acquisition technologies for the detailed 3D mapping of sites, including the generation of point clouds, are based on imaging devices (photogrammetry) and Lidar sensors (laser scanning). Accuracy – or to use a more general term, *data quality* – is an essential component of geodata in general and point clouds in particular. There are four essential ingredients determining the quality of raw point clouds. These are the sensors, the software, the platform and, above all, the survey plan. The design of the survey plan requires thorough knowledge, skills and expertise, and some of the peculiarities the surveyor will encounter are addressed in Chapter 7. This is where the geomatics specialist comes in. Given the strong societal needs for geoinformation in general and point clouds more specifically, outlined earlier, it is odd that universities in so few countries offer bachelor-level geomatics degrees; at best, the subject is usually on offer at master's level only. There is a serious risk that society will pay the price for this in the future and be forced to increasingly depend on the less specialised knowledge of the multinational informatics industry.

It is my sincere hope that my book helps to spread the essential knowledge about the core, idiosyncrasies and unparalleled potential of point cloud technology, a technology for which I have coined the term *pointcloudmetry*. This book is also intended to make ordinary and occasional point cloud users aware of the manipulations and conversions that a point cloud may have undergone from raw acquisition to the data sets which land on one's computer screen. For example, when the user buys or downloads a DEM for free under open data, a key question to ask is: "What processing steps did the data set go through before it landed on my hard drive or on the cloud server?" Is it the original point cloud as provided by the contractor who conducted the survey, or has the data set undergone processing steps such as ground filtering to contain only the bare ground surface? Has the point cloud been manually edited to remove erroneous ground points or have automated filtering techniques been used? Indeed, point clouds have undergone a series of processing steps before becoming visible on the analyst's computer screen. It is important to be aware of this. This book is an indispensable aid to arrive at the required levels of knowledge.

References

Arapakis, T., van Heerden, N., Rodriguez-Mon Barrera, G., Wang, Q., Wang, X. (2018) *Digitizing Real-World Scenes from Images. Geomatics Synthesis Project.* Delft University of Technology.

Bomford, G. (1980) *Geodesy.* 4th Edition, Clarendon Press, Oxford.

Burkholder, E.F. (2008) *The 3D Global Spatial Data Model: Foundation of the Spatial Data Infrastructure.* CRC Press, Taylor & Francis Group, London.

Heiskanen, W.A., Moritz, H. (1967) *Physical Geodesy*. Freeman and Co., San Francisco, London.

Heiskanen, W.A., Vening-Meinesz, F.A. (1958) *The Earth and Its Gravity Field*. McGraw-Hill, New York.

Hirschmüller, H. (2008) Stereo Processing by Semiglobal Matching and Mutual Information. *IEEE Transactions on Pattern Analysis and Machine Intelligence*, 30(2), pp. 328–341.

Janssen, V. (2009) Understanding Coordinate Reference Systems, Datums and Transformations. *International Journal of Geoinformatics*, 5(4), pp. 41–53.

Lemmens, M. (2014) Point Clouds (1): The Functionalities of Processing Software. *GIM International*, 28(6), pp. 16–19.

Lemmens, M. (2018) Mobile Laser Scanning Point Clouds: Status and Prospects of Automatic 3D Mapping of Road Objects. *Geomatics World*, 26(2), pp. 16–19.

Lemmens, M. (2019) Smart Parking: App Based on Mobile Mapping Point Clouds and Imagery. *GIM International*, 33(6), pp. 11–13.

Moritz, H. (1990) *Advanced Physical Geodesy*. Wichmann, Karlsruhe.

Sansó, F., Rummel, R. (Eds.) (1989) *Theory of Satellite Geodesy and Gravity Field Determination*. *Lecture Notes in Earth Sciences*, Vol. 25. Springer, Berlin, New York.

Seeber, G. (1993) *Satellite Geodesy*. Walter de Gruyter, Berlin.

Zheng, M., Wu, H., Li, Y. (2019) An Adaptive End-to-End Classification Approach for Mobile Laser Scanning Point Clouds Based on Knowledge in Urban Scenes. *Remote Sensing*, 11, pp. 186.

Electromagnetic Energy

CHAPTER 2

Having insight into the particularities and peculiarities of geodata is a first key step in the beneficial exploration of the data. The use of geodata may be directed towards arriving at meaningful information for exploiting oil, gas, metallics and other natural resources; preventing flooding; avoiding the collapse of levees and dikes; or limiting the effects of soil erosion. Indeed, without knowledge about the mechanisms and physics underlying the generation of the geodata, any manipulation of the data becomes vain, a gambling game and manoeuvring on slippery ice. How sophisticated the software running on your fast computer may be, when data is treated as garbage you get garbage out. The keyword with respect to the acquisition of point clouds is *electromagnetic* (EM) energy.

The point clouds considered in this book are obtained from capturing EM energy emitted either by the sun, the earth or the sensing system itself. Sensor systems which capture EM energy emitted by the sun or by the earth are called passive systems. Sensing systems which emit themselves EM energy and next capture the backscattered energy, also called returns, are called active systems. From the characteristics of the points, one wants to extract geometric and thematic (some call thematic also semantic) information about the objects present in the scene. The EM signals are somehow the messengers who tell the story about the scene objects, where they are, what they are and how they evolved over time. EM signals are the cradle source determining which objects and phenomena can be detected in a scene, classified and mapped.

Some basic understanding of the interaction of EM energy with the atmosphere and matter on or near the surface of the earth is thus essential for readers who want to use point clouds to detect or map objects present in a scene. This means that understanding the appearances, properties and particularities of a point cloud requires insight into how EM energy is transmitted through the atmosphere and how it interacts with the air, the ground surface and man-made and natural objects present on this thin skin on top of the lithosphere.

In this chapter, we use the generic term *matter* to indicate solid objects such as buildings, paved roads, bridges and other man-made objects, bare ground and water bodies, as well as semi-permeable objects such as trees, scrubs and other vegetation. Matter thus appears on or near the surface of the earth. We use the overarching term "substance" to indicate matter and air molecules – that is, particles present in the atmosphere, together (Figure 2.1).

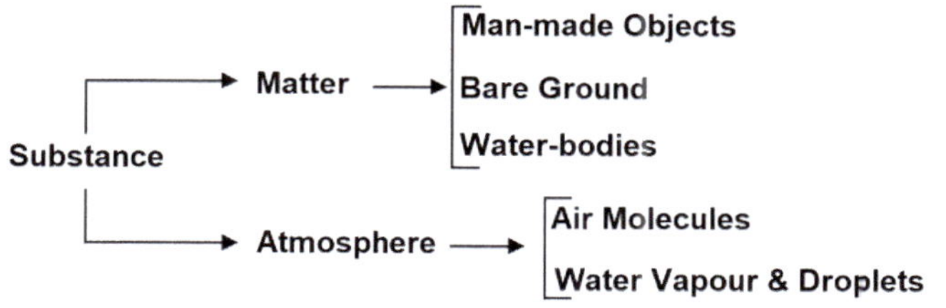

Figure 2.1 Different substance types interact differently with EM energy

The basics and principles of EM energy and radiation are excellently documented in a large body of text books, including Born and Wolf (1999), Boyd (2009), Jensen (2006), Lillesand et al. (2015) and Wynne and Campbel (2011). Compared to these textbooks, the depth of elaboration in this chapter is modest as it primarily aims at providing essential basic knowledge on EM energy and EM radiation indispensable for understanding the characteristics of point clouds. First, we will have a closer look at the characteristics and modelling of EM energy. Next, we focus on the sources of EM energy. Section 2.3 outlines how EM energy interacts with substances. The interaction with the atmosphere causes disturbances in the EM signals. By introducing models about the atmosphere or placing reference panels in the field, these disturbances can be (partly) eliminated or their effects on the final results reduced. Mathematical correction models are considered in the section on radiometric corrections.

A special form of EM energy is laser light. This is light artificially created by devices which are able to convert electric power into concentrated beams of light. Laser is extensively exploited in the creation of point clouds. The laser sensors are usually laser scanners and operate from ground-based (fixed or mobile) or airborne platforms. Because of its importance – not to say indispensability – for creating point clouds we outline the basics of laser light in a separate chapter (Chapter 3).

2.1 Electromagnetic Energy

The EM energy captured by the receiver of a sensor has first travelled through the atmosphere. If the sensor is a Global Navigation Satellite System (GNSS) receiver – commonly used in the creation of point clouds for georeferencing purposes – the EM energy is emitted by GNSS satellites and directly captured by the receiver preferable without interaction with matter. In other applications, the sensor captures sunbeams or beams emitted by the sensors themselves which – after having travelled through the atmosphere – bounced back from building roofs, roads, façades and nearly all other objects on or near the surface of the earth. The earth itself also emits EM energy, which may be captured by a sensor after having travelled some distance through the atmosphere. The strength of the EM energy received by the sensor, the shape of the return beam and the travel time through the atmosphere depend on the characteristics of the atmosphere at the time of data capture and on the properties of the objects. The properties of objects concern, in particular, the materials the object is made of and surface roughness.

First, this section looks at the modelling of EM energy. There are two ways of modelling: the wave model and particles model in which photons play a central role. For understanding the behaviour of EM energy and how it affects the characteristics of point clouds, the wave model is particularly crucial. Next, we will have a closer look at the wave model and the relationship between wavelength and frequency. This section continues then by describing the relationship between photon energy, wavelength and frequency, followed by an outlining of the EM spectrum, which is the partitioning of the wide range of EM wavelengths into spectral bands. These bands consist of a range of wavelengths, with each band indicated by a name label.

2.1.1 Waves versus Particles

The behaviour of EM energy can be modelled in two ways. The first way is to interpret EM energy as waves which propagate through the atmosphere and bounce on matter. This way of approaching the EM energy and its propagation through space has resulted in modelling in the form of waves, called *wave theory* (Figure 2.2). EM waves cover a broad range of wavelengths including radio waves, thermal infrared (TIR), reflected infrared, visible light, ultraviolet and X-rays.

Humans have learned to explore parts of the wavelengths lying in the range from X-rays, which are able to penetrate human tissue and used in medical diagnoses, amongst others, to radio waves, used for broadcasting sound and moving pictures. Already in prehistorical times, humans discovered how to create EM energy in an artificial way. Using animal fat or pitch enabled humans to bring light in the darkness, and by burning firewood TIR was created to bring warmth on cold days or nights. The ability of man to control fire is often called the greatest invention of mankind.

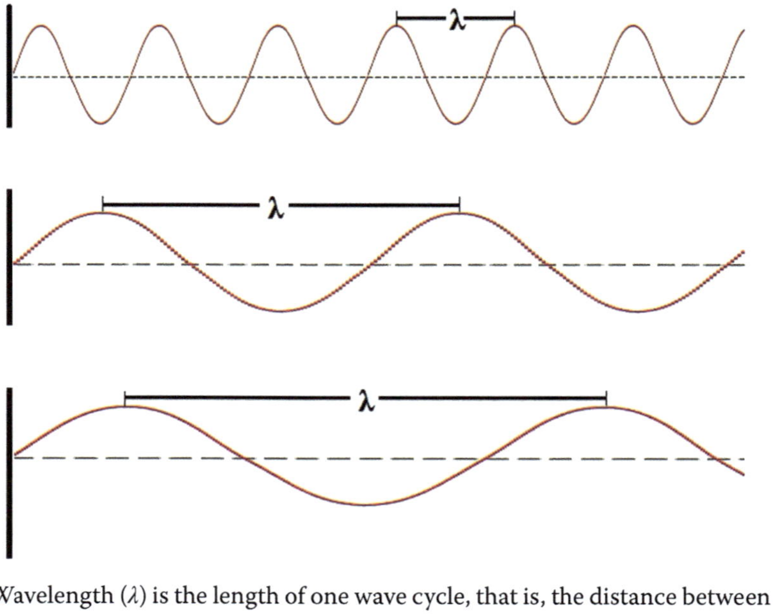

Figure 2.2 Wavelength (λ) is the length of one wave cycle, that is, the distance between successive wave tops

X-rays have a wavelength which is measured in micrometres (1 µm = 0.001 mm) or nanometres (1 nm = 0.001 µm), while radio waves have a wavelength which is expressed in kilometres. The range of wavelengths is called the EM spectrum. The EM spectrum thus consists of a continuum of wavelengths stretching from nanometres to kilometres. The interaction of EM energy with substances depends on the wavelength for the major part.

The second way to model EM energy is as particles of which the energy is transmitted. These particles are called *photons* resulting in the *particle theory*. This way of modelling EM energy is particularly useful when analysing the brightness of an image or the strength of the return beam of a laser scanner or a radio detection and ranging (Radar) system.

2.1.2 Wavelength and Frequency

In the wave-theoretic modelling of EM energy, one often uses the term *frequency* instead of *wavelength*. The frequency terminology is particularly preferred for use in the microwave and radio part of the EM spectrum. The microwave part covers the wavelength range from approximately 1 mm up to around 1 m. The microwave bands are specifically exploited in Radar remote sensing. The signals emitted by GNSS satellites, including the US Global Positioning System (GPS), Russian GLONASS, European Galileo and Chinese BeiDou, lie also in the microwave part of the EM spectrum. Indeed, also the microwave in your kitchen exploits this part of the EM spectrum to heat your frozen meal. Microwaves are also used by the police for speed checks of cars.

The relationship between wavelength and frequency is straightforward. The frequency measures the number of waves – more exactly cycles – passing a fixed point per unit of time. The longer the wavelength, the fewer cycles will pass the fixed point provided that the waves have the same propagation velocity. Hence, the number of cycles that pass a fixed point per unit of time depends on the wavelength and the propagation velocity. The speed of EM waves – usually called the speed of light – depends on the properties of the medium through which the waves travel and is for vacuum determined at 299,792.458 km/s, which is often rounded to 300,000 km/s for practical cases. The speed of light through a vacuum, usually symbolically represented by the character "*c*," differs slightly from the speed through the atmosphere.

Waves with a wavelength λ and travelling with speed c will pass a fixed point c/λ times. So the relationship between the frequency (f) of an EM wave and wavelength is given as:

$$f = c/\lambda \tag{2.1}$$

Let, for example, $\lambda = 23$ cm, then $f \approx 30\ 10^9/23 \approx 1{,}304{,}347{,}826$ cycles per second. One cycle per second is denoted as 1 Hz, 1,000 cycles per second as 1 kHz, one million cycles per second as 1 MHz, and one billion cycles per second as 1 GHz. So the frequency of the above example can be denoted – in rounded form – as 1.3 GHz.

Just to get an impression of the travel time dimensions of EM energy, let us have a look at the time it takes for EM energy emitted from the sun to travel to the earth.

The average distance of the earth to the sun is 149,600,000 km. This means that it takes sunlight nearly 500 s or 8 min and 20 s to travel from the surface of the sun to the surface of the earth. So the reflection of sunlight on matter captured with a digital camera has already gone a long way before the snapshot is taken. The GNSS satellites, which are exploited by your smartphone to determine your location, orbit at an altitude of approximately 20,000 km above the earth's surface. A small calculation learns that it will take GNSS signals 67 ms to bridge the distance between its source to your smartphone.

2.1.3 Wavelength and Photon Energy

Each photon carries energy. The amount of energy depends on the wavelength of the EM beam; the longer the wavelength the lower the energy. This means that the amount of energy of the photon (E) is inversely proportional to the wavelength. Mathematically this can be expressed as:

$$E = \frac{h \cdot c}{\lambda} = h \cdot f \tag{2.2}$$

The numerator in the first part of the above equation consists of a multiplication of two constants: Planck's constant ($h = 6.626 \times 10^{-34}$ J s) and the speed of light (c) discussed in the previous section. Substituting Equation 2.1 into Equation 2.2 gives the multiplicative relationship between Planck's constant and frequency, shown as the last part of the above equation. This equation shows that the higher the frequency the higher the energy. Indeed the photons of X-rays and ultraviolet, which have very short wavelengths and thus very high frequencies contain so much energy that X-rays can penetrate skin, tissue and muscles and ultraviolet may burn your skin. Two units of energy are commonly used to quantify photon energy. These are electronvolt (eV) and Joules (J). The relationship between energy and frequency is proportionate and just connected by Planck's constant, which is a fundamental constant of physics and forms the cornerstone of quantum mechanics. It is astonishing that the relationship between two fundamental physical quantities can be expressed in such simple terms and that the relationship is valid for the entire domain of wavelengths (frequencies) covered by the EM spectrum. It quite often happens that the relationship between parameters is only valid within a limited domain of the parameter values. Of course, the value of the constant has to be determined. Today, it is possible to determine Planck's constant in a classroom experiment using LEDs fairly accurately (around 1% of its actual value). The effort lies in the exact determination of the value.

The above equation, often called Planck's equation or the Planck–Einstein relation, shows that the amount of energy of a photon solely depends on the wavelength of the photon. This means that two photons having the same wavelength will carry the same amount of energy; it does not matter whether the photon is emitted by a powerful laser gun, a light bulb or a candle at night. This also means that the energy of a beam entirely depends on the number of photons present in that beam. Increasing or decreasing the intensity of a beam emitted by an active sensor system can thus only be tuned by changing the number of photons.

2.1.4 Spectral Bands

It is convenient to group ranges of wavelengths together and give a unique name to each range. Earlier, we already mentioned a few spectral bands. In this section, we will discuss spectral bands in a more systematic manner. The ranges of wavelengths are called spectral bands. For example, the near-infrared (NIR) band starts at the wavelength where the red band ends, which is around 0.7 μm and stretches out to a wavelength of around 1.7 μm. The blue, green and red bands together are called visible bands and cover the range from 0.4 μm to 0.7 μm. Adjacent to the NIR band are two TIR bands (3–5 μm and 8–14 μm). The spectral band ranging from around 5 μm to 8 μm is blocked by the atmosphere.

Just like the visible spectral band can be captured by the red, green and blue (RGB) cameras, so can the two TIR bands be captured by sensors specifically developed for recording emitted heat. Let us have a look at the possible use of such sensors and what they look like. The smaller images in Figure 2.3 show TIR images of Mount Etna, Italy, taken with a video sensor and a handheld TIR camera. The video sensor was mounted on a hexacopter – a rotary-wing unmanned airborne system (UAS) with six propellers. The aim of the UAS survey was to arrive at a better understanding of the signs that prelude volcano eruptions and temperature rise is one of the key indicators for detecting seismic activities underneath the surface (Amici and Turci, 2014). Thermal imaging from a UAS is a promising technique for acquiring data in volcanic areas which are difficult to access. The large image is taken with a handheld TIR scanner capturing the 8–14 μm thermal spectral band and sensing temperatures from 0°C to 350°C. This image was

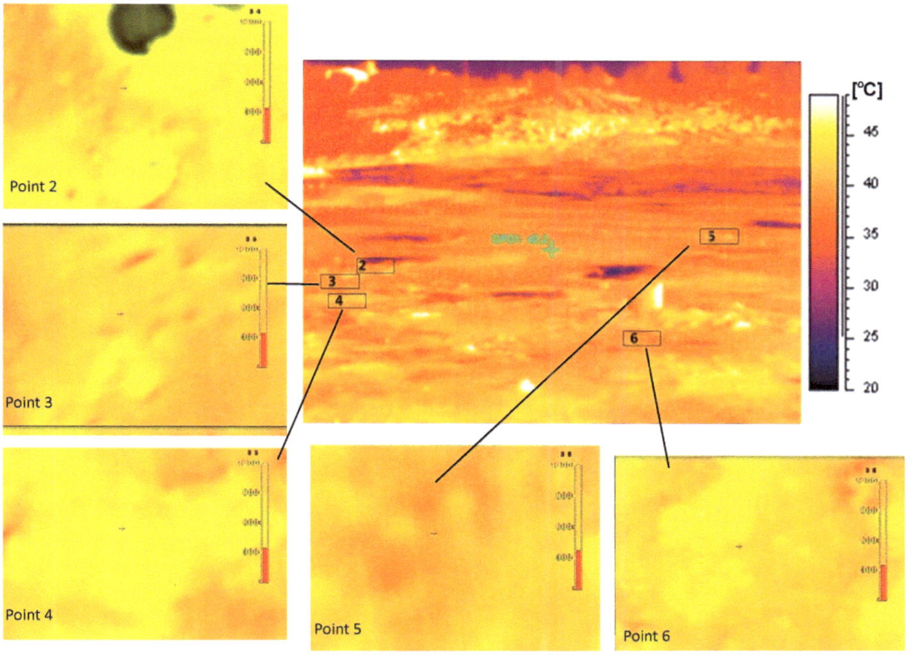

Figure 2.3 TIR images of Mount Etna; the large image is taken with a handheld TIR scanner and the smaller images are captured from an unmanned hexacopter (source: Amici and Turci, 2014; modified)

Table 2.1 Radar spectral bands

Band	Wavelength (cm)	Frequency (GHz)
K	0.8–2.4	37.5–12.5
X	2.4–3.8	12.5–7.9
C	3.8–7.5	7.9–4.0
L	15–30	2.0–1.0
P	30–100	1.0–0.3

taken for comparison purposes. Note the temperature of the surface of the volcano lies around 27°C, that is, 300°K. According to Wien's displacement law (Equation 2.3), the wavelength which contains the maximum amount of energy is ~10 μm. This has been the reason to choose a sensor which captures the 8–14 μm thermal spectral band.

For Radar remote sensing applications, the microwave part between 1 cm and 1 m is usually divided into five spectral bands, each indicated by a character from the alphabet (Table 2.1). The choice of the letters is based on the historical development of Radar.

2.2 Sources of EM Energy

Within the framework of the creation of point clouds, three sources of EM energy are important to distinguish (Figure 2.4). The first source of EM energy is emitted by the sun, the bearer of all life on earth. This is also the oldest source of EM energy exploited for photogrammetric mapping applications. In 1837, Louis Daguerre, by profession a painter of dioramas, obtained the first practical photograph using a process named after him. The camera based on the daguerreotype process of photography became a commercial success and contributed much to the popularity of photography. It resulted also in the birth of a new mapping discipline called *photogrammetry*. Since World War I, photogrammetry has been intensively exploited for mapping remote areas. For example, in 1936, New Guinea was mapped from aerial images using photogrammetric techniques for oil exploitation purposes. Till the invention of false colour photography, films captured the same part of the EM spectrum as the human eye catches.

The second source of EM energy is the earth itself. From Wien's displacement law, it can be computed that the earth's surface radiates the highest amount of energy at the wavelengths around 10 μm. Wien's displacement law states that the wavelength which contains the maximum amount of energy (λ_{max}) is inversely proportional to the temperature (T in degree Kelvin) of the body which emits the energy (Figure 2.5):

$$\lambda_{max} = 2898/T \tag{2.3}$$

An active volcano may have a temperature of around 800°K, meaning the wavelengths, which contain the highest energy, lie in the range of about 3–5 μm. So, to monitor volcanoes but also bush fires, a sensor operating in this range can be fruitfully exploited.

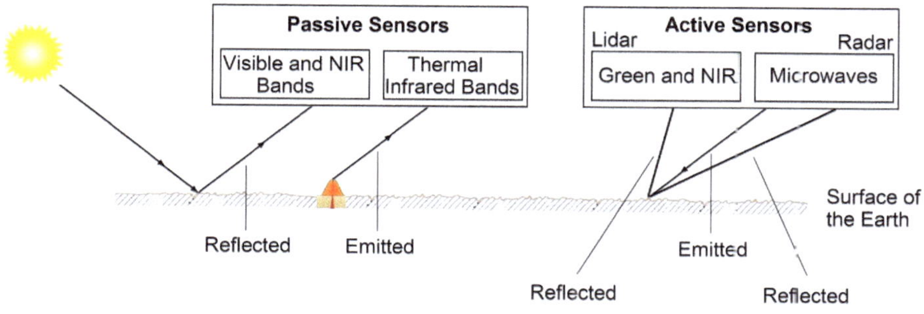

Figure 2.4 Sources of EM energy and the corresponding parts of the EM spectrum captured by active and passive sensors

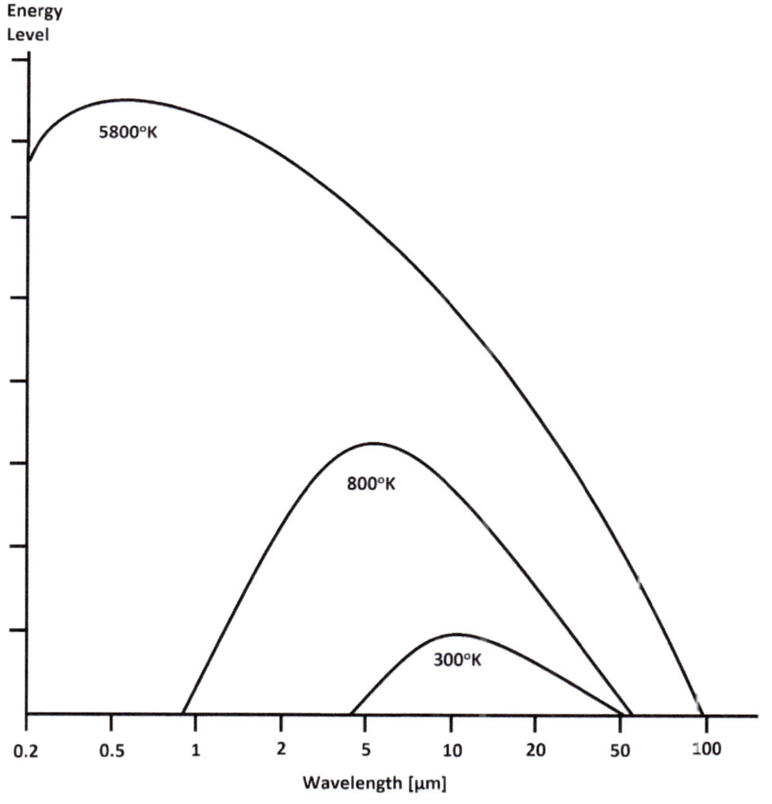

Figure 2.5 The temperature of a body emitting EM energy determines the level of energy in the diverse wavelengths

The surface temperature of the Sun lies around 5,800°K. Wien's displacement law states that – at this temperature – the wavelength, which contains the highest energy, lies around 0.5 µm. This is the part of the EM spectrum which is visible to the human eye. Obviously, the eye has evolved through the ages of evolution to capture the sunlight which contains the maximum amount of energy.

The third source of EM energy is delivered by the sensors themselves. These sensors consist of transmitters, which generate and emit themselves EM energy. The transmitter is accompanied by a receiving sensor which captures the backscattered energy. Since they emit themselves EM energy, these types of sensors are called *active sensors.* The most important active sensors are, in order of appearance, Radar and Light Detection and Ranging (Lidar). Radar reached maturity during World War II, while Lidar became an operational remote sensing technique in the early 1990s and is experiencing impetuous developments since the turn of the millennium.

Whether emitted by external sources or by the sensor itself, the EM energy, which passes through the atmosphere and bounces back from matter located on or near the surface of the earth, is captured by the receiver part of the sensor. The most important component of EM energy which is captured by both active and passive sensor systems is the amount of energy which enters the receiving part of the sensor. The level of energy is also called signal strength. In addition to the signal strength, active systems can capture other components, which are as follows:

- Travel time through the air, that is, the time between the emission of the EM energy and the receival of the backscatter.
- Phase of the signal, important for GNSS surveying and laser scanners based on the phase-shift principle.
- Frequency of the signal, particularly important for Interferometric Synthetic Aperture Radar (InSAR).

One may argue that the above-mentioned components are the most important part of the signals of an active system providing the information one is looking for and that is partly true. However, without sufficient signal strength, the components listed earlier can't be determined.

2.3 Reflection, Absorption and Transmission

When EM radiation meets substance, three types of interactions may occur: The entire energy or part of it is reflected, absorbed or transmitted. The degree of reflection, absorption and transmission depends on wavelength, the materials the matter is made of, the

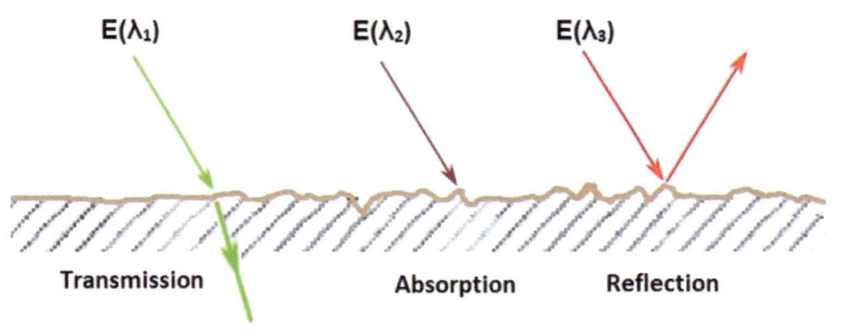

Figure 2.6 Whether an incident EM beam is (partly) reflected, absorbed or transmitted depends on wavelength

surface roughness of the matter and the incident angle. In other words, when EM energy incidents on surfaces with identical properties, a part of the EM energy will be reflected, another part absorbed and a further part transmitted (Figure 2.6).

Reflection is the part of incident EM energy that is bounced back from the surface. This EM energy is also called backscatter. In the visible part of the EM spectrum, the human eye observes reflected light as colour. For example, humans see the leaves of deciduous trees as green in spring and summer. This is caused by chlorophyll, a chemical component present in the leaves of plants. Chlorophyll reflects green wavelengths much more than red and blue wavelengths, which are absorbed for the major part. However, the reflection of NIR light – invisible to the human eye – is stronger than the reflection of green. This is clearly visible in false colour images which represent NIR light in red.

Absorption is the part of the incident EM energy that is immersed by substances and transformed into another form of energy. For example, the body of a pale blue car absorbs green and red light and transfers it into heat, which may transform the bonnet of the car into a cooker during hot summer days. As discussed earlier, plants absorb red and green lengths and other parts of the EM spectrum and use the energy for photosynthesis to create organic material from inorganic substances, particularly carbon dioxide and water.

Transmission is the part of the incident EM energy that passes substances. Sun energy that passes the atmosphere, or penetrates water to reach the bottom of a waterbed, belongs to this part. Let us call the energy of an incident EM beam with wavelength λ, $E_i(\lambda)$. When this beam meets a substance, it will be decomposed into three components, $E_r(\lambda), E_a(\lambda), E_t(\lambda)$. These three components sum up to $E_i(\lambda)$ (Figure 2.7):

$$E_i(\lambda) = E_r(\lambda) + E_a(\lambda) + E_t(\lambda) \tag{2.4}$$

The rate of reflection – usually called reflectance – absorption or transmission is strongly dependent on, in addition to the wavelength, the properties of the substance and the incident angle of the beam. Reflection is the most important component for remote sensing applications since this is the part which reaches the sensor and thus determines the amount of energy captured.

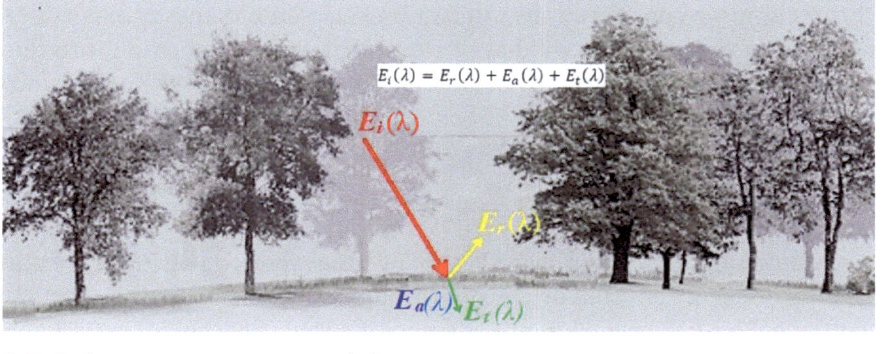

Figure 2.7 Reflection, transmission and absorption

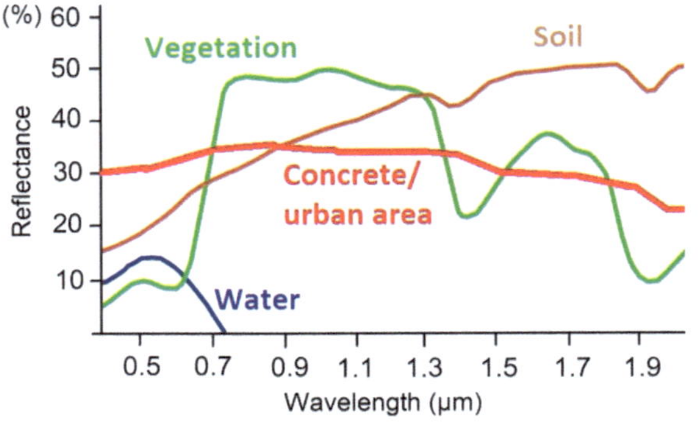

Figure 2.8 Spectral signature of water, vegetation, soil and concrete/urban area

The description of the EM interaction properties of matter is often done in the form of reflectance (*R*), absorption (*A*) and transmission (*T*) measures, which are defined as:

$$R(\lambda)=\frac{E_r(\lambda)}{E_i(\lambda)}$$
$$A(\lambda)=\frac{E_a(\lambda)}{E_i(\lambda)} \tag{2.5}$$
$$T(\lambda)=\frac{E_t(\lambda)}{E_i(\lambda)}$$

These measures are usually expressed as a percentage by multiplying by 100. Of course, the sum of reflectance, absorption and transmission should be 100%.

Reflectance can be determined using a spectrometer. As a result, for each type of matter, reflectance curves can be determined with the wavelength within a range of interest on the horizontal axis and the reflectance in percentage along the vertical axis. Reflectance curves for four types of matter – water, vegetation, building materials and bare ground – are shown in Figure 2.8. Water absorbs most of the wavelengths. This knowledge is essential when interpreting point clouds acquired by laser scanners operating in the NIR bands. This means that clear water bodies will be "visible" in the point cloud as pointless gaps.

In this section, we considered the interaction between EM energy and substances in general. In the next section, we will have a closer look at the interaction with the atmosphere and in a subsequent section at the interaction with the earth's surface and objects present on this surface.

2.4 Interaction with Atmosphere

For the creation of point clouds as well as many other remote sensing applications, the transmission of energy emitted by the sun through the atmosphere is important. Many wavelengths of the EM spectrum are absorbed by the atmosphere, entirely or partly. The atmospheric particles which cause absorption are above all water droplets, carbon dioxide, oxygen and ozone. The spectral bands, which pass for the major part through the atmosphere, are called atmospheric windows (Figures 2.9 and 2.10). For some, the term atmospheric window sounds confusing, since it seems to refer to a window in the atmosphere. However, the term refers to ranges in the EM spectrum which are largely unaffected by the atmosphere.

Reflection on air molecules and water droplets in the atmosphere is called scattering. The rate of scattering depends on the wavelength in relation to the size of the particles. When the size of a particle is 1 mm, it will form an enormous obstacle for a wavelength of 0.001 mm (1 μm). This is because the particle appears as a giant for a wavelength of minuscule size. But for a wavelength of 100 mm (1 dm), the particle appears like a germ, much the same as a children's autoped may get stuck in a fist-sized hole in the pavement, while the man-high wheels of a tractor will not notice its presence. So, when the waves are much

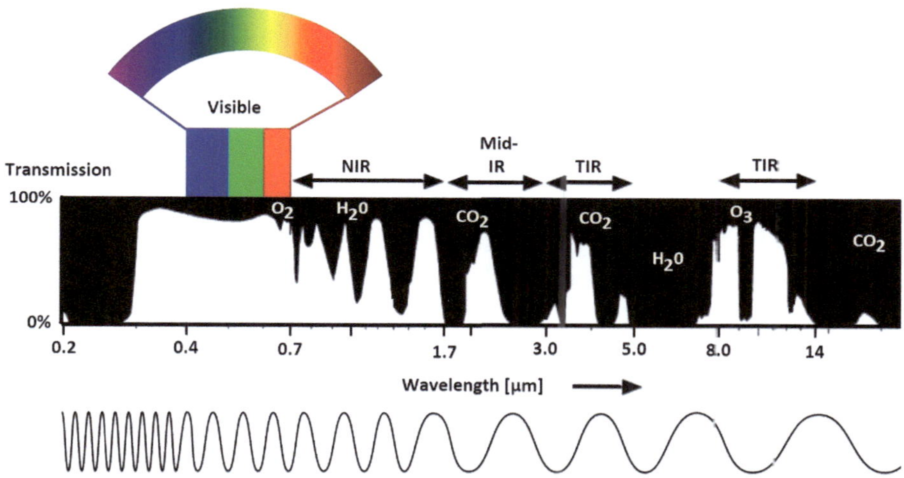

Figure 2.9 Atmospheric windows of the visible, NIR and TIR part of the EM spectrum

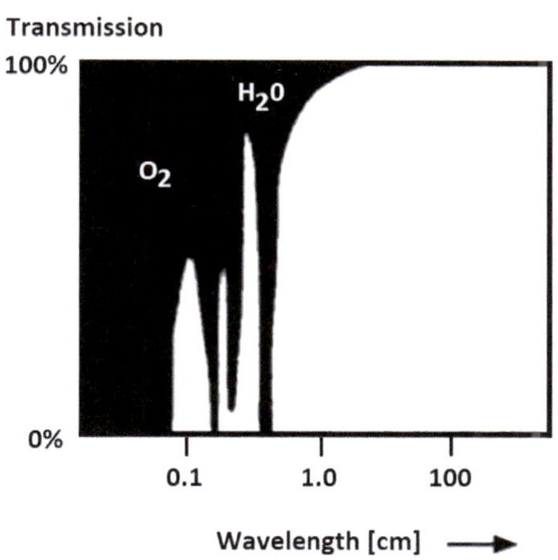

Band	Wavelength [cm]
K	0.8 – 2.4
X	2.4 – 3.8
C	3.8 – 7.5
L	15 – 30
P	30 – 100

Figure 2.10 Atmospheric windows of the microwave part of the EM spectrum in the range of 0.1–100 cm

larger than the particles, the EM beam will be not or only slightly affected. The beam waves or undulates, so to say, around the particle and does not care about its presence.

The range of EM spectrum between 1 mm and 100 cm – microwaves – is only slightly or not at all hindered by the presence of water vapour and water droplets in the air – heaped together in clouds. The larger the wavelength, the less they will be impeded. As a result Radar can penetrate clouds. Therefore, it is often said that Radar has an all-weather capacity. Since Radar is an active system, its operation is also independent of the presence of sunlight. This means that Radar can operate day and night in addition to being weather independent. These are favourable properties for capturing tropical regions from the air or from space by Radar systems.

When the size of particles, for example water droplets, is much larger than the wavelength, the EM energy is severely hampered. Therefore, the visible bands – blue, green and red – are scattered. This is the reason that clouds are observed as white by the human eye or more specifically the human visual system.

2.5 Interaction with Surfaces

When an EM beam survives its travel through the atmosphere, it will meet matter present on or near the surface of the earth. The part of the beam, which is reflected, is of particular interest for most remote sensing applications. It should be noted that in bathymetric Lidar, which uses green laser pulses, the pulses pass the water surface and reflect on the bottom of the water body (Figure 2.11). In contrast, NIR used in topographic or land Lidar systems is absorbed by water. The pattern of reflection depends on the roughness of the surface and the components matter is made of.

A smooth surface will act as a mirror resulting in specular reflection: The direction in which the beam is reflected depends on the incident angle. A smooth surface is a surface where the roughness or the sizes of the particles the matter consists of are small compared to the wavelength. Only a few surfaces will act as an ideal specular reflector and they are usually man-made for the major part. The other extreme is that

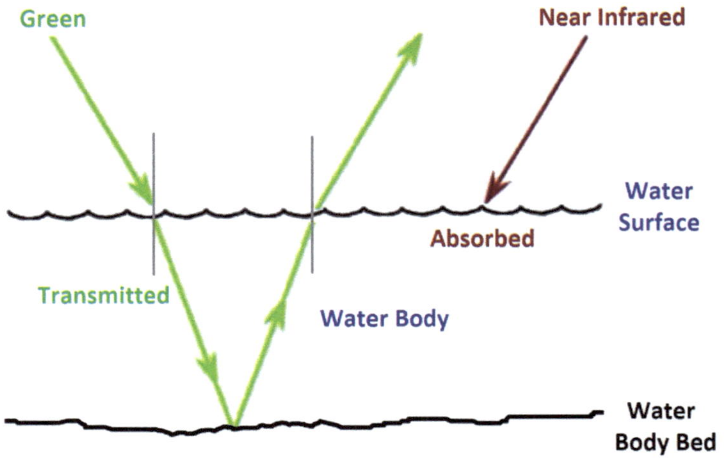

Figure 2.11 The green band is transmitted through water while the NIR band is absorbed

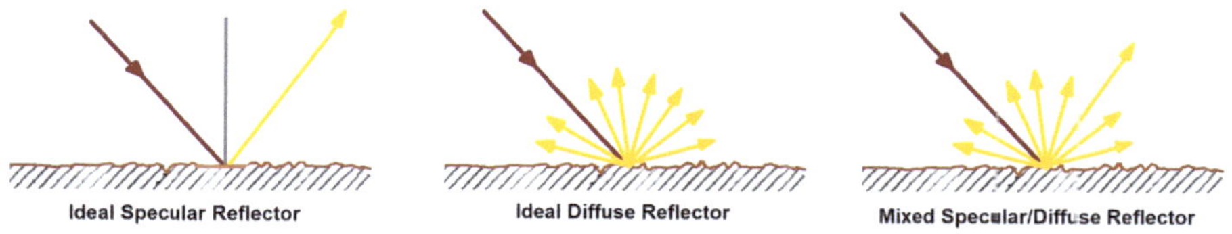

Figure 2.12 Ideal specular reflection (left), ideal diffuse reflection (middle) and partly specular and partly diffuse reflection, indicated as mixed (right)

a surface scatters the beam in all directions equally. This is called ideal diffuse reflection, and the type of surface is called a Lambertian surface (Figure 2.12). In reality, the type of reflection will lie somewhere in between the two extrema of ideal specular reflection and ideal diffuse reflection. Diffuse reflection will dominate if the surface roughness or the sizes of the particles the surface consists of are large compared to the wavelength. The definition of small and large is relative in this case. This means that the definitions of smooth and rough will vary depending on the type of sensor used. For example, in dune areas, specular reflection will be dominated if the sensor is a Radar system since Radar exploits the microwave part of the EM spectrum. In contrast, dune areas act diffusely for a camera since it captures the visible part of the EM spectrum. The amount of the EM energy captured by both passive and active sensors depends on the mixture of specular reflection and diffuse reflection. With the increasing exploitation of Lidar and Radar two other types of reflection have emerged eminently: fuzzy reflection and multiple reflection. We outline these two types of reflection in the sequel.

2.5.1 Fuzzy Reflection

Many objects have a crisp boundary. The surface where the object starts and ends can be defined sharply. Examples are plenty and include building façades, pavements of roads and parking lots, bridges and many other man-made objects. However, there exists also matter, particularly vegetation, which exhibits a more fuzzy appearance; it is not clear where the object starts or ends. This characteristic can be fruitfully exploited when mapping forests using airborne Lidar or Radar.

Fuzzy reflection can be fruitfully exploited by Lidar by using the beam divergence of the laser pulse which results in a footprint of a certain size when the pulse hits matter. Foliage cover is usually not completely impermeable but rather semi-porous; a part of the laser pulse may reflect on leaves but gaps between the leaves allow parts of the pulse to reach the ground and be scattered back to the receiver (Figure 2.13). In other words, the first return reflects off the foliage and the last return off the ground. In this way, the bare ground of forest areas can be measured and thus digital elevation models (DEM) be generated. The difference between the last and first returns provides a measure of the tree heights and ultimately of the forest stand as a whole. To allow more information to be derived from the data, returns also started to be recorded between the first and last return, finally resulting in full-waveform digitisation (FWD). So fuzzy reflection enables the sampling of the entire return from its leading edge to its trailing edge at regular intervals, resulting in, for example, 64, 128 or 256 samples per return.

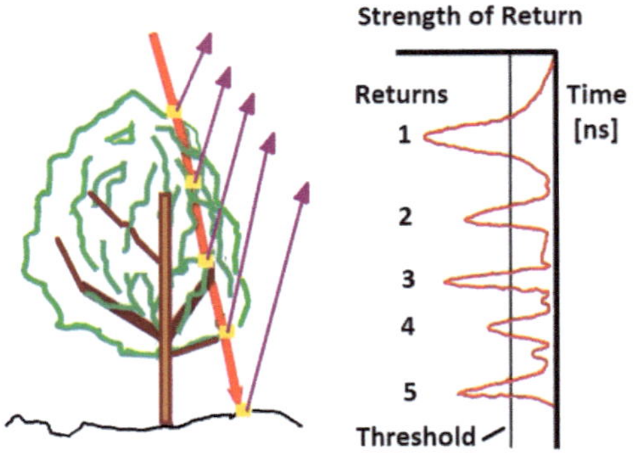

Figure 2.13 Fuzzy reflection of a tree resulting in multiple returns, which can be captured by Lidar when applying full-waveform digitisation

2.5.2 Multiple Reflection

The measurement of the pulse travel time or phase is the most important asset of active sensors, which have been designed for collecting point clouds of scenes. The following discussion is limited to time-of-flight systems. Since the travel time determines the distance between sensor and matter – and thus indirectly the 3D coordinates of the point – faults in the time measurement will cause errors in the range calculation. One such fault is introduced when the beam reflects multiple times before reaching the receiver.

EM energy may be reflected on more than one surface before it reaches the receiver of the sensor. Having, for example, been reflected by a horizontal surface such as the pavement of a road, the beam may be reflected a second time on a building façade. Multiple reflection affects all active systems, not only Lidar and Radar but also GNSS measurements. Surveyors who conduct positioning with the help of GNSS receivers experience multiple reflection as an error in the determination of the coordinates of the point on which their GNSS receiver is positioned. In GNSS surveying, multiple reflection is usually called multipath. Multipath is primarily due to the fact that the receiver does not only pick up the direct signals emitted by the GNSS satellites but also signals which first hit a surface in the vicinity of the receiver. In other words, multipath is caused by the fact that the same satellite signal has travelled along different paths – at least two – before the resulting riven signals are received by the GNSS receiver (Figure 2.14, left). So the reflections from nearby objects such as road surfaces, buildings, vehicles or trees affect the precision of GNSS positioning. Multipath is a GNSS error source which requires precautionary measures of the surveyor and attentiveness while conducting the measuring process.

In airborne Lidar, multiple reflection results in longer pulse travel times which ultimately appears as a dip in the elevation model (Figure 2.14, right). Such errors have to be eliminated through outlier detection, which is an important part of the 2.5D mapping of the earth's surface. Since these are usually individual events, they often can be successfully removed by spatial filtering. Similarly, in Radar imagery, multiple reflection causes objects to produce returns of greater signal strength than might be expected from the size and reflection properties of the object. Multiple reflection results in bright spots in Radar imagery. The Radar term for multiple reflection is corner reflection or, sometimes,

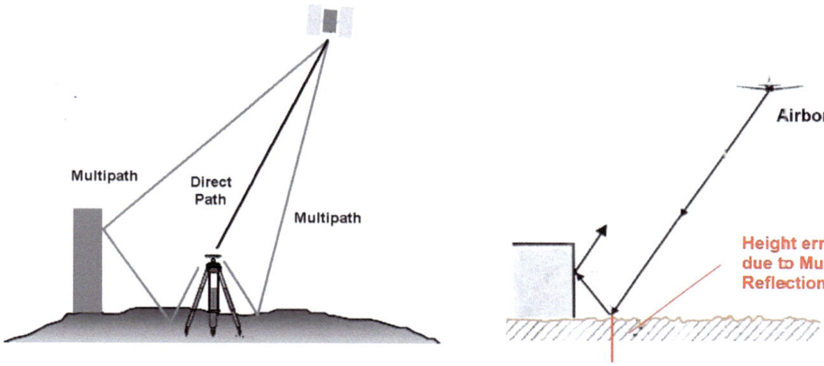

Figure 2.14 Multiple reflection; some signals reach directly the GNSS receiver, others along indirect paths (left); for airborne Lidar multiple refection causes elevations dips (right), which are outliers (source: Lemmens, 2011; modified)

double bounce. Whatever the name may be, the physical cause is the same for laser scanning, Radar and GNSS: A system emits beams with the aim to determine – from travel time – the distance from the receiver sensor to a surface or satellite in the case of GNSS. But due to multiple reflection, the range is measured erroneously causing errors in the final products.

2.5.3 Double Bounce

What may be a deficit, our shortcoming of a system for one application may be a gift for another application. In airborne Radar, multiple reflection is exploited for creating DEMs of tropical forest areas using the P-band. The wavelength of the P-band ranges from 30 cm to 100 cm (Table 2.1), which is large enough to pass leaves. The capability of P-band Radar to penetrate clouds as well as canopy has been exploited to map an area of 1.2 million km^2 of the Brazilian Amazon between 2008 and 2013. The survey used a dual Radar system mounted on aircraft using X- and P-bands. This was the world's largest systematic aerial-mapping project using X- and P-bands (Sambatti and Lübeck, 2014). The aerial survey took 5,230 flight hours using two aeroplanes carrying the dual-band Radar. The P-band used had a wavelength of 75 cm, which is large enough to penetrate the foliage and reach the terrain surface underneath the vegetation. P-band signals reflect off the terrain and only return as backscatter to the Radar sensor when being reflected a second time in particular by nearby tree trunks. This is called double bounce (Figure 2.15). In forests, where tree trunks are more abundant and/or larger than in open areas, more P-band signals return to the sensor.

The X-band used had a wavelength of 3 cm. This band reflects on the vegetation canopy. Radar X-band interferometry can be used to create a digital canopy model (DCM), while P-band interferometry provides the data for modelling the terrain underneath the vegetation (DEM). If no vegetation is present, DCM and DEM coincide. The difference between the DCM and DEM gives an estimate of the vegetation height. Assuming that the ground surface does not change over time, changes in DCM are caused either by vegetation removal or vegetation growth.

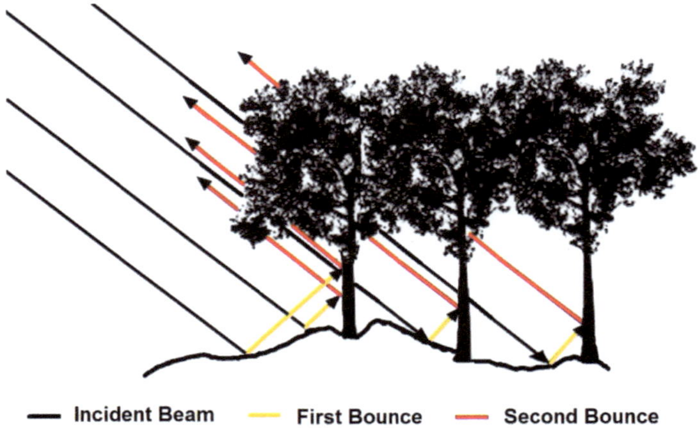

Figure 2.15 Double bounce effect of the Radar P-band capturing a forest region

2.6 Radiometric Corrections

Both the atmosphere and the sensor systems themselves introduce aberrations. Sensor systems introduce aberrations since they are physical devices which – operating in physical reality – do not exactly behave as objected by the designer. If the type and extent of aberrations are known, which means a priori information about the aberrations is available, the distortions on the sensor outputs can be partly or entirely eliminated. The deviations can concern both the geometry and the radiometry. Geometric deviations are related to the where component of geodata, which means the position. Radiometric, sometimes also called photometric, distortions concern the EM energy captured by the sensor. This section considers some of these radiometric distortions and how they can be compensated for. Geometric distortions are considered in Part II which focuses on modelling the geometric sensor behaviours of cameras and laser scanners.

The atmosphere affects the speed of the EM beams as well as the signal strength which is captured by the sensor. Deviations from a model atmosphere cause distortions which appear as delays in the propagation of the EM beams and as attenuations of the signal strengths. If, for example, a NIR laser pulse is emitted and propagates through the atmosphere, some photons will be absorbed by water vapour while other photons may be scattered by oxygen or carbon-dioxide molecules, resulting in a weakening of the EM energy of the pulse (Figure 2.16).

2.6.1 Absolute Reflectance Values

In some applications, one wants absolute reflectance values, rather than relative values. One such application concerns the comparison of multispectral images taken at different times. Such a comparison is important to judge whether measures have resulted in improvements or whether the natural conditions of an area have further deteriorated in course of time. Another application is to use the strength of returns of point clouds generated by mobile laser scanning as a feature in the classification process. Due to beam divergence, atmospheric attenuations and incident angle, the same object, if

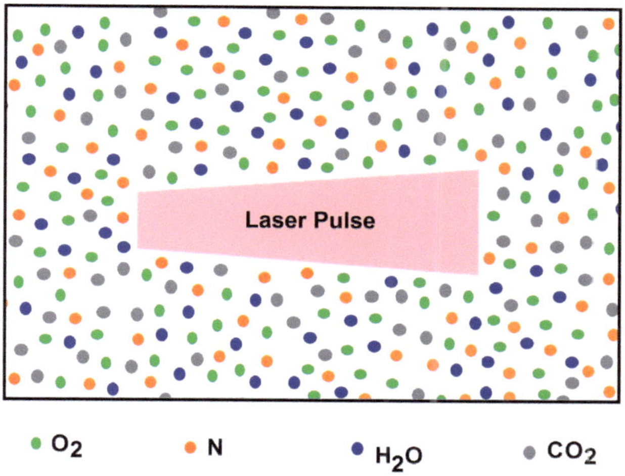

Figure 2.16 While propagating through the atmosphere with approximately the speed of light laser, pulse energy may be weakened by scattering and absorption by air molecules

located close to the sensor, would give a higher reflectance value than when located farther away. In these cases, it is worth carrying out a correction of the strength of return.

2.6.2 Determining Correction Values

Basically, corrections can be conducted along two lines: (1) measuring at sampling locations the actual reflectance values and (2) modelling the distortions in a mathematical way. The measuring of the actual reflectance values means that reference values are collected which provide information for correcting all the values measured by the sensor. This can be done along two lines. The first line is by travelling to the scene with an accurate spectrometer and measuring at sampling locations the reflectance values of the spectral bands of interest for the task at hand. Prerequisites are that the sensor system and the spectrometer operate simultaneously and that the locations of measuring reference values or placing reference panels are well distributed over the site.

A second method is to place reference panels in the scene (Jackson et al., 1987). The reflectance values in the diverse spectral bands of these panels are exactly known through calibration. Both methods provide reference values enabling correction of the actual values measured by the sensor. By introducing a mathematical relationship between the actual (R_a) and reference values (R_r), transformation parameters can be computed. If a linear relationship:

$$R_r = \alpha R_a + \beta \tag{2.6}$$

appears to be suitable, the two parameters α and β can be adjusted in a least-squares sense if reference values are available at three or more locations. The parameters of non-linear relationships can also be determined using least squares. Least-squares adjustment is based on redundancy, which means more observations are available than unknowns. The method enables the determination of the precision of the measurements and to detect possible outliers. Once the transformation parameters are adjusted, all actual values can be corrected:

$$R_c = \alpha R_a + \beta \tag{2.7}$$

This approach assumes homogeneity of the distortions, which means they should be space invariant.

2.7 Concluding Remarks

This chapter has focused on the interaction of EM energy with the atmosphere and objects on or near the surface of the earth. In the present context, interaction is a collective term, which assembles reflection, absorption and transmission. For the creation of point clouds, the EM energy reflected by objects is most important. The EM energy captured by the receiver of an active system depends not only on the properties of the materials an object is made of and its surface roughness, but also on sensor characteristics, the distance between sensor and object and the incident angle. When the reflectance values of a point cloud acquired under the same conditions are used for analysing the properties of objects, it is necessary to normalise the reflectance values.

The next chapter forms a further and focused prelude to Part II of this book. This part elaborates on the sensors able to acquire point clouds. Chapter 3 outlines the characteristics of a particular type of EM energy: laser light. Since the early 1990s, laser light has played a crucial role in remote sensing and its importance is steadily growing. Particularly, this chapter outlines in greater detail how laser interacts with the air, the surface of the earth and objects present on this surface.

References

Amici, S., Turci, M. (2014) Monitoring Volcanoes with UAS: Testing the Suitability of UAS Equipped with Thermal Infrared Sensor. *GIM International*, 28(6), pp. 27–29.

Born, M., Wolf, E. (1999) *Principles of Optics: Electromagnetic Theory of Propagation, Interference and Diffraction of Light.* 7th Edition, Cambridge University Press, Cambridge.

Boyd, D.S. (2009) Physical Basis of Remote Sensing. In Atkinson, P.M. (Ed.) *Geoinformatics*. Eolss Publishers Col. Ltd., Oxford, UK.

Jackson, R.D., Moran, M.S., Slater, P.N., Biggar, S.F. (1987) Field Calibration of Reference Reflectance Panels. *Remote Sensing of Environment*, 22(1), pp. 145–158.

Jensen, J.R. (2006) *Remote Sensing of the Environment: An Earth Resource Perspective.* Pearson Education, Upper Saddle River, NJ.

Lemmens, M. (2011) *Geo-information: Technologies, Applications and the Environment.* Springer, Dordrecht, Heidelberg, London, New York.

Lillesand, Th., Kiefer, R.W., Chipman, J. (2015) *Remote Sensing and Image Interpretation.* 7th Edition, Wiley, Hoboken, NJ.

Sambatti, J., Lübeck, D. (2014) Mapping the Amazon, Radar X- and P-Band For Detecting Illegal Logging and Assess Biomass. *GIM International*, 28(5), pp. 17–19.

Wynne, R.H., Campbel, J.B. (2011) *Introduction to Remote Sensing.* 5th Edition, Guilford Publications, New York.

Laser Light

CHAPTER 3

Laser scanners are active sensors which emit laser beams for measuring the distances to objects without human contact with the objects; "active" means that the sensors themselves emit electromagnetic (EM) energy. The distance measured is usually called range. Exploiting laser beams enables the construction of contact-free measurement devices from which 3D coordinates of points in a preferred reference frame can be computed. This technology has become vital for the 3D mapping of outdoor and indoor scenes, activities which belong to the major tasks of geomatics.

The previous chapter highlighted the general characteristics of EM energy and its behaviour while travelling through the atmosphere. The chapter also outlined the interaction of EM energy with objects on or near the surface of the earth. This chapter outlines and models the characteristics of a particular type of EM energy called laser. The term *laser* is an acronym for *l*ight *a*mplification of *s*timulated *e*mission of *r*adiation. To create a laser beam, a device is required that transfers electrical power into concentrated monochromatic EM energy.

This chapter preludes Chapters 5 and 6, which focus on airborne Light Detection and Ranging (Lidar) and laser scanners mounted on ground-based static or moving platforms, respectively. The first section of this chapter focuses on the basics of laser, including stimulated emission, the characteristics of laser beams and the safety aspects of using laser in scenes, where people are around. Next, Section 3.2 outlines three principles of exploiting laser beams for accurately measuring distances between the laser sensor and objects, also called ranges. Section 3.3 focuses on the effects of the heterogeneous composition of the atmosphere on range determination and path curvature. Section 3.4 briefly looks at the effects when a green laser beam leaves the atmosphere and meets a water body. Section 3.5 considers the interaction with object surfaces on which laser beams bounce; in particular, reflection characteristics are described and modelled. In Section 3.6, a mathematical modelling of beam divergence is provided. Section 3.7 concludes this chapter.

3.1 Basics of Laser

Basically, a laser sensor used for generating point clouds emits EM energy in the visible or near-infrared (NIR) part of the EM spectrum. The visible part commonly concerns the green band, which is able to penetrate water and therefore often used for bathymetric purposes, including the mapping of the sea, rivers and lakebeds. After emission of the EM energy, a receiver records the energy which an object in front of the scanner reflects in the direction of the scanner. The determination of the distance between sensor and object – range – is the main aim of the majority of laser sensors. From the range, the coordinates of points can be calculated when the position and orientation of the sensor, as well as the scan angle, are known in a (local) reference system.

3.1.1 Stimulated Emission

A laser emitter creates a narrow, intense beam of coherent and monochromatic light generated by particles (atoms or molecules) that emit radiation. The particles have only two energy levels, an upper state and a lower state, separated by some energy difference. Electrons in the lower state can be excited to the upper state, usually by heat. Upon returning to a lower state, the particle emits light. Usually, the particles behave independently of one another and the resulting light can be of any wavelength (colour). However, the emission of radiation can be stimulated to produce light of a desired wavelength. This occurs when a particle in the upper state interacts with a photon matching the energy separation of the levels. This particle may decay, resulting in the emission of another photon with the same phase and wavelength as the incident photon, leading to the emission of two photons. This process is known as stimulated emission.

The output beam consists of light of a single wavelength; the light is said to be monochromatic. The wavelength depends on the amount of energy released when the electron falls to a lower state. The beam is also coherent, meaning that the photons move in step with each other; they have the same phase. The divergence of the beam is small, which means the beam is very concentrated, narrow and cone-shaped. A flashlight, on the other hand, releases weak and diffuse light in many directions. Compared to other light sources (Siegman, 1986; Wilson and Hawkes, 1987; Koechner, 1992; Silfvast, 1996), laser beams have the following characteristics:

- High spectral purity
- High intensity, which means they contain a lot of energy
- Very monochromatic
- Small beam divergence; that is, the width of the beam and thus the footprint on the object is small

The monochromatic property, combined with small beam divergence and high energy level, enables laser beams to bridge relatively long distances through the atmosphere. Added to this, it enables reflectorless measurements which is a key prerequisite when gathering dense point clouds of any scene type, indoors or outdoors.

Albert Einstein recognised the existence of stimulated emission in 1917, but not until the 1950s were ways found to transfer the theory into practice (Bertolotti, 1983). In 1954, Charles Townes, a researcher at Bell Labs US and, independently, Basov and Prokhorov in Russia suggested a practical method of achieving laser using ammonia gas, which produced amplified microwave radiation instead of visible light (called a Maser). For their achievements, the scientists shared the 1964 Nobel Prize for Physics.

3.1.2 Laser Safety

A price to be paid for the beneficial properties discussed above is that laser beams can harm living tissue. As laser scanners often are used for capturing urban scenes, plants, dunes and other sites in which people are around safety is a key issue. Since the early 1970s, lasers have been subject to safety classification based on their potential for causing damage to the eye and skin. The degree of damage depends on the maximum output power, laser principle and wavelength. The initial classification system of six classes was slightly revised and extended to include seven classes in 2002. In the old system, the United States indicated classes by Roman numerals, sometimes appended with letters (I, II, IIA, IIIA, IIIB and IV), whereas the European Union used Arabic numerals (1–4), also appended by one letter; today the latter system is used worldwide in the revised version. The higher the numeral, the more injury the laser can cause. According to the European Norms (EN), maintained by CEN (European Committee for Standardization), the seven safety classes of lasers are, in ascending order from safe to hazardous, as follows:

- 1: Safe.
- 1M: Safe provided optical instruments are not used.
- 2: Visible lasers, safe for exposures taking no longer than a quarter of a second.
- 2M: Visible lasers. Safe for accidental exposure (<0.25 s) providing optical instruments are not used.
- 3R: Low risk of damage.
- 3B: Direct contact with the laser beam is hazardous; indirect contact through reflection on walls and other surfaces is safe.
- 4: Direct and indirect contact is hazardous; there is also a fire risk.

Most terrestrial laser systems are assigned to safety class 3R or lower.

3.2 Measurement Principles

Two main principles are in use for measuring longer ranges: emission of pulses and modulation of a carrier wave. For measuring close ranges in indoor scenes, triangulation-based measurement systems are often used. These systems are able to calculate directly 3D coordinates of points in a device-centred coordinate system by exploiting trigonometric triangulation principles.

After emission, a laser beam backscatters from the first non-transparent object it meets. The part of the backscatter which reaches the receiver is called return. In addition to measuring the distance from the sensor to the object, forth and back, the energy level of the return is usually captured. Measuring with laser sensors for geomatics applications may be based on three principles:

- Phase shift: laser beams are amplitude modulated, which means signals of varying wavelengths are superimposed on the laser carrier beam.
- Pulse measurements, also coined time-of-flight (ToF): pulses are emitted and their travel time forth and back from the object is measured.

- Triangulation-based aims at creating point clouds for short-range applications and small objects.

The first two laser principles – phase shift based on frequency modulation and pulse measurements – are commonly used in terrestrial laser scanners (TLSs) for collecting point clouds in outdoor scenes, while airborne Lidar systems are based on the pulse measurement principle. The latter principle is also called time-of-flight (ToF). Overall, phase-shift and pulse-based systems are called roundtrip systems because they measure ranges forth and back. The triangulation-based approach doesn't measure distances but angles. In the sequel, we will outline the principles of ToF laser sensors, the phase-shift principle and the triangulation-based principle.

3.2.1 Pulse (ToF)

The emitter and receiver are both parts of the same laser device. A laser scanner, based on the pulse measurement principle, emits single laser pulses at a high frequency, which are bounced off from an object. The roundtrip travel time of the laser pulse forth and back is measured. Since the path is followed twice, the travel time has to be divided by two. By multiplication with the speed of light, the range is calculated. ToF sensors are the only type of laser system which can bridge long distances and therefore are used in airborne Lidar systems, which operate at a flight height of hundreds of metres up to 5–6 km. The bridging of a distance of 50 m forth and back is done in just 333 ns (1 ns = 10^{-9} s). So ToF systems require very accurate sub-nanosecond time measurement.

The length of the pulse varies from 1 ns to 10 ns. A pulse of 1 ns has a length of 30 cm. The laser beams propagate through the atmosphere until they meet a surface from which they backscatter. The material the object is constructed from and the incident angle determine the type of reflection. In general, the more a surface behaves as a diffuse reflector, the higher the intensity of the signal reflected in the direction of the laser sensor (Figure 3.1).

The range (r) is defined as the distance between the sensor and the object surface hit by the pulse. So range measurement is based on the dependency of the path followed by the laser pulse and the time necessary to complete this path. Since air pressure, temperature and humidity of the atmosphere differ along the path also the velocity (v) of the pulse will vary within the time the pulse is propagating through the atmosphere. As a result, the formula to compute the range from the travel time reads as follows:

$$r = \frac{1}{2}\int_{t_0}^{t_1} v(t)\,dt \tag{3.1}$$

with $v(t)$ the velocity of the pulse at time t and t_0 and t_1 the time of emission and the time of receipt, respectively. The travel time is: $\Delta t = t_1 - t_0$. If the velocity of the pulse is independent of the position within the path, then $v(t)$ becomes v everywhere, with v a constant value which is close to the speed of light through a vacuum (Figure 3.2). Equation 3.1 reduces accordingly to:

$$r = v\frac{\Delta t}{2} \tag{3.2}$$

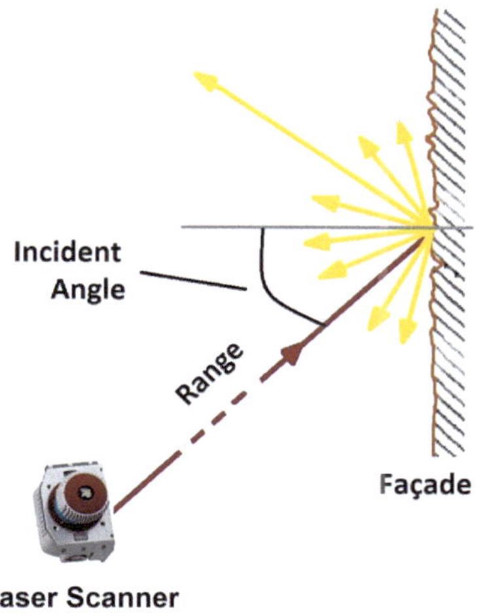

Figure 3.1 Within the same survey, objects made of identical materials and having identical surface roughness may deliver returns with different strengths, depending on the range and incident angle

Figure 3.2 Time-of-flight measurement; the instrument shown is a Riegl VZ-400 terrestrial laser scanner

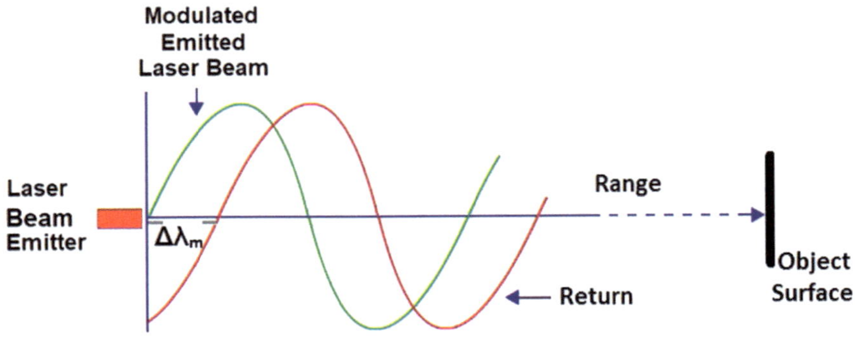

Figure 3.3 Diagram showing the phase-shift principle; the emitted laser beam is the carrier for another signal superimposed on it

3.2.2 Phase Shift

The EM energy may be emitted as laser beams which act as carriers for sine waves with a much larger wavelength. The generation process of such waves is called modulation resulting in laser beams of which the intensity is modulated as a sinusoidal signal with wavelength λ_m. The emitted, modulated laser beam reflects on the object, and a part of the reflected energy – return – is captured by the receiver of the laser system. Here, the phase of the returning modulated signal is compared to the phase of the emitted signal. This can be done with precision at the millimetre level. This means that laser phase-shift systems modulate the power of the laser beam, and the receiver compares the phase of the laser sent out and returned to the sensor.

In principle, the counting of the number of wavelengths gives a measure of the distance forth and back between the sensor and the object. It would, of course, be just by accident that the distance exactly equals the number of wavelengths. Rather, a portion of a wavelength should be added to the integer number of wavelength. This portion of a wavelength is called phase shift, which has given the name to this type of laser system. Indicating the phase shift related to the wavelength λ_m as $\Delta\lambda_m$, the travel time forth and back becomes:

$$\Delta t = \frac{1}{v}\left(N_m \lambda_m + \Delta\lambda_m\right) \tag{3.3}$$

where N_m is an integer, which is equal to the number of full wavelengths along the forth and back trajectory. Substituting Δt in Equation 3.3 into Equation 3.2 results in (Figure 3.3):

$$r = \tfrac{1}{2}\left(N_m \lambda_m + \Delta\lambda_m\right) \tag{3.4}$$

The issue is that the number of full wavelengths can't be counted. There are no means available to do so. To resolve this problem, waves with differing wavelengths are emitted in a systematic manner. This means that to resolve the ambiguous number of cycles, several waves whose wavelengths increase in a systematic manner have to be exploited (Figure 3.4).

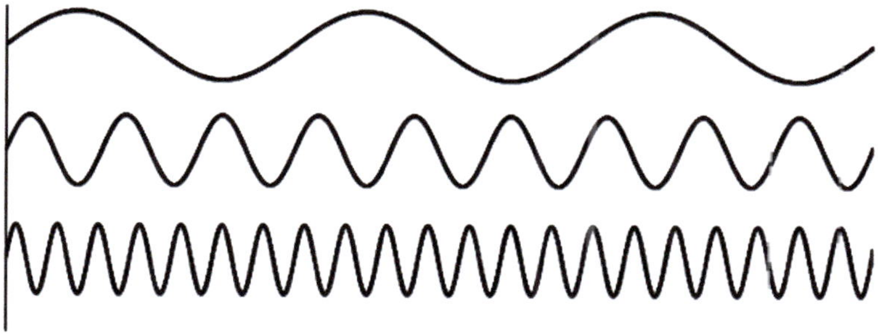

Figure 3.4 Phase-shift laser systems exploit multiple modulated waves with diverse wavelengths varying from small to large

The modulation of a laser beam enables the determination of ranges with high precision. This is because the phase shift ($\Delta\lambda_m$) can be determined with a resolution of 0.01% of the wavelength of the modulated signal. This means when we want to measure at the millimetre level, the smallest usable wavelength is 10 m (0.01% of 10,000 mm is 1 mm). When using only this wavelength of 10 m, objects up to a distance of 5 m can be measured (10 m/2). If we want to measure distances up to 50 m, two wavelengths are required and when the round trip distance is larger than 50 m but smaller than 1,000 m, three wavelengths are needed. Suppose that a roundtrip distance is 198.186 m, then the 10 m wavelength will give a phase shift of 8.186 m (there are 19 full wavelengths, which can't be counted, however). To determine the total number of wavelengths, a second signal with a wavelength of 100 m is necessary. In addition, the phase shift of this wavelength can be measured with a precision of 0.01%. Given the roundtrip distance of 198.186 m, the phase shift of this wavelength will be measured as 98.19 m (there is one full wavelength). The third signal with a wavelength of 1,000 m will provide a phase shift of 198.2 m, taking the measurement precision of 0.01% into account (there are zero full wavelengths). The roundtrip distance is computed by combining the three phase shifts as follows: the range is $198 + 0.1 + 0.086 = 198.186$ m. By dividing this number by 2, the range is found to be 99.093 m.

How do we know that the roundtrip distance is 198.186 m and not 298.186 m? If the latter would be the case, the phase shift ($\Delta\lambda_m$) of the third signal with a wavelength of 1,000 m would have been 298.2 m.

3.2.3 Triangulation-Based Laser

In addition to roundtrip laser systems, there are sensors exploring the trigonometric principles of triangulation. These instruments have typically been designed for scanning objects at close range usually under indoor conditions. These types of sensors are often used in industrial settings but increasingly also for 3D mapping of indoor scenes. The operating range of these short-range scanners is usually 5 m or less. So they are not suited for the 3D mapping of the outdoor environment. They are most suited for capturing small objects with complicated shapes. Laser triangulation is one of the simplest and cheapest scanning methods.

Triangulation-based systems combine the emission of laser beams with imaging the reflection. The advantages of these systems are that no accurate time measurements or phase measurements have to be carried out. Usually, the laser source is a semiconductor, which is of low cost and small. The wavelength of the emitted beam lies around 0.65 µm and thus is visible to the human eye as red. The position of (1) the mirror which distributes the laser beam in object space and (2) the projection centre of the camera are accurately calibrated. The angular position of the mirror with respect to the Z-axis (α) is measured (Figure 3.5). The incident angle of the reflected beam (β) with respect to the optical axis of the camera can be computed from the image coordinates of the spot on the charge-coupled device (CCD) or CMOS chip of the camera. This indicates that the point P on the surface of an object is imaged as point p in the digital image. Since the length of the base (B) of the triangle as well as the two angles (α and β) are known, the 3D coordinates of point P can be computed along with well-known trigonometric calculations:

$$X_p = Z_p \tan\alpha$$
$$Y_p = Z_p \tan\gamma \tag{3.5}$$
$$Z_p = \frac{B}{\tan\alpha + \tan\beta} = \frac{Bf}{\left(x_p - x_0\right) + f\tan\alpha}$$

where

$$\tan\beta = \left(x_p - x_0\right)/f \text{ and } \tan\gamma = \left(y_p - y_0\right)/f$$

In Equation 3.5, f is the focal length of the camera and (x_0, y_0) is the position in the image plane where the optical axis intersects with the image plane. Note that in the field of photogrammetry, the focal length is called camera constant and abbreviated to c. In this chapter, we abbreviate focal length to f in order to avoid confusion with the speed of light through a vacuum (c). Ideally, the optical axis would intersect the image plane exactly at its origin. However, it is virtually impossible to construct an optical-electronic device

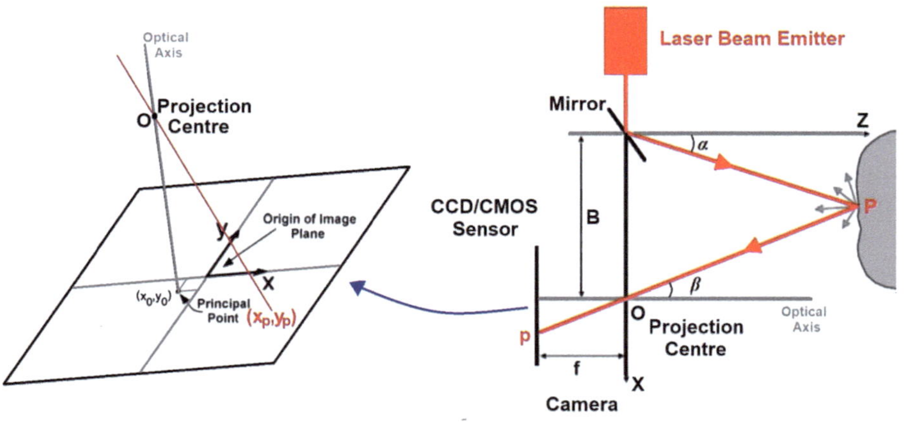

Figure 3.5 Sketch demonstrating the principles of triangulation-based laser measurements

such that the projection centre lies exactly perpendicular to the origin of the image plane. The triple (x_0, y_0, f) can be determined by calibration.

Compared to roundtrip systems, which aim at accurately determining ranges by measuring time or phase, triangulation has the following advantages:

- Simple concept and low price
- High acquisition speed
- High precision (up to 0.01 mm)
- Complex shapes of small objects can be captured in high detail

The major limitation of this method concerns the small distance between the sensor and objects. The range is limited to a maximum of 5 m.

3.2.4 Concluding Remarks

Phase-shift scanners can capture more than one million points per second while the range of precision lies in the order of a few millimetres. Phase-shift scanners produce higher point densities compared to ToF scanners. Pulse scanners can bridge longer distances than phase-shift scanners – therefore they are used in airborne Lidar systems – but the rate of laser beams emission is lower. Triangulation-based systems measure with high precision but have a low maximum range.

3.3 Range and Path

In order to create a point cloud, laser beams are fired into the atmosphere. The air particles present in the atmosphere affect the speed of the laser beam, which is lower than the speed of light through a vacuum. In addition, the path of the laser beam will deviate slightly from a straight line – the path will be curved. This section aims at modelling the impact of path curvature as well as the varying speed of laser beams due to the varying composition of the atmosphere.

After emission, laser beams travel through the atmosphere before they reach a surface and bounce back. To determine the range, it is necessary to know the speed of the EM beam through the atmosphere (see Equation 3.1). The speed of light through a medium depends on the presence of particles, particularly on the density and size of these particles. Since the atmosphere consists of particles of which nitrogen and oxygen are the major components, the speed of light (v) slightly deviates from the speed of light through a vacuum (c). The density of air at sea level is about 1.3 kg/m^3, which drops by half to 0.65 kg/m^3 at an altitude of 5.6 km. The different densities of air along the path of the laser beam also cause patch curvature: the laser beam doesn't propagate along a straight line but runs along a slightly curved path. So the varying density of the atmosphere, particularly with attitude, affects the speed of light as well as the path followed by the beam which may deviate from a pure straight line. The path length (range) and the deviations introduced by path curvature can be computed from measurable quantities, particularly air pressure, temperature and humidity, when introducing assumptions about the composition of the lowest part of the atmosphere, called troposphere. First, this section considers the modelling of the speed of light through the atmosphere and next models the planar displacement caused by the path curvature.

3.3.1 Range and Speed of Light

It is impractical to determine the various speeds along the path, in Equation 3.1 indicated as $v(t)$. The speed varies depending on air density: the closer to the surface of the earth, the denser the atmosphere. Around 50 km above the surface, the atmosphere becomes very thin and here the EM speed will be close to c, the speed of light through a vacuum. If no dynamics would be present in the atmosphere, the decrease of density would mainly depend on the height above the ground surface level. However, there is plenty of convection and turbulence and other air flow which affects the air density. Air flow in the lower atmosphere, called the troposphere, is mainly determined by the heating effects of the sun.

In principle, v can be determined by measuring the range between the point of emission and the point where the laser beam hits the terrain. The position of the aircraft in which the laser sensor is mounted and the point on the ground can be determined by Global Navigation Satellite System (GNSS) measurements. By identifying the point on the ground in the laser point cloud and using its travel time through the atmosphere, which has been measured, the average speed of the EM beam can be computed and used to calculate the ranges of other points. Another way to obtain the average speed is by modelling using the refractive index.

Refractive index or index of refraction (n) describes how EM beams propagate through a medium and is defined by the ratio of the velocity of light through a vacuum (c) and the velocity through the medium (v):

$$n = c/v \tag{3.6}$$

Obviously, n is a dimensionless number. The refractive index of water is 1.333, meaning that the speed of light in water slows down by 25% compared to a vacuum. For optical wavelengths, to which laser light used in Lidar systems belongs, the refractive index is determined by the following four quantities:

- Wavelength of the laser beam
- Air pressure
- Air temperature
- Humidity

If a mathematical relationship between the refractive index and the above four variables can be developed, the velocity of the EM beam through the atmosphere can be mathematically modelled. The wavelength of the laser beam is a defined characteristic of the laser device. Air pressure, temperature and humidity can be measured with proper instruments. The velocity can be computed from these measurable quantities when introducing assumptions about the composition of the troposphere. The principal assumption to arrive at a practical solution is that the change of v along the path is gradual and can be modelled in a linear way. Under this assumption, the change of refractive index can be described by measuring air pressure, temperature and humidity at the position where the laser beam is emitted, which means the aircraft, and at ground level. In course of time, many empirical equations have been developed to describe the relationship between refractive index and air pressure (P), temperature (T) and humidity (E). One of these relationships is given by the equation of Kohlrausch (Bolsakov et al., 1985; p. 64):

$$n_a - 1 = \frac{98.7 \cdot 10^{-5}(n-1)P - 4.1 \cdot 10^{-8} E}{1 + 36.61 \cdot 10^{-4} T} \tag{3.7}$$

where n_a is the actual refractive index at a certain location in the air, E the partial water pressure depending on the air humidity, P in hPa and T in °C. Here, n is the refractive index of the standard atmosphere. According to international agreements, the above-mentioned three parameters have been defined for a standard atmosphere at mean sea level as follows:

$$T_0 = 288.16°\text{K} \approx 15°\text{C}$$

$$P_0 = 1013.25\,\text{hPa}$$

$$E_0 = 0$$

The refractive indices per wavelength for a standard atmosphere have been tabulated (see e.g. Penndorf, 1957). The refractive index of a standard atmosphere for wavelengths around 1 µm – these are the wavelengths of laser beams – is $n \approx 1.00028$. Since the atmospheric refractive indices are only slightly larger than 1, i.e. in the order of tenths of permille, often refractivity (N) is used. If $n = 1.00028$, $N = (1.00028 - 1)10^6 = 280$.

The three atmospheric quantities – pressure, temperature and humidity – do not influence the refractive index to the same extent. Obviously, air pressure is the most important quantity. This is because the density of the atmosphere depends for a major part on pressure. On the other hand, the influence of air humidity is minimal. For temperatures lower than 30°C and a relative air humidity of less than 60%, the influence is less than 1 ppm, which means 1 mm/km (Kahmen and Faig, 1988). Omitting the influence of air humidity, Equation 3.7 reduces to:

$$n_a - 1 = \frac{98.7 \cdot 10^{-5}(n-1)P}{1 + 36.61 \cdot 10^{-4} T} \tag{3.8}$$

Since we assume that the change of v along the path can be modelled in a linear way, the computation of the mean of the velocities at the position where the laser beam leaves the sensor in the aircraft (v_1) and at ground level (v_2) suffices. Furthermore, exploiting Equation 3.6, the mean velocity along the path (v_m) becomes:

$$v_m = \frac{v_1 + v_2}{2} = \frac{1}{2}\left(\frac{c}{n_1} + \frac{c}{n_2}\right) = \frac{c(n_1 + n_2)}{2n_1 n_2} \tag{3.9}$$

where n_1 and n_2 are the refractive index at the position of the aircraft and at ground level, respectively. Both refractivity indices can be determined using Equation 3.8.

3.3.2 Path Curvature

Because of the composition of the atmosphere, the laser pulse emitted from an aircraft doesn't usually follow a straight line; the path is bent or curved. To arrive at a modelling of the path curvature, we assume that the composition of the atmosphere is layered. This is the same assumption as exploited to determine the speed of EM beams. This means that when going vertically up into the air the refraction index changes gradually while the refractive index gradient is constant. Under these assumptions, an EM beam follows a circle arch. The arch starts where the laser beam is emitted, which means the position of the aircraft at the moment the pulse leaves the emitter. The parameter which describes the bending of the arch is the radius of the circle. This radius is called the radius of curvature and depends on the refraction coefficient (n), the angle with the zenith under which the EM beam is fired and de refraction index gradient (dn/dH). The angle with the zenith corresponds approximately to the scan angle (φ), which is a measurable quantity. The relationship is as follows (Lemmens and Fortuin, 1997):

$$\rho = \frac{n}{\sin\varphi \dfrac{dn}{dH}} \tag{3.10}$$

Assuming that the refractivity varies gradually along the beam path from sensor to ground so that one may suffice with determining the refractive index at the position of the aircraft with flight height H and at ground level, n_1 and n_2, respectively ($\Delta n = n_1 - n_2$), the curvature ($1/\rho$) becomes:

$$\frac{1}{\rho} = \frac{(n_1 - n_2)\sin\varphi}{H} = \frac{\Delta n \sin\varphi}{H} \tag{3.11}$$

where we have neglected the term n since its number is close to 1.

From the geometric relationships shown in Figure 3.6, the following can be derived:

$$\begin{aligned}(\rho + \Delta\rho)^2 &= \rho^2 + r^2 \rightarrow \rho^2 + 2\rho \cdot \Delta\rho + \Delta\rho^2 \\ &= \rho^2 + r^2 \rightarrow 2\rho \cdot \Delta\rho + \Delta\rho^2 = r^2\end{aligned} \tag{3.12}$$

Compared to ρ, the value of $\Delta\rho$ will be small. So, without loss of accuracy, we may neglect the term $\Delta\rho^2$, resulting in:

$$\Delta\rho = r^2 \cdot \frac{1}{2\rho} \tag{3.13}$$

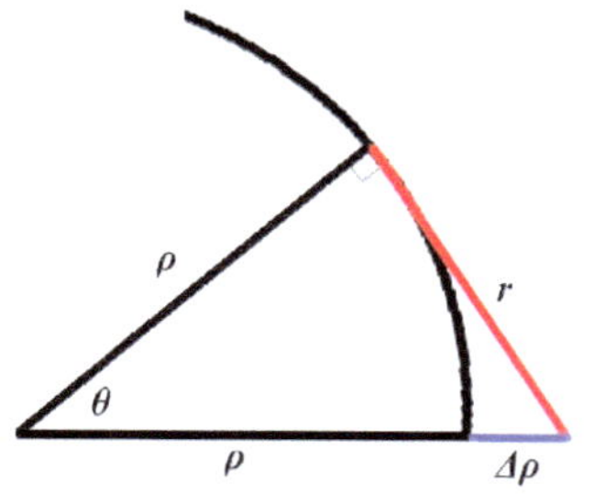

Figure 3.6 The planar displacement $\Delta\rho$, caused by path curvature, depends on the radius of curvature and distance travelled by the laser beam

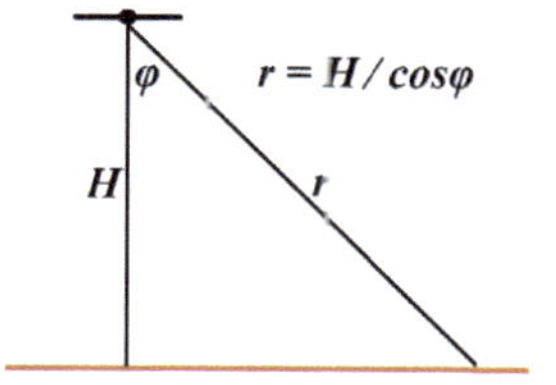

Figure 3.7 Relationship between flight height, range and scan angle

Substituting Equation 3.11 in Equation 3.13 and observing from Figure 3.7 that $r = H/\cos\varphi$ gives the following expression:

$$\Delta\rho = \frac{H}{2}\frac{\sin\varphi}{\cos^2\varphi}\Delta n \tag{3.14}$$

Table 3.1 shows an example of displacement of point location due to path curvature for three heights above ground level (500 m, 1,000 m and 2,000 m) for scan angles φ of 5°, 10°, 20°, 30° and 45°.

With Equation 3.14, the effect of the path curvature can be computed if the following quantities have been measured:

- Air pressure and air temperature at the position of the aircraft as well as on the ground from which Δn can be computed
- Scan angle (φ), which is a measurable quantity
- Flight altitude (H), which is part of the flight planning of the survey

Table 3.1 Displacement of point location due to path curvature

	500	**1,000**	**2,000**	***H* (m)**
φ/Δn	**−18**	**−35**	**−68**	Δn (10^6)
5	0	−0.2	−0.6	(cm)
10	−0.1	−0.3	−1.2	(cm)
20	−0.2	−0.7	−2.6	(cm)
30	−0.3	−1.2	−4.5	(cm)
45	−0.6	−2.5	−9.6	(cm)

Note: for notational convenience Δn is denoted as $(n_1 - n_2)10^6$

3.4 Angle of Refraction

The refractive index not only describes the ratio of the speed of light through a vacuum and through a medium containing particles but also how the direction of an EM beam deflects when the beam leaves a medium and enters another medium. As discussed in Section 3.3, the refractive index, also called the index of refraction, is a dimensionless number that compares the speed of light in a vacuum to the speed of light in a medium (v) as $n = c/v$. The beam deflection and the change of velocity are essential parameters for bathymetric Lidar. The refractive index of water (n_w) is 1.333, so travel time of EM beams is 25% slower in water than in a vacuum. The beam deflection can be computed from Snell's law when the beam leaves the atmosphere and enters the water (Figure 3.8):

$$n_a \sin\theta_1 = n_w \sin\theta_2 \tag{3.15}$$

where n_a, θ_1, and θ_2 are the refractive index of the atmosphere, the angle of incidence and the angle of refraction, respectively. As an example, we compute the angle of refraction for several angles of incident assuming that n_a is 1.00027, which is the value of the refractive index of green light for a standard atmosphere at mean sea level and n_w is 1.333 (Table 3.2). The velocity of EM beams through water becomes c/n_w. So bathymetric Lidar surveys are confronted with a chicken and egg problem, which is solved by including a height model of the water surface in the solution.

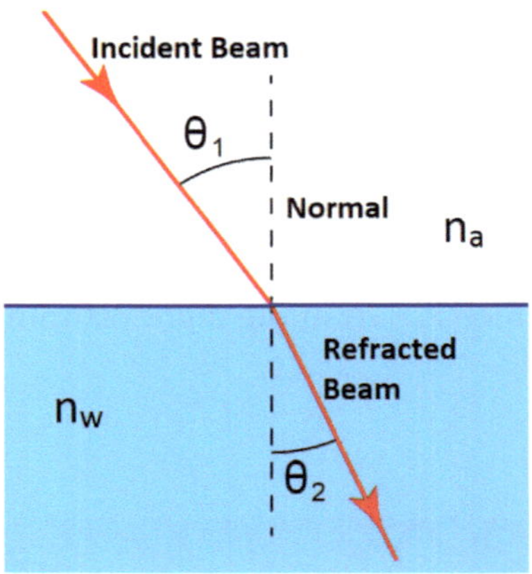

Figure 3.8 Illustration of Snell's law relating angle of incidence and angle of reflection at the transition zone between two media and their refractive indexes

Table 3.2 Angles of refraction (θ_2) for given angles of incident (θ_1) at the interface of atmosphere and water

θ_1	0	1	5	10	20	30	40	50	60	70	80	90
θ_2	0	0.75	3.75	7.49	14.87	22.04	28.84	35.09	40.53	44.84	47.65	48.62

3.5 Reflection

The return energy caught by the receiver part of the laser sensor depends on the reflection characteristics of the surface of the object, mathematically expressed as the reflection coefficient, that is the ratio between reflected energy and incident energy. The reflection coefficient of smooth and bright surfaces lies between 70% and 100% but reduces to 20% when surfaces become more rough and dark. A measurement is only usable if the return energy lies above a certain threshold. The measurement precision depends on distance between laser sensor and object, beam divergence, shape and reflectivity of the object and environmental conditions.

Sometimes, the reflectance values of Lidar systems are calibrated in advance using reference targets with known reflectance; such a procedure enables rigorous radiometric correction of forthcoming surveys (Kashani et al., 2015). Also the suitability of natural targets, such as beach sand, concrete, asphalt and gravel for airborne Lidar radiometric calibration, has been studied (Vain et al., 2009).

The second approach of correcting the received EM energy is by modelling without using reference values. Usually, this approach is exploited when it is hard, impossible or too expensive to collect reference values. The reflectance value is actually the total amount of EM energy (E_r) received by the receiver part of the laser sensor. The factors which influence the amount of received energy can be grouped into four categories: (1) sensor parameters, (2) data acquisition geometry, (3) atmospheric effects and (4) object surface properties. As a result, the total amount of EM energy received by the receiver of the sensor depends on a sequence of parameters, including:

- Energy of the emitted pulse (E_e)
- Atmospheric transmission factor (P_a)
- Area of the footprint of the pulse (A_f)
- Diffuse reflection characteristics of the patch covered by the footprint (B)
- Area of the detector opening (A_d)
- Sensitivity of the detector (s)
- Incident angle – that is, angle between laser beam direction and the surface normal φ

The relationship can be expressed in a mathematical equation as follows (Tradowsky, 1979):

$$E_r = E_e \cdot P_a^2 \cdot s \cdot B \frac{A_d}{A_f} \cdot \cos\varphi \tag{3.16}$$

This equation relates the received EM energy to the emitted EM energy, receiver parameters, acquisition geometry, atmospheric conditions and object properties. The atmospheric permeability or attenuation depends on the distance travelled through the atmosphere from the sensor to the object (r) and the absorption of emitted laser pulse:

$$P_a^2 = e^{-2\gamma r} \tag{3.17}$$

The factor γ is primarily determined by the amount of water present in the atmosphere. When conducting a survey with the same sensor and platform, a number of parameters in Equation 3.17 are the same for all points in the point cloud. These parameters are energy of the emitted pulse, area of the detector opening, sensitivity of the detector and γ. Furthermore, the footprint is determined by the beam divergence and distance between the sensor and the object. The beam divergence is also a constant value for the same system. Since it is the aim to obtain a normalised reflectance value, these parameters can be omitted. The surface normal can be computed from adjacent points. So the normalised reflectance value for each point R_n is a function F of the energy of the received pulse, range and incident angle. Written in a symbolic representation:

$$R_n = F\left(E_r, r, \varphi\right) \tag{3.18}$$

where E_r and r are measured while φ has to be computed.

Many functions and methods have been developed by researchers for normalising or correcting the received EM energy to reflectance values, for example, Höfle and Pfeifer (2007) and Yan and Shaker (2014). The current situation is complicated by a lack of standardisation, as well as confusing and inconsistent use of terminology (Kashani et al., 2015). If the reflectance value is used in the analysis of the point cloud acquired through mobile mapping or TLS, it may be beneficial by placing, in addition to markers and ground control points, reflectance reference panels in the scene.

3.6 Beam Divergence

A laser beam is not a ray with a diameter of zero and the laser spot hitting the surface is not a point in the mathematical sense but will have a footprint with a certain size; the greater the distance, the larger the footprint. The reason for this is the presence of particles in the atmosphere causing diffraction, resulting in divergence of the laser beam. As a result, EM beams expand when propagating away from the emitter. This phenomenon is called beam divergence. In this section, we have a look at how the EM energy distribution across the beam can be modelled and the mathematical description of beam divergence. A suited model for the drop-off of the EM energy across the beam is the Gaussian bell-shaped function. In addition to the modelling of a laser beam and its divergence as a Gaussian beam, this section will have a closer look at measurement errors introduced by beam divergence.

3.6.1 Gaussian Beam

At the exit of the emitter, the laser beam has a certain width. Rather than using a model in which the cross-section of the beam has no dimension, a better model is a capped, narrow cone, which means the diameter of cross-sections increases with increasing distance from the emitter. However, it appears that the energy distribution across a cross-section of the beam is not everywhere the same – the energy level does not drop suddenly to zero. In contrast, the energy level drops gradually from the centre of the beam to the edges. So a better model for describing the energy drop across the beam is the two-dimensional Gaussian function (Alda, 2003). When leaving the emitter, which means at a distance zero from the laser device, the beam has a certain width, called beam waist. While moving away from the collimator, the beam widens and gets the shape of a cone (Figure 3.9). The rate of the expansion of the beam width with distance from the laser source depends on the wavelength of the laser beam, the beam waist width and the distance from the source (Alda, 2003):

$$w(r) = \sqrt{w_0^2 + \frac{\lambda^2 \cdot r^2}{\pi^2 \cdot w_0^2}} = w_0 \sqrt{1 + \left(\frac{\lambda \cdot r}{\pi \cdot w_0^2}\right)^2} \tag{3.19}$$

It should be emphasised that the Gaussian beam is a model of how a laser beam diverges through space. This means that the Gaussian beam model describes an ideal world. However, reality never behaves ideally and never obeys a model superimposed on a part of the world for the full 100%. One may even argue that the shape of a laser beam propagating through the atmosphere resemblances more the shape of a biscuit tube than the Gaussian bell-curve (Jutzi and Stilla, 2006). This means that everywhere the energy distribution is circular and flat. The cut-off edges are abrupt: The energy level drops suddenly from maximum to zero. Nevertheless, in the majority of cases, the circular symmetric Gaussian 2D bell-shape model is adequate enough for explaining laser beam behaviour. In the Gaussian beam model, the beam diameter is usually defined as the circle (or ellipse), where the energy drops to a certain fraction of the total energy of the beam. Several drop-off radii have been proposed (Pfeifer and Briese, 2007). A generally accepted drop-off radius is where the energy ends up to $1/e^2$ of the maximum energy (Figure 3.10). Since $e = 2.7183$, $1/e^2$ defines the radius which encircles 86.5% of the total beam energy. Although the beam width is commonly defined as where energy drops to $1/e^2$ of the maximum energy; others may prefer slightly other

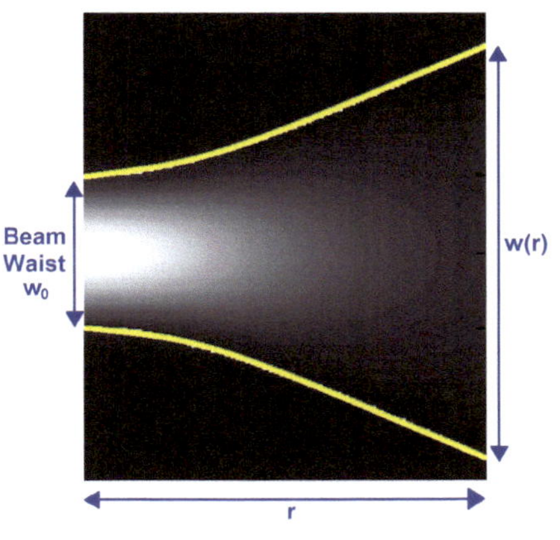

Figure 3.9 Gaussian beam

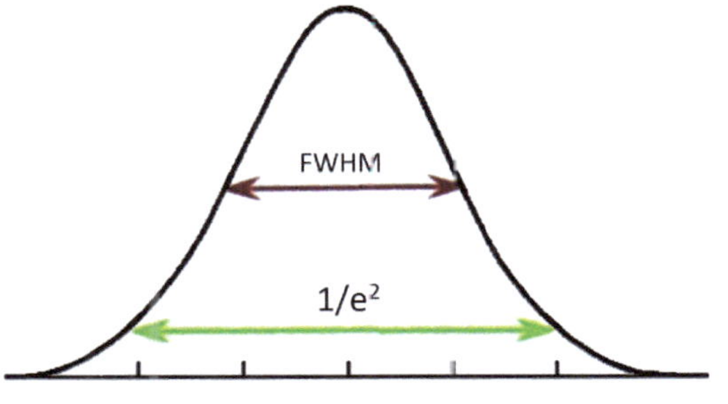

Figure 3.10 Several definitions of beam width are in use; outlined here are $1/e^2$ and Full Width at Half Maximum (FWHM)

definitions, including $1/e$ and full width at half maximum (FWHM), which corresponds to a radius which encircles 50% of the energy – that means the beam width corresponds to a radius of $1/e^{0.7}$.

The widening of the laser beam while moving away from the emitter can be described by an angular measure, called beam divergence. This means that the angular measure defines the increase in beam diameter as the beam moves away from the laser beam emitter. The footprint, also called spot size, is the area covered by the laser beam when it hits a surface. The diameter of the circular footprint, where the beam hits perpendicularly an object surface, depends on the range (r), the beam divergence and the pulse width at the exit of the laser beam emitter (Figure 3.11):

$$d(r) = w_o + 2r\tan(\theta/2) = w_o + r \cdot \theta \tag{3.20}$$

The tan operation in Equation 3.20 can be omitted (right part of Equation 3.20), since the beam divergence angles are small. As known from trigonometry, the tangent of a small angle is equal to the angle itself.

Beam divergence is sensor specific and usually calibrated by the manufacturer of the laser device. It is provided to the user in terms of beam width at the exit, beam divergence in milliradian (mrad) and spot size at a certain distance from the laser beam emitter, for example, 50 m. In Figure 3.11, the beam divergence is demonstrated by showing an Optech Polaris TLS, which has been introduced on the market in 2016. The manufacturer provides the following numbers about this instrument: The beam waist (beam diameter at exit) is 10 mm and the beam divergence is 0.35 mrad. The range measurements are done using pulses; the maximum range is 2 km and the minimum range is 15 m (source: geo-matching.com).

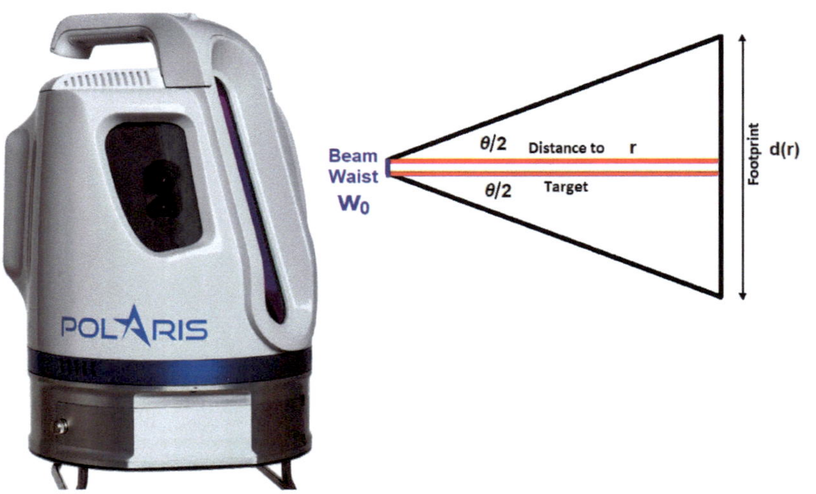

Figure 3.11 Diagram showing the geometric modelling of beam divergence; the laser scanner shown is an Optech Polaris TLS introduced on the market in 2016

3.6.2 Errors Due to Beam Divergence

Since the signal may reflect upon any surface present in the line of sight, blunders may easily occur. For example, the signal may be reflected on a leaf that whirls through the line of sight. Another concern is beam divergence, that is, the increase in beam diameter with distance from the source, discussed earlier. This may result in three types of errors (Lemmens, 2016a,b).

The first error type is introduced by the target surface. Differences in the surface characteristics of the target and its vicinity may cause the waveform of the backscattered part of the pulse to be a distorted version of the emitted one, which impedes the resolve of the exact time the return enters the instrument and thus affects the accuracy.

A second error type stems from a target that is located on a plane that is not perpendicular to the line of sight; the part of the plane that is closest to the instrument will reflect the beam first and the part farthest away last. The effect will be negligible when the reflectance characteristics of the plane are homogeneous: the reflected pulse will be elongated but will have everywhere the same signal strength enabling the resolve of the exact time the backscatter enters the instrument. When the reflectance changes within the footprint, the return will be distorted and it will be hard to resolve the exact arrival time.

Discontinuities in the vicinity of the target will induce time differences between the reflections of the same laser beam, introducing a third error type. If the depth differences do not exceed one-half of the pulse length, the reflected laser beam will be treated as stemming from one surface. Since the return will be elongated with fluctuating signal strengths, the ToF cannot be detected accurately.

Related to the above-mentioned effect is micro-relief, which plays a role in airborne Lidar. Micro-relief is the degree of height variability within the footprint of the laser beam. The reflected laser pulse will be smeared and change shape due to micro-relief. The change of shape depends on the spatial structure of height variation and can't be determined in advance (Gardner, 1992). However, it may be assumed that the effect on range measurement will be directly proportional to the micro-relief. A suited measure for the micro-relief is the height variance within the footprint: σ_h^2. The influence of the micro-relief can be expressed as a root mean square error ($RMSE_{mr}$):

$$RMSE_{mr} = \sigma_h / SNR \tag{3.21}$$

where SNR is the signal-to-noise ratio of the laser sensor system, which is a measure of the quality of the return. SNR is the ratio between the EM energy of the return stemming from the emitted beam and the EM energy due to noise. Further discourse on the effects of micro-relief and other terrain characteristics on the quality of digital elevation models is provided in Section 7.5.

3.7 Concluding Remarks

Laser has several beneficial properties as a means for accurately measuring ranges. These properties include that a laser beam is narrow and coherent. Furthermore, laser beams contain a lot of energy and consist of monochromatic light. The determination of ranges depends on accurate knowledge of the speed of EM beams through the atmosphere. The composition of the atmosphere causes EM beams to propagate slightly slower through the atmosphere than through a vacuum. The speed of EM beams varies mainly with attitude because of the layered heterogeneous composition of the atmosphere. Therefore, airborne Lidar will be affected most by variations in the speed of light. The ratio between the speed of EM beams propagating through a vacuum and a medium containing particles is called the refractive index (Equation 3.6). The heterogeneous composition not only affects speed but also causes a laser beam to not follow a straight line – its path is slightly curved. The path curvature can be modelled for attitudes up to a few kilometres above ground level – these are the attitudes laser scanners usually operate – the impact of path curvature is relatively small, particularly at small scan angles (Equation 3.14). Other effects of the atmosphere which impact the measuring capabilities of laser are beam divergence and the angle of refraction when an EM beam, consisting of green light, leaves the atmosphere and penetrates a water body. Added to this, NIR laser beams are absorbed by water bodies and the properties of the surfaces of objects, together with the incident angle, determine the strength of the return. Another behaviour of laser beams propagating through the atmosphere, which influences the quality of the return, is beam divergence. Although the product of a laser scanning survey is usually called a point cloud, these points are non-dimensionless. This is because the laser spot hitting an object's surface is not a point in the mathematical sense but will have a footprint with a certain extension; the larger the distance, the larger the footprint. Section 7.4 provides a further discussion on the effects of the "points aren't points" characteristic of laser points.

References

Alda, J. (2003) Laser and Gaussian Beam Propagation and Transformation. In *Encyclopedia of Optical Engineering*, pp. 999–1013, CRC Press, Boca Raton, Florida.

Bertolotti, M. (1983) *Masers and Lasers: An Historical Approach*. Adam Hilger, Bristol.

Bolsakov, V.D., Deumlich, F., Golubev, A.N., Vasilev, V.P. (1985) *Electronische Streckenmessung*. VEB Verlag für Bauwesen, Berlin, Verlag Nedra, Moscow.

Gardner, C.S. (1992) Ranging Performance of Satellite Laser Altimeters. *IEEE Transactions on Geoscience and Remote Sensing*, 30(5), pp. 1061–1072.

Höfle, B., Pfeifer, N. (2007) Correction of Laser Scanning Intensity Data: Data and Model-driven Approaches. *ISPRS Journal of Photogrammetry and Remote Sensing*, 62(6), pp. 415–433.

Jutzi, B., Stilla, U. (2006) Range Determination with Waveform Recording Laser Systems Using a Wiener Filter. *ISPRS Journal of Photogrammetry and Remote Sensing*, 61, pp. 95–107.

Kahmen, H., Faig, W. (1988) *Surveying*. De Gruyter, Berlin.

Kashani, A.G., Olsen, M.J., Parrish, Ch. E., Wilson, N. (2015) A Review of LIDAR Radiometric Processing: From Ad Hoc Intensity Correction to Rigorous Radiometric Calibration. *Sensors*, 15(11), pp. 28099–28128, doi: 10.3390/s151128099.

Koechner, W. (1992) *Solid-state Laser Engineering*. 3rd Edition, Springer, Berlin.

Lemmens, M. (2011) *Geo-information: Technologies, Applications and the Environment*. Springer, Science+Media Business.

Lemmens, M. (2016a) EDM without Reflectors. *The American Surveyor*, September 2016, pp. 1–3.

Lemmens, M. (2016b) Total Stations: The Surveyor's Workhorse. *GIM International*, 30(10), pp. 20–25.

Lemmens, M., Fortuin, E. (1997) *Fouten-Analyse van Vliegtuig-Laseraltimetrie*. Ministerie van Verkeer en Waterstaat, Rijkswaterstaat Meetkundige Dienst, Afdeling Geo-Advisering, Delft.

Penndorf, R. (1957) Tables of the Refractive Index for Standard Air and the Rayleigh Scattering Coefficient for the Spectral Region between 0.2 and 20.0 μ and Their Application to Atmospheric Optics. *Journal of the Optical Society of America*, 47(2), pp. 176–182.

Pfeifer, N., Briese, C. (2007) *Laser Scanning–Principles and Applications*. Vienna University of Technology Institute of Photogrammetry and Remote Sensing, Austria.

Siegman, A.E. (1986) *Lasers*. University Science Books, Mill Valley, CA.

Silfvast, W.T. (1996) *Laser Fundamentals*. Cambridge University Press, Cambridge, UK.

Yan, W.Y., Shaker, A. (2014) Radiometric Correction and Normalization of Airborne LIDAR Intensity Data for Improving Land-Cover Classification. *IEEE Transactions on Geoscience and Remote Sensing*, 52, pp. 7658–7673.

Tradowsky, K. (1979) *Laser: Grundlagen, Technik, Basisanwendungen*. Vogel, Würtzburg.

Vain, A., Kaasalainen, S., Pyysalo, U., Krooks, A., Litkey, P. (2009) Use of Naturally Available Reference Targets to Calibrate Airborne Laser Scanning Intensity Data. *Sensors*, 9(4), pp. 2780–2796, doi: 10.3390/s90402780.

Wilson, J., Hawkes, J.F.B. (1987) *Lasers: Principles and Applications. International Series in Optoelectronics*. Prentice Hall, Upper Saddle River, NJ.

Part II
Acquisition of Point Clouds

Photogrammetry and 3D Computer Vision

CHAPTER 4

This chapter aims at providing the basics and concepts of automatically extracting point clouds from overlapping images. The disciplines dealing with information extraction from images are photogrammetry, remote sensing and computer vision. Once the images, which are acquired without contact with the objects to be measured, have been captured with cameras mounted on a diversity of platforms such as aircraft, tripod or human hand, the processing and measuring are done in the office using software. Photogrammetry is a measurement technique mainly concerned with the geometric aspects of images: *where* are points, features and objects of interest, such as buildings and roads, located. Remote sensing, on the other hand, focuses on the extraction of thematic, also called semantic, information: This implies measuring *what* is present at certain locations. The importance of the *when* component, which means at what time the geodata has been captured, is steadily growing. This is because many geo-managers want to know how urban fringes, sea-level changes, dune erosion and many other earth-related phenomena evolve over time. The origins of photogrammetry and remote sensing are in topographic mapping and land cover mapping

Computer vision has a broader background and originates in the desire to simulate the human visual system using digital cameras and computers and to automate tasks that the human visual system does (Marr, 1979; Ullman, 1979; Marr and Poggio, 1979; Ballard and Brown, 1982; Shapiro and Stockman, 2001). Such tasks vary from inspection of industrial products, detection of tumours in X-ray images, to replacement of the human visual system in self-driving cars. Many achievements of computer vision have been applied in photogrammetry and remote sensing. There is an ongoing cross-pollination between these disciplines and computer vision. Compared to photogrammetry, remote sensing and computer vision are relatively young techniques, which emerged in the 1960s and 1970s, respectively.

Photogrammetry has been in existence since the early 1860s. The main endeavour during all of its long history has been the reduction of human involvement in the process of geometric information extraction by automating parts of the complex process to arrive from the planar (2D) coordinates of image points to 3D coordinates of salient object points in a common coordinate system, such as a national geodetic reference system. The complexity appeared so massive that photogrammetry evolved as an autonomous discipline with dedicated scientific and professional organisations. Today, the extraction of geometric information from images is done along fully digital paths. After capturing the images, photogrammetric products such as digital elevation models (DEMs) and digital landscape models can be generated along (semi-)automatic pipelines.

A brief but broad overview of how photogrammetry evolved from a camera held in the hands of a balloonist to its present stage as an advanced mapping technology is given in the *Manual of Photogrammetry*. Today, the mathematical relationships between image points and their corresponding object space points are well understood and documented in diverse excellent textbooks, including the *Manual of Photogrammetry* published by the American Association of Photogrammetry, of which the sixth edition was released in 2013 and which serves as the comprehensive reference for photogrammetrists and others working on the extraction of information from aerial images.

Textbooks by Kraus (2007), Mikhail et al. (2001), and Wolf and Dewitt (2000) provide a general introduction to the basics of (aerial) photogrammetry. The principles of digital photogrammetry are treated by Schenk (1999), while an introduction to digital photogrammetry from a practical point of view is provided by Linder (2006). The basic essentials of aerial digital photography, emphasising the survey aspects, are presented by Graham and Koh (2013). The principles of extracting information from images taken from ground-based platforms – close-range photogrammetry – are described by Atkinson (2003) and Luhmann et al. (2019). A volume specially dedicated to the calibration and orientation of cameras in computer vision has been written by various authors brought together by Gruen and Huang (2001). Hartley and Zisserman (2004) describe the geometric aspects of images from a computer vision perspective. Some of the above-mentioned books acted as a source for writing the present chapter, together with our own lecture notes and publications.

We will describe the basics from the viewpoint of aerial photogrammetry. The main reason for doing so is the growing importance of acquiring images from cameras mounted on unmanned airborne systems (UASs). However, the basics are equally valid for images captured from other platforms. Figure 4.1 shows the creation of overlaps during an aerial photogrammetric survey. A special characteristic of aerial images is that the image planes are intentionally captured approximately parallel to each other. An additional aim of this chapter is to provide the basics of correlation, which is a statistical method, and least-squares adjustment (LSA), which is an optimisation method, in the context of their applications in image matching, that is, the automatic search for corresponding points in the overlap of images. Points in overlapping images are called corresponding or conjugate if they are generated by the same point in the object space.

Figure 4.1 Creation of image overlaps during an aerial photogrammetric survey

4.1 Basics of Photogrammetry

The key sensor in photogrammetric data acquisition is the image camera. The camera types may vary from highly specialised, high-definition metric cameras, developed as workhorses for topographic mapping of large parts of the territory of countries, to consumer-grade cameras mounted on a UAS or handheld. The main difference between metric and non-metric cameras is the stability over time of their interior orientation parameters, their imaging capacity and their image quality.

4.1.1 Central versus Orthogonal Projection

The basic task of photogrammetry is to transfer 2D image coordinates to 3D coordinates in an object space–based coordinate system, in the sequel to be called ground-based reference system or simply reference system. The fundamental assumption is that each and every point in the image uniquely corresponds to one point in the object space, which is valid since the geometry of a camera can be modelled as a central projection. Although an aerial image showing a city resembles a map of that city, the differences between these two planar representations are significant – with respect to not only the content but also the geometric projection. The projection of a map is orthogonal – between the mapped area and the map, there is only a scale difference (Figure 4.2). Knowing the scale allows to measure distances in a map fairly accurately; the calculation of the distance in the object space just requires multiplication with the scale factor. Obtaining sound geometric information from images is much more complicated.

The central projection established by the lens system ensures that the rays of sunlight reflected by each point in the scene pass the camera exactly through one point, called the projection centre. Next, the ray of light meets the image plane and leaves its mark there. The intensity of the ray, which depends on the reflection properties of the object and the scattering and absorption properties of the atmosphere, determines whether the

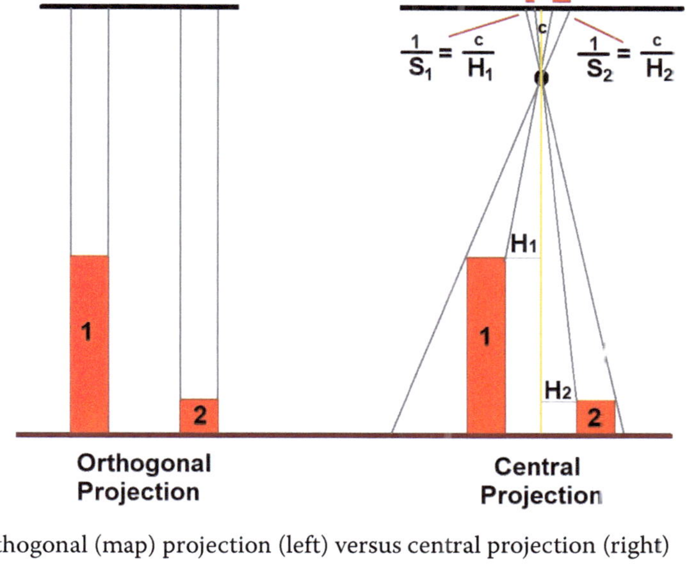

Figure 4.2 Orthogonal (map) projection (left) versus central projection (right)

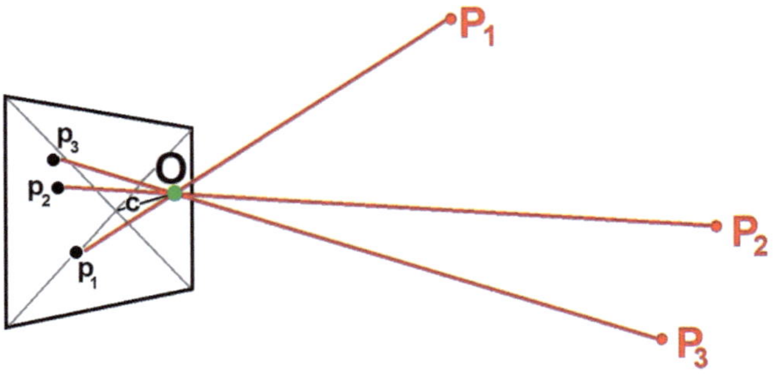

Figure 4.3 Collinearity of central projection: the object point, image point and projection centre lie on one straight line

image spot is dark, bright or somewhere in between. The basic notion on which all the mathematics in photogrammetry is founded is that the object point, projection centre and image point constitute one straight line; they are said to be *collinear* (Figure 4.3).

4.1.2 Scale

In the geometric modelling of a camera, or in other words defining the interior orientation parameters, one of the most important parameters is the focal length (f) – in photogrammetry, this is usually called camera constant (c), and we will adopt this term from now on. The camera constant determines the scale of the image and field of view (FoV). The *nominal* scale is defined by 1: S, where S is the scale factor (Figure 4.4):

$$1/S = d/D = c/H \text{ or } S = H/c \tag{4.1}$$

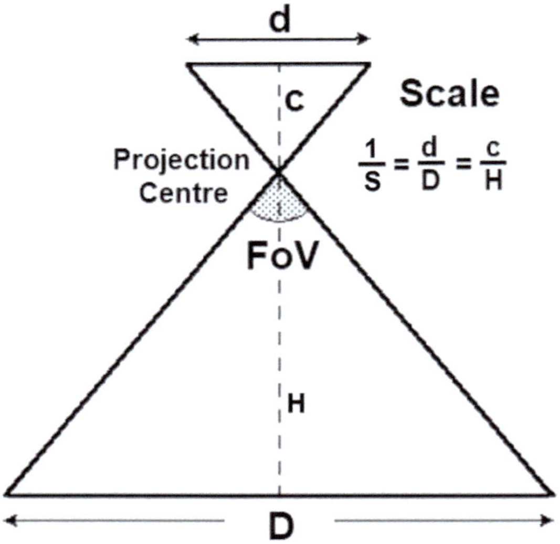

Figure 4.4 Scale is the ratio between the camera constant (focal length) and the distance of the camera to the object

Or, in words, the nominal scale factor is equal to the flight height divided by the camera constant. The flight height is measured with respect to the ground surface or, more generally, the distance of the camera to the object. If the camera constant is 21 cm and the average flight height is 840 m, then the nominal scale is 1:4,000, which means 1 cm in the image is 40 m in the object space.

In very rare cases, the scale will be the same throughout the entire image. This will be when (1) the terrain is flat, which means no relief or height differences are present and (2) the image plane is exactly parallel to the terrain. This will hardly ever occur in practical circumstances. The dynamics of a moving platform through the air cause turbulence and vibrations. As a result, the image plane will be slightly tilted with respect to the terrain surface. To give an image the scale characteristics same as those of a map, which means the scale is the same everywhere, the image has to be transformed from a central projection to an orthogonal projection. So, compared to maps, which do have an orthogonal projection, raw images show scale differences, and these are caused by the following:

1. Tilt of the camera with respect to the vertical
2. Height differences of buildings and other objects
3. Fluctuations of the bare ground

The tilt of the camera with respect to the vertical, also called the nadir direction, results in oblique images. The area nearest to the flight track will have the largest scale. The scale gradually decreases with distance to the flight track (Figure 4.5). If the horizon is visible, the image is called high oblique; otherwise, the image is said to be low oblique. If the photogrammetric survey aims at vertical images, oblique images deviate from the

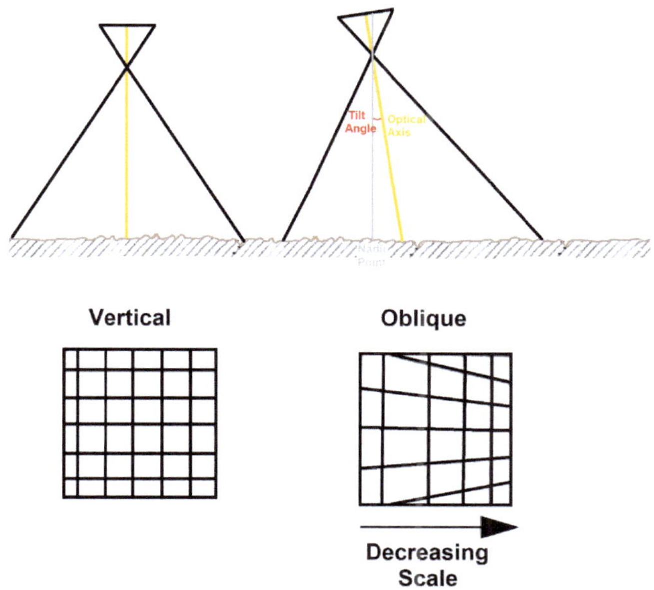

Figure 4.5 The scale difference in an oblique image

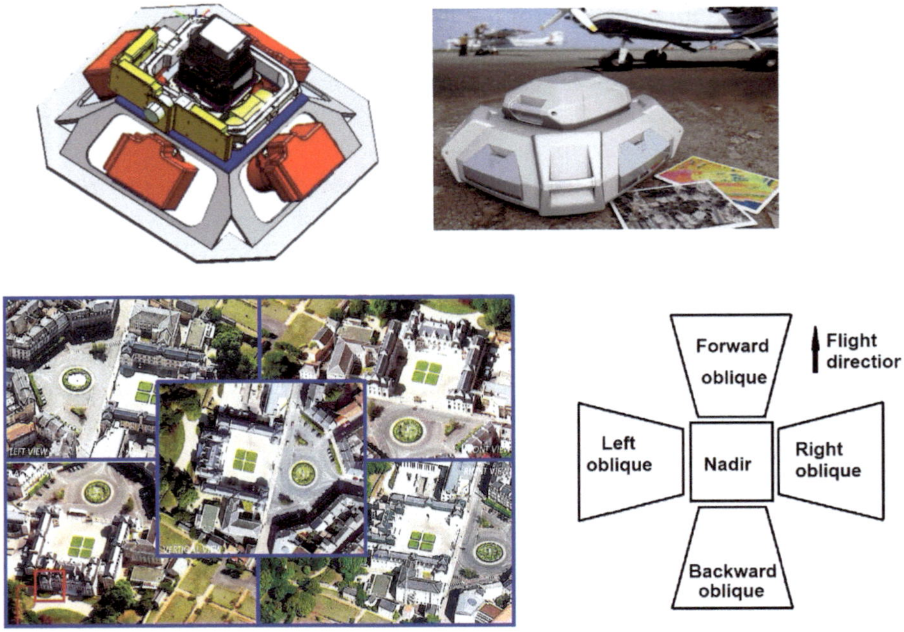

Figure 4.6 An example of an oblique camera system, the IDM 1000, which consists of a nadir camera system extended with four oblique cameras configured in a Maltese cross configuration (top) and a composite of the five Maltese cross images recorded by the Dimac oblique camera system (source: Lemmens, 2014a,b)

intended imaging configuration. Capturing of oblique images while aiming at vertical images will often occur because of dynamic forces such as air turbulence. But oblique images may also be captured willingly using multiple lens cameras, because they enable easier analysis which is beneficial for many non-topographic mapping purposes. Oblique camera systems consist of two or more lenses mounted in the same camera body or two or more cameras mounted in an array so as to keep the angles between their optical axes fixed. Figure 4.6 shows an example of an oblique camera system and a composite of the resulting images taken at the same instant.

4.1.3 Lean-Over and Occlusion

The camera constant also determines the FoV. The smaller the camera constant, the larger the FoV (Figure 4.7) and thus the larger the area captured by the camera. Therefore, these cameras are called wide-area cameras. This camera type enables the capture of large areas and requires fewer flight lines and images than cameras with larger FoV. This reduces fuel consumption and airtime and thus saves surveying costs. Because of the limited number of images needed for capturing an area, processing time is also limited. However, buildings and other objects extending in the vertical direction will introduce excessive lean-over and occlusion (Figure 4.8), which is a general characteristic of every camera because of the central projection. Due to the central projection, building facades are pronouncedly visible in wide-area imagery – this effect increases from the centre of

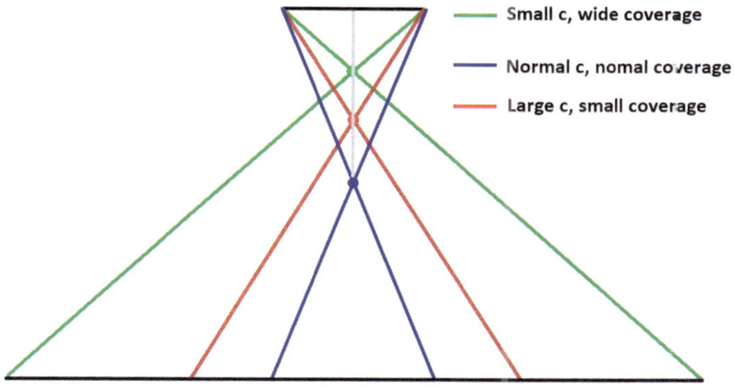

Figure 4.7 Relationship between camera constant and field of view (FoV)

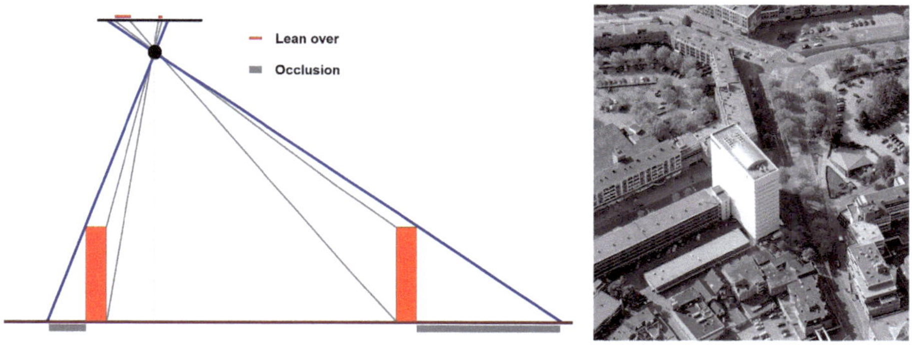

Figure 4.8 Vertical height differences cause lean-over and occlusion (left) demonstrated by an oblique pictometry image (right) (source: Lemmens et al., 2007)

the image to the borders. The opposite facade side of the building will be invisible and occlude objects which are located close to the facade. Consequently, wide-area cameras are suited for recording large areas which are relatively flat. For urban areas, cameras with a small FoV, and thus a large camera constant, should be preferred.

4.1.4 Orthoimages

Rectified images and images corrected for relief displacement are called orthoimages. The geometric properties of ortho-rectified images approximately resemble those of a map. So an orthoimage is an image on which geometric transformations have been conducted to convert it from perspective projection to orthogonal projection by performing corrections for tilt and relief displacement. Tilt correction is done using ground control points (GCPs). Correction for relief displacement requires a DEM which can be automatically extracted from overlapping images by image-matching techniques. Draping an orthoimage over a DEM enables extracting 3D coordinates of points from a digital landscape model and adding red, green and blue (RGB) values to these points. Usually, this type of photogrammetric 3D mapping is carried out by a human operator. The in-office surveyor can easily detect objects and move around and jump over houses,

Figure 4.9 Conventional orthoimage (top) and true orthoimage in which the effects of lean-over and occlusion have been largely reduced of a site in Utrecht, the Netherlands, GSD 5 cm (courtesy: Slagboom en Peeters, Teuge, the Netherlands)

trees, hedges and other obstructions; therefore, some call this type of photogrammetric mapping *virtual surveying* (Op't Eyndt and Volkmann, 2013). Since lean-over and occlusion increase with increasing distance from the centre of the image, they will never be absent. Using high along-track (e.g. 80%) and across-track (e.g. 60%) overlaps enables exploiting the centre part of images, which significantly reduces lean-over and occlusion (Figure 4.9). Since the resulting images show a near-real orthogonal projection, they are called true orthoimages. The price to pay is a strong increase in the number of images to be captured during the aerial survey. In addition, the corrections for relief displacements require a dense DEM.

4.2 Camera Coordinates

The previous section discussed the camera constant as the most important parameter for geometric camera modelling. However, more parameters are necessary to describe the geometry of a camera. The computation of object space coordinates from points identified in the image requires the definition of a coordinate system related to the camera. The planar image coordinates (x,y) are defined by introducing two orthogonal axes, which usually cross at the centre of the image. The beginning and end of the axes may be identified by fiducial marks on the film or by the row and column number of a digital image. The third dimension is defined by the optical axis, of which the camera constant is the key parameter. This means that all image coordinates have the same z-coordinate, indicated by c. Ideally, the optical axis would cross the image plane at its origin. Due to

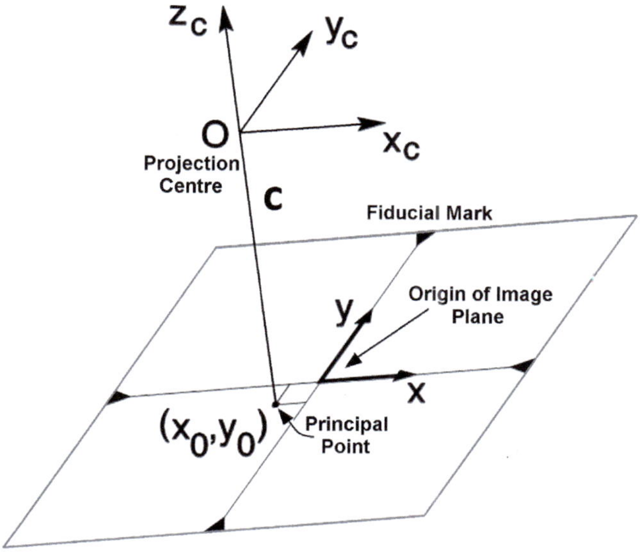

Figure 4.10 Schematic overview of the elements of the interior orientation of a camera; X_c,Y_c,Z_c indicate the axes of the camera coordinate system

construction limitations, a slight shift will be present in the order of a few to hundreds of microns depending on whether the camera is metric or non-metric (Figure 4.10). The point where the optical axis crosses the image plane is called the principal point. The coordinates of the principal point (x_0,y_0) with respect to the origin of the image plane and the camera constant (c) are the three basic elements of the interior orientation.

A camera is an optical-mechanical construction. As a result, the actual path of light rays through the camera will slightly deviate from a central projection. To turn a camera into a high-precision measurement device, these deviations have to be modelled and described mathematically. Sources which introduce major deviations are radial lens distortions and image plane unflatness, which means slight deviations of the image plane from a perfectly flat plane. Fraser (1997) provides a further survey of the deviations of the camera model in the framework of self-calibration. As an example of the modelling of distortions, we give here the formula of radial lens distortion (Δr), modelled as an odd power polynomial, based on the lens design theory (*Manual of Photogrammetry*, 2013, p. 342):

$$\Delta r = K_0 r + K_1 r^3 + K_2 r^5 + K_3 r^7 \tag{4.2}$$

where K_i, $i = 0, \ldots, 3$ are the radial distortion coefficients and r is the radial distance from the principal point.

The parameters describing the geometry of the camera are called interior orientation parameters. High-precision applications require accurate calibration of the coefficients of these parameters. The conventional calibration method is done under well-conditioned circumstances in a laboratory. Temperature and humidity are standardised, and an accurate measurement frame is imaged (Figure 4.11). To obtain coefficients under

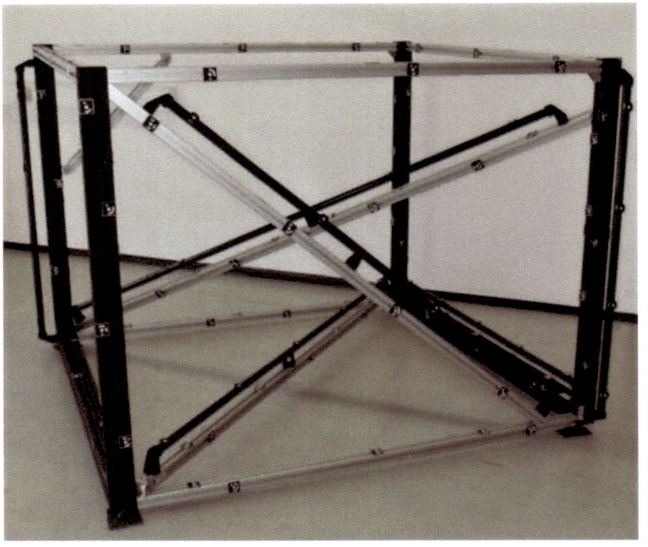

Figure 4.11 High-accuracy calibration frame in a conditioned laboratory environment for determining the interior orientation parameters of cameras for high-precision measurements (courtesy: AICON 3D Systems, Brunswick, Germany, which is part of Hexagon Manufacturing Intelligence since 2016)

image capture conditions in aerial photogrammetry, test fields have been established in the terrain containing known points. After take-off, the aircraft is flown over the test field, and after the aerial survey, the test field is imaged again to improve the precision of the calibration. It is also possible to carry out a calibration of the camera without using known geometry in the object space, called self-calibration, which is extensively exploited when using a UAS as the camera carrier (see Section 4.3.5).

4.3 From Camera Coordinates to 3D Coordinates

Obtaining 3D object space coordinates from camera coordinates also requires information about the relationship between the camera and the ground-based coordinate system (Figure 4.12). This means that the coordinates of the projection centre at the instant of exposure have to be known in a reference system. Also, the orientation of the image plane (or, alternatively formulated, optical axis) with respect to the three axes of the reference system has to be known. The 3D position coordinates and the three orientation angles together are called the exterior orientation parameters. They are six in total. So, in addition to the parameters of the interior orientation, the position and orientation of the camera in space have to be known.

4.3.1 Collinearity Equation

We denote the three coordinates of the position of the projection centre by (X_0,Y_0,Z_0). Three angular parameters (ω, φ, and κ; omega, phi, kappa) describe the direction of the

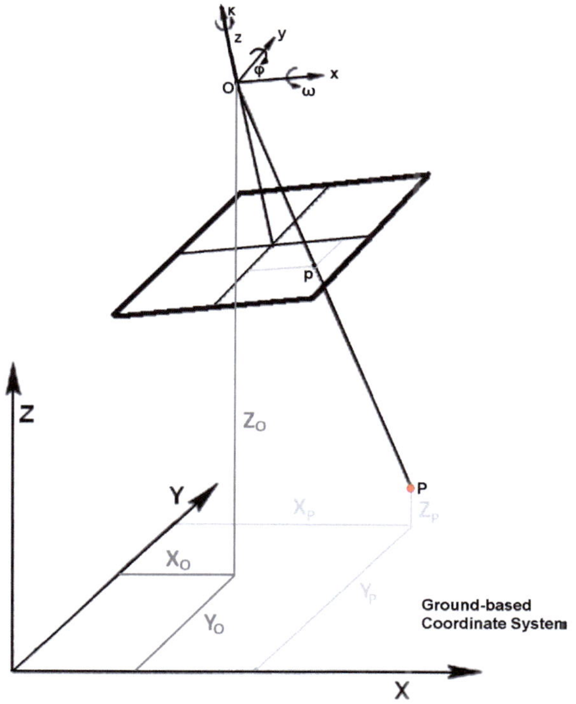

Figure 4.12 Tying the camera coordinate system with the ground-based coordinate system through the six parameters of exterior orientation: three coordinates of the projection centre (O) and ω, φ and κ

optical axis and thus the orientation of the image plane with respect to the three axes of the reference system. The direction of the optical axis is described by sine and cosine functions of ω, φ, and κ; in total, there are nine such functions, three for each axis of the reference frame. The rotation, often called camera transform, is given by a 3×3 matrix, containing the elements r_{ij}, $i = 1, \ldots, 3; j = 1, \ldots, 3$.

The relationship between the camera coordinates (x,y,c) and the coordinates (X,Y,Z) of the corresponding object space point is given by the collinearity equation, which can be derived from the projective geometry after many, but straightforward, steps:

$$
\begin{aligned}
x - x_0 &= -c\frac{r_{11}(X - X_0) + r_{21}(Y - Y_0) + r_{31}(Z - Z_0)}{r_{13}(X - X_0) + r_{23}(Y - Y_0) + r_{33}(Z - Z_0)} \\
y - y_0 &= -c\frac{r_{12}(X - X_0) + r_{22}(Y - Y_0) + r_{32}(Z - Z_0)}{r_{13}(X - X_0) + r_{23}(Y - Y_0) + r_{33}(Z - Z_0)}
\end{aligned}
\tag{4.3}
$$

4.3.2 Spatial Intersection

If the X,Y,Z-position of a point in the object space and the elements of interior and exterior orientation are known, the image coordinates can be calculated using the

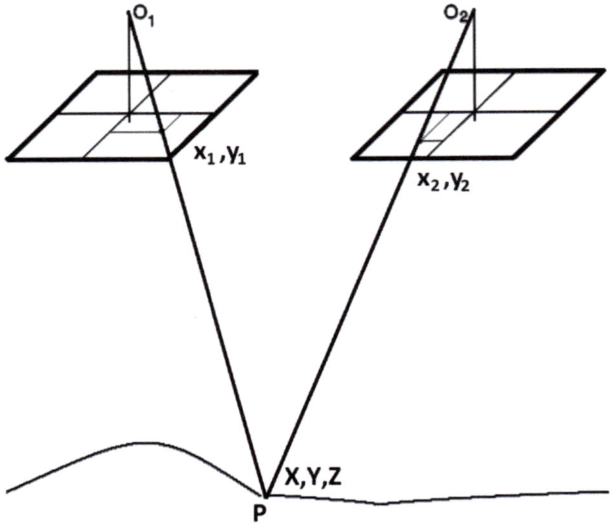

Figure 4.13 If the interior orientation parameters and the six parameters of exterior orientation are known, 3D coordinates of points in the object space can be computed from two overlapping images (stereopair)

collinearity equation (Equation 4.3). However, the other way around is not true. From measuring the image coordinates (x,y) of a point, it is impossible to calculate the X,Y,Z-position of the corresponding point in the object space. Algebraically, this can be understood by considering that the collinearity equation consists of two equations and it is impossible to solve three unknown parameters from two equations. Geometrically, this can be understood as follows. The known image coordinates and the projection centre determine the ray on which the object point lies. This provides a straight line in the space on which the object point may lie anywhere. Determining where the point exactly lies requires a second image which has captured the same part of the object space from another viewpoint. This second image also provides a straight line, and its spatial intersection with the straight line of the first image results in a unique solution (Figure 4.13).

4.3.3 Stereo Viewing

The two overlapping images covering the same part of the object space and taken from different viewpoints are called stereo images. The capability of the human visual system to fuse two overlapping images into a 3D representation is exploited for 3D mapping using digital photogrammetric workstations, often abbreviated to DPW. When the two overlapping images are presented to the human eyes – one image to the left eye, the other image to the right eye – the human visual system is able to combine the two images into one representation as if one is floating through a landscape. The complete separate presentation of the two images allows the brain to create a realistic 3D model of the scene, and this ability of the human visual system is extensively exploited for simultaneous measurement of the image coordinates of corresponding points in stereo images. This is demonstrated by the predecessor of the DPW, called the analytical plotter, shown in Figure 8.12 in Chapter 8.

4.3.4 Georeferencing

The determination of (X,Y,Z) coordinates – georeferencing – requires known coefficients of the parameters of interior and exterior orientation. The coefficients of the interior orientation parameters are determined by calibration, whereas those of exterior orientation can be determined directly, indirectly or by a combination of both. Direct measurement is done by using an onboard global navigation satellite system (GNSS) and inertial measurement unit (IMU) – GNSS provides the position coordinates of the projection centre of the camera at the instant of exposure and IMU the orientation of the image plane. Indirect determination is done by using GCPs. Usually, the coefficients measured with GNSS and IMU are treated as approximate values. To increase precision, four or more GCPs, which are well distributed over the area captured by the images, are used. Whatever method is used, georeferencing starts with the relative orientation of the images with respect to each other, a process which is called registration or co-registration in computer vision. Relative orientation requires the determination of corresponding or conjugate points in the overlaps. These points, called tie points or key points, need to be well distributed over the overlaps to obtain high precision and reliability. In the pre-digital era, human operators selected and measured tie points – a time-consuming effort. Consequently, minimising the number of tie points, while preserving photogrammetric accuracy and stability, was a major challenge. The abundant research on the best distribution of tie points in the overlaps resulted in a standardised and well-accepted pattern where the tie point should be measured to obtain the highest precision and reliability – the so-called von Gruber positions (Figure 4.14). Today, tie points can be determined automatically using image-matching techniques. High precision is possible when using matching methods which trace corresponding points at the subpixel level. After relative orientation, the coordinates in the local system have to be transformed to

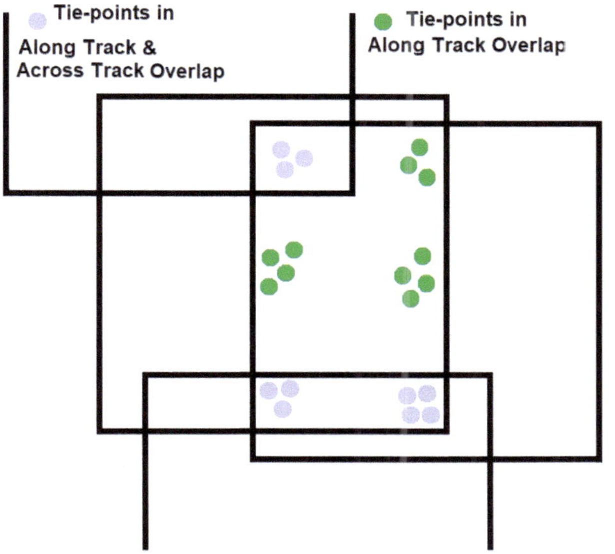

Figure 4.14 Von Gruber positions: optimal distribution of tie points in image overlaps for relative orientation (registration or co-registration)

a ground-based reference system, using a 3D similarity transformation as a geometric transformation model. This georeferencing step, called absolute orientation in photogrammetry, requires GCPs to compute the seven parameters of the transformation.

It is possible to carry out the two-step procedure of relative orientation and absolute orientation in one single step by incorporating the image coordinates of tie points, the interior orientation parameters, the approximate values of exterior orientation parameters and GCPs in one rigorous LSA process. This approach, which results in coefficients of the six exterior orientation parameters for each image in a ground-based reference system, avoids that all the *X,Y,Z* coordinates of the measured points have to go through a 3D similarity transformation. As said before, today, the approximate values of the exterior orientation parameters are obtained from onboard GNSS and IMU measurements.

4.3.5 UAS Photogrammetry

Although pilotless aircraft steered by radio signals broadcasted from the ground started their development during World War I, it was nearly a century before unmanned airborne vehicles (UAVs) became a preferred platform for photogrammetric surveys for multiple applications (see Chapter 13). Usually, lightweight copters and fixed wings – some with vertical take-off and landing (VTOL) abilities – are deployed because of costs and licence restrictions. Figure 4.15 shows a picture of a copter during an aerial survey. The maximum payload of these tiny platforms is usually a few kilogrammes. Metric cameras and their peripheral devices, which are used for mapping extensive areas, are bulky and heavy. Lightweight means relatively small size and a maximum payload of a few kilogrammes, which limits the payload to non-metric and often consumer-grade cameras, small GNSS antennas and IMUs based on microelectromechanical systems (MEMS) technology. The elements of the interior orientation of non-metric cameras are not stable. This means the coefficients of the interior orientation parameters may change during flight. Thus, high-precision surveys with UASs require calibration procedures other than those prevailing in manned aerial surveys. The calibration method should be able to estimate the interior orientation of the camera separately for each image captured during the survey. Would it be possible to compute the coefficients of

Figure 4.15 A UAS equipped with a camera in operation

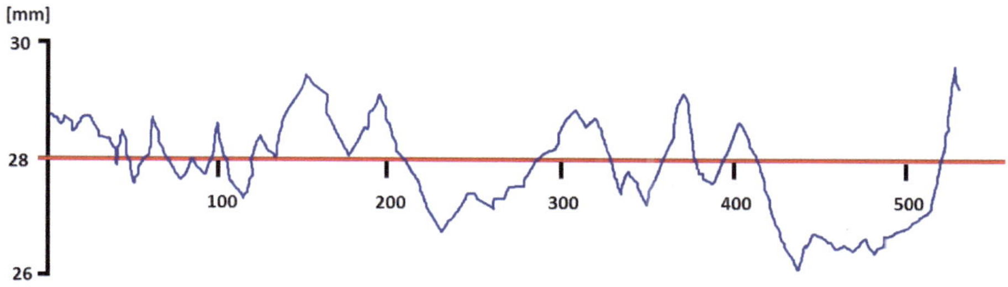

Figure 4.16 Variation of the camera constant of a non-metric camera with a nominal focal length of 28 mm during a UAS photogrammetric survey (graph based on research conducted by Rosnell and Honkavaara, 2012)

the interior orientation parameters using a mathematical extension of the collinearity equation (Equation 4.3)? This would enable calculating the camera constant and principal point and also the coefficients of lens distortions and other deviations from the central projection model without calibration under laboratory conditions or using a test field in the terrain. The many research efforts to find an answer to this question revealed that it is possible, providing that there is sufficient overlap between the images. The direct computation of the interior orientation elements allows the gathering of precise coordinates from non-metric cameras and even consumer-grade cameras. In addition to many developments in electronics and lightweight materials, self-calibration and its implementation in software have given a boost to UAS photogrammetry. Various research efforts on the self-calibration of non-metric cameras mounted on UASs have shown that the interior orientation coefficients may considerably vary during one flight (Figure 4.16).

4.4 Introduction to Image Matching

Storing the content of an image in the computer in the form of pixels enables the automatic detection of corresponding points in the overlap of images and the computation of the 3D object space coordinates from spatial intersection. Image matching aims at tracing corresponding points automatically. Tracing of corresponding points is called image matching in photogrammetry and structure from motion (SfM) in computer vision. Image matching has two main aims:

1. Detection of tie points in the relative orientation step, which means the (co-)registration of images
2. Generation of a dense point cloud, for example, a digital surface model (DSM) or digital terrain model (DTM)

Although the categorisation of image-matching techniques into two groups, area-based and feature-based, introduced by Lemmens (1988), is still valid today, the matching algorithms implemented in today's commercial and open-source software are often quite complex. Solutions may use a sequence of different matching techniques to obtain robust and accurate matches.

Due to the illumination and reflectance properties of surfaces in the object space, some points will result in more reliable matches than other points. In conventional image-matching approaches, the first step is to find image patches with sufficient contrast – distinctive points – because these are better suited for matching than other patches. Area-based matching detects distinctive points in one image and predicts the position in the other image using the information on the perspective differences and the terrain relief. Next, an image patch, usually a square window, is defined around the distinctive point, called here target patch, and the intensity values are compared within a search patch in the other image.

Feature-based matching traces distinctive points in the overlap of both images, assigns features to the distinctive points and conducts the matching process based on the similarity between the features.

Area-based matching allows the determination of corresponding points at the subpixel level and is particularly applied for finding tie points in overlapping images for determining the exterior orientation parameters. Usually, area-based matching is subdivided into two main groups:

1. Correlation-based methods
2. Least-squares matching

Actually, once corresponding points have been detected at pixel-level precision, least-squares matching acts as a refinement method to determine the position of the points with high-subpixel accuracy.

The intensive research on image-matching techniques, both by photogrammetric researchers and by computer vision researchers, since the early 1980s, has resulted in a diversity of solutions. The different approaches and their refinements are a result of the many appearances of stereo images, object spaces and intended tasks. Notably are the following:

- Required accuracy – one of the main consideration criteria in photogrammetry
- Point density – should be high and homogenously distributed for DEMs
- Tie point detection in overlapping images for registration purposes – requires high robustness, reliability and precision
- Image content – presence of sufficient contrast, texture and distinctive points such as corners of streets
- Illumination differences between overlapping images
- Perspective differences
- Computational and storage demands

In many applications, more than two images may have overlaps. This is, for example, the case for UAS photogrammetry where the along-track overlap may be 80% or even 90% and the across-track overlap 60% to 70%. This has resulted in the approach of multi-image matching (Hartley and Zisserman, 2004. To improve the robustness of the match, a multi-resolution approach may be adopted. In this approach, an image pyramid is created with the original, full-size image at its base and images generated by uniting

2×2 pixels at subsequent higher levels. This may be repeated until a feasible series of images have been created covering the same part of the object space but with ground sample distances (GSDs), which become larger in a systematic way. In the treatment of the diverse methods in the following sections, we focus on the matching of stereopairs at the same level of spatial resolution. We commence with matching methods which compare the intensity values of two image patches on similarity. We call these methods correlation-based image matching. We continue with least-squares matching. This section is supported by a side bar introducing non-linear LSA, which builds upon the linear least-squares solution presented in Chapter 11. Next, we step over to feature-based image matching, which extensively exploits features derived from small image patches, as described in Chapter 11. Finally, we consider a matching technique called semi-global matching (SGM), developed around 2005 for dense image matching (DIM) purposes and proven to be very successful in a broad range of applications for creating dense point clouds.

4.5 Correlation-Based Image Matching (CBM)

In the area-based method exploiting correlation, correspondence is sought using the grey values of pixels in a regularly shaped patch. A square target patch with a size of, for example, 5×5 pixels is defined in one image and shifted over a search patch in the other image. The size of the search patch depends on how well the approximate location of the conjugate point is known, and this, in turn, depends on the accuracy of the relative orientation and the height differences. This process of shifting the target patch pixel by pixel over the search patch is schematically outlined in Figure 4.17. For each position, a similarity measure is computed, resulting in a connected set of similarity measurements stored in a similarity map or matrix. If the target patch covers 5×5 pixels and the search patch 14×13 pixels, the size of the similarity matrix will be 10×9. The highest similarity value determines the corresponding point, and the centre pixels are selected as corresponding points. Acceptance of the match depends on whether the similarity value exceeds a predefined threshold; the value can be selected in a heuristic way or, when using normalised cross-correlation, on a statistically sound basis using the Student's

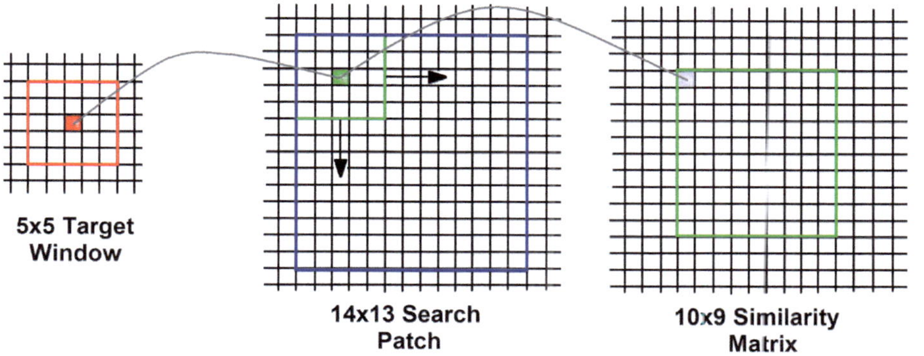

Figure 4.17 Area-based matching by shifting a target patch (left) over a search patch resulting in a similarity matrix

t-test. Subpixel accuracy can be achieved by fitting a function, for example, a second-order polynomial, through the correlation values and then determining the position where the function reaches its maximum.

4.5.1 Similarity Measures

When applying correlation for image matching, several similarity measures can be exploited. Most of these measures are based on statistics, including mean values, variances and covariances. The selection may depend on various user requirements, including computation time and required accuracy. In addition, measures derived from Lp-norms are suited. This means a power function of the absolute values of the differences of the grey values per pixel in the two image patches is calculated and is summed up next. A general expression of the Lp-norm is:

$$\|x\|_p = \left(\sum_{i=1}^{n} \lceil x_i \rceil^p \right)^{1/p} \tag{4.4}$$

In particular, the L1-norm and L2-norm are suited to act as similarity measures. The differences can be normalised by the sample means. Other suited measures are as follows.

Covariance of the grey values in the target patch and the search patch. Dividing the covariance by the product of the square roots of the variances of both patches results in normalised cross-correlation coefficient, which is usually indicated by the symbol R. The nice property of normalised cross-correlation is that the measure is normalised: Its value lies between −1 and + 1. The terms in R can be rearranged to obtain a computationally more convenient expression. To take account of the geometric differences between the target and search areas, a weight function (w_i) can be introduced, which favours the central parts of the patches (Mori et al., 1973).

For the sake of completeness, Table 4.1 lists ten similarity measures, discussed earlier; g_i defines the grey values in the target patch and h_i the grey values in the search patch. If the size of the target patch is 5×5 pixels, the number of grey values, n, will be 25.

The similarity measures in Table 4.1 become increasingly computationally demanding from top to bottom, with a few exceptions. The measures do have different properties. None of them is robust against geometric differences between the two patches, except for a 2D shift, although weighted normalised cross-correlation can partly cope with geometric differences. The difference of means is not invariant to intensity differences, and the same is true for the non-normalised L1-norm and L2-norm. This doesn't mean that these measures would be useless. They may work well for indoor scenes with a limited number of objects in which the lighting can be tuned. Because these measures are obtained fast, they are often used in real-time applications. However, for outdoor scenes and high-precision applications, more sophisticated measures are necessary.

Another criterion is that the setting of the match acceptance threshold should be uniformly applicable and should not be determined by trial and error on the images of a particular photogrammetric survey. The last three measures in Table 4.1 are normalised, meaning the similarity measures range from −1 to +1, with −1 meaning perfect

Table 4.1 Similarity measures suited for area-based matching

Measure	Formula
Difference of means	$\frac{1}{n}\sum_{i=1}^{n} g_i - \frac{1}{n}\sum_{i=1}^{n} h_i$
Difference of variances	$\frac{1}{n-1}\sum_{i=1}^{n}(g_i-\bar{g})^2 - \frac{1}{n-1}\sum_{i=1}^{n}(h_i-\bar{h})^2$
L2-norm, or Euclidean distance in n-dimensional space: sum of squared differences (SSD)	$\sqrt{\frac{1}{n}\sum_{i=1}^{n}(g_i-h_i)^2}$
L1-norm: sum of absolute differences (SAD)	$\frac{1}{n}\sum_{i=1}^{n}\lceil g_i-h_i\rceil$
L2-norm, or Euclidean distance in n-dimensional space normalised to the mean	$\sqrt{\frac{1}{n}\sum_{i=1}^{n}\left((g_i-\bar{g})-(h_i-\bar{h})\right)^2}$
L1-norm, normalised to the mean	$\frac{1}{n}\sum_{i=1}^{n}\left\|(g_i-\bar{g})-(h_i-\bar{h})\right\|$
Covariance	$\frac{1}{n}\sum_{i=1}^{n}(g_i-\bar{g})(h_i-\bar{h})$
Normalised cross-correlation (NCC or R)	$\frac{\sum_{i=1}^{n}(g_i-\bar{g})(h_i-\bar{h})}{\sqrt{\sum_{i=1}^{n}(g_i-\bar{g})^2}\cdot\sqrt{\sum_{i=1}^{n}(h_i-\bar{h})^2}}$
Alternative formulation of NCC	$\frac{\sum_{i=1}^{n} g_i\cdot h_i - 1/n\sum_{i=1}^{n} g_i\cdot\sum_{i=1}^{n} h_i}{\sqrt{\sum_{i=1}^{n} g_i^2 - 1/n\left(\sum_{i=1}^{n} g_i\right)^2}\cdot\sqrt{\sum_{i=1}^{n} h_i^2 - 1/n\left(\sum_{i=1}^{n} h_i\right)^2}}$
NCC with weight function	$\frac{\sum_{i=1}^{n} w_i(g_i-\bar{g})(h_i-\bar{h})}{\sqrt{\sum_{i=1}^{n} w_i(g_i-\bar{g})^2}\cdot\sqrt{\sum_{i=1}^{n} w_i(h_i-\bar{h})^2}}$

opposite correlation and +1 perfect correlation. The setting of the threshold does not depend on the conditions of the particular scene. Furthermore, the setting of the threshold on the NCC coefficient can be derived from sound reasoning of statistical hypothesis testing theory. In addition to the various similarity measures, the image patch, usually a quadrangular-shaped window, may vary depending on the image content. For example, Levine et al. (1973) introduced an adaptive window, of which the size was adjusted to the variance of the grey values: The larger the variance, the more reliable the match can be established, justifying the choice of a small window size.

4.5.2 Threshold

Shifting the target patch – pixel by pixel – over the search patch results in a similarity matrix (Figure 4.18). Let R_{max} be the maximum value after shifting the target patch over the search patch, using the normalised cross-correlation as a similarity measure. Based on R_{max}, a decision has to be made as to whether the target patch and the search patch cover the same part of the scene. Also, patches with relatively constant grey values will yield a high correlation, although the point in the search patch cannot be located in a reliable way, since the discrete cross-correlation function will have nearly the same high value everywhere. Since the discrete cross-correlation function is rather flat, the location of the match cannot be determined in a precise and reliable way. So checks have to be conducted on whether R_{max} exceeds a prespecified threshold and on the steepness of the similarity matrix. The check on the significance of R_{max} can be formulated in the following decision rule: If $R_{max} > T_R$ then the corresponding pixel of (I_t,J_t) in the target image is located at R_{max}, say, (I_0,J_0) in the search image. Which value should be assigned to T_R? This can be done in a heuristic way by trial and error. Another way is by formulating the decision rule as a statistical testing problem: Can we conclude at a certain significance level, for example, 5%, that the correlation coefficient differs from zero? A significance level of 5% means that there is a probability of 5% to reject the null hypothesis, given that it were true. This means we have to decide, on the ground of the value of R_{max}, to reject the null hypothesis that the hypothesised correlation coefficient is zero in favour of the alternative hypothesis that the hypothesised correlation coefficient is larger than zero. In this formulation of the test statistics, the hypothesised correlation coefficient R_H is zero. Assuming that the grey values in both target and search patches are normally distributed, then for $R_H = 0$, the distribution is symmetric and a statistic involving a one-tailed Student's t-test can be carried out. The critical threshold T_R depends on the number of observations and the significance level. The number of observations is determined by the number of pixels in the target patch: n. The relevant window sizes are 3×3 pixels ($n = 9$); 5×5 pixels ($n = 25$); 7×7 pixels ($n = 49$); and 9×9 pixels ($n = 81$). The levels of significance that are appropriate in area-based matching based on cross-correlation

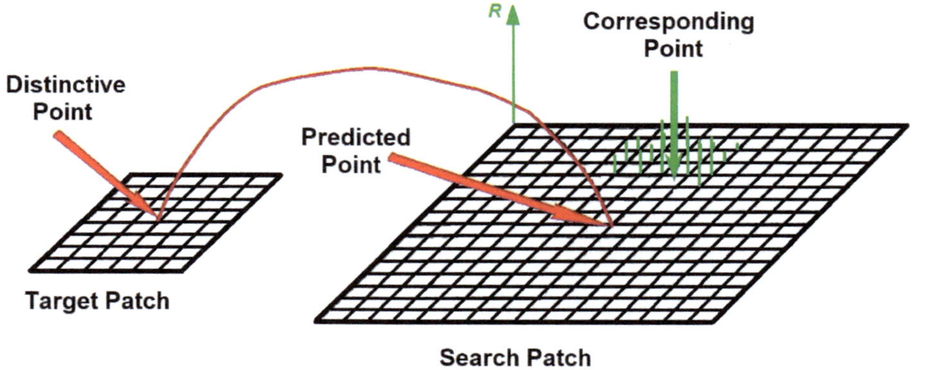

Figure 4.18 After a distinctive point has been found, its corresponding point is predicted and determined by shifting the patch of adjacent pixels over the search patch; green arrows indicate the similarity measures (R); the pixel with the largest R is further scrutinised on correspondence

Table 4.2 List of critical values for R_{max} based on Student's t-test

n	$t.999$	$t.995$	$t.975$	$t.95$
9	0.88	0.80	0.67	0.58
25	0.59	0.51	0.40	0.34
49	0.43	0.36	0.28	0.24
81	0.34	0.29	0.22	0.18

are 0.1%, 0.5%, 2.5% and 5%, respectively. Table 4.2 lists the critical threshold for R_{max} for the aforementioned significance levels and window sizes.

Note that aforementioned t-test states that there could be a relationship between the target patch and the search patch. A high significance level is most appropriate. For a significance level of 0.1%, R_{max} computed over a 5×5 target window should exceed 0.59. Testing the value of R_{max} by the t-test assumes that the grey values in both patches are normally distributed. However, violations of this assumption are important only when the sample size is less than about 10, a situation which will rarely occur in the image-matching problem. Added to this, the test assumes that the relationship between the grey values in the target patch and the search patch is linear. When the images are recorded during the same survey, there will usually be a time difference of seconds or minutes between the exposures of overlapping images. In such short time spans, the illumination differences may be modelled by a linear relationship.

4.5.3 Steepness Check

After accepting a match based on the value of R_{max} exceeding a threshold, a second test has to be conducted on the steepness of the discrete NCC function. This can be done by fitting a surface through the function and computing the two principal curvatures. If both curvatures are close to zero, the cross-correlation values are quite similar. If one curvature is large and the other is small, the correlation has been conducted along an edge, resulting in multiple high-correlation values along this edge. This means that the exact location of the corresponding point in the search patch can't be reliably determined, and the match should not pass the local consistency check. Only when both curvatures have high values and lie approximately within the same value range, a reliable match has been established. To compute the principal curvatures, a two-dimensional polynomial can be used as a surface and fitted through the discrete cross-correlation functions. In general terms, a two-dimensional polynomial of order n can be written as:

$$f(x, y) = \sum_{i=0}^{n} \sum_{j=0}^{n} a_{ij} x^i y^j \; ; i + j \leq n \tag{4.5}$$

For the steepness check on the discrete NCC function, a second-order polynomial appears to be a suitable option:

$$R(x, y) = a_0 + a_1 x + a_2 y + a_3 xy + a_4 x^2 + a_5 y^2 \tag{4.6}$$

The (x,y) coordinates refer to the similarity matrix. The principal curvatures κ_1 and κ_2 of a curved surface in 3D space can be calculated by solving the quadratic form:

$$\kappa^2 - \kappa\left(R_{xx} + R_{yy}\right) + R_{xx}R_{yy} - R_{xy}^2 = 0 \tag{4.7}$$

resulting in the solution:

$$\kappa_{1,2} = \frac{R_{xx} + R_{yy}}{2} \pm \sqrt{\left(\frac{R_{xx} + R_{yy}}{2}\right)^2 - R_{xy}^2} \tag{4.8}$$

with $R_{xx} = \frac{\delta^2 R}{\delta x^2} = 2a_4; R_{yy} = \frac{\delta^2 R}{\delta y^2} = 2a_5; R_{xy} = \frac{\delta^2 R}{\delta x \delta y} = a_3$.

The solution of Equation 4.8 becomes:

$$\kappa_1 = a_4 + a_5 + D; \kappa_2 = a_4 + a_5 - D \tag{4.9}$$

with $D = \sqrt{\left(a_4 - a_5\right)^2 + a_3^2}$.

The steepness condition is now that $\kappa_1 \approx \kappa_2$, while both should exceed a prespecified threshold. The parameters $a_0, a_1, a_2, a_3, a_4, a_5$ can be computed from an LSA.

4.5.4 Subpixel Location

The actual match is not located where the discrete NCC function reaches its maximum but where the continuous NCC function has its maximum. Considering the discrete NCC function as samples of the continuous function, the latter can be reconstructed by fitting a second-order polynomial (Equation 4.6) through the samples. So the same parameters used for the steepness check can also be used for determining the subpixel location. An extremum (maximum or minimum) of a function is located at the position where its first partial derivatives are zero. To determine the maximum, the partial derivatives, R_x and R_y of the 2D second-order polynomial, given by Equation 4.6, should be set to zero:

$$\begin{aligned} R_x &= a_1 + a_3 x_0 + 2a_4 y_0 = 0 \\ R_y &= a_2 + a_3 y_0 + 2a_5 x_0 = 0 \end{aligned} \tag{4.10}$$

resulting in the solution:

$$\begin{aligned} x_0 &= \frac{a_2 a_3 - 2a_1 a_5}{4a_4 a_5 - a_3^2} \\ y_0 &= \frac{-a_2 + a_3 x_0}{2a_5} \end{aligned} \tag{4.11}$$

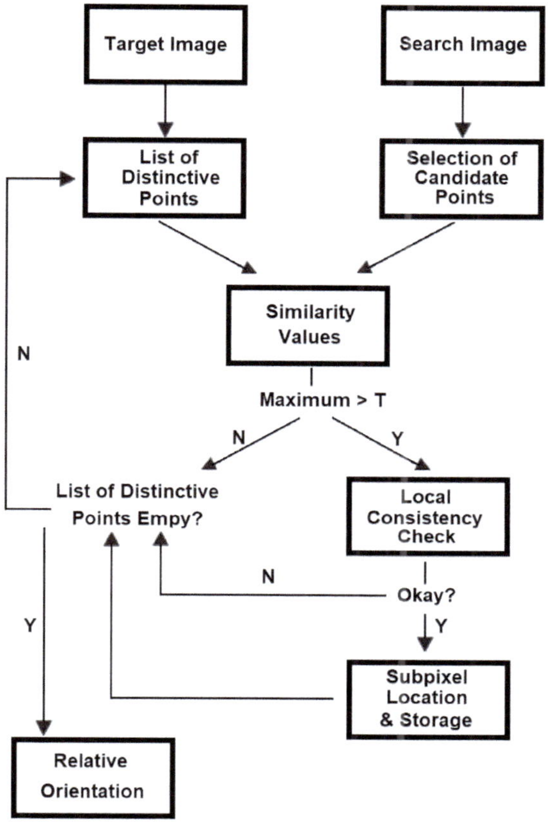

Figure 4.19 Schematic overview of image matching for establishing image co-registration (relative orientation)

where (x_0, y_0) is the subpixel location of the match with respect to the centre of the pixel within the search patch where $R(k,l)$ reached its maximum. Indicating the centre of this pixel by (I_0, J_0), coordinates of the subpixel location of the match in the search image are $(I_0 + x_0, J_0 + y_0)$.

Correlation can handle only shifts, which means correlation cannot cope with height discontinuities and differences in the relative orientation of the images. Furthermore, the finding of reliable matches requires relatively large patches, which causes sharp height continuities such as corners of buildings being smeared and leaves structures, which cover only a few pixels, undetected. Small patches enable the detection of small structures but provide unreliable results. Figure 4.19 shows a schematic overview of applying correlation for detecting tie points at the subpixel level aimed at relative orientation (registration) of overlapping images.

4.5.5 Census

The similarity measures discussed earlier are based on the first and second moments of the grey values. Another approach, called census and introduced by Zabih and Woodfill

Table 4.3 Hamming distance between two bit vectors of size 8

	1	2	3	4	5	6	7	8
Bit vector	1	0	0	0	1	0	1	0
Bit vector	0	0	0	0	0	1	0	1
Difference	1	0	0	0	1	1	1	1

(1994), doesn't exploit the grey values themselves but is based on non-parametric statistics in the correlation process. The approach relies on the ranking of local intensity values and not on the intensity values themselves. Within each of the two image patches, each pixel is compared with the centre pixel. If the grey value is lower than the centre pixel, the outcome is 1; if the grey value is higher, the outcome is 0. If the patches are 7×7 windows, then the 48 outcomes are stored in a bit vector of size 48 in a prespecified order. The same is done for the other image patch. Next, the bit vector of the candidate pixel in the search patch is compared to the bit vector in the target patch. The similarity is quantified through computing the Hamming distance. The Hamming distance between two bit vectors of equal length is the number of positions at which the corresponding symbols are different. To illustrate the Hamming distance, consider the example in Table 4.3. In a 3×3 window, the grey value of the centre pixel is compared to the grey value of each of the eight neighbour pixels. The same is done for the 3×3 search patch, resulting in a Hamming distance of 5.

The centre pixels of the target window and search window, which show the smallest Hamming distance, are chosen as corresponding pixels, provided that the distance is smaller than a certain threshold. The census algorithm allows only a shift in intensity values between the two patches but not a linear transformation. The algorithm is also not robust against geometric changes. On the other hand, the method is robust against the presence of a substantial number of outliers and intensity differences and has a low computational complexity, which is important for robotic applications (Banks and Corke, 2001). The robustness against outliers also improves performance near object boundaries, where correlation methods which exploit the intensity values themselves may fail.

NON-LINEAR LEAST-SQUARES ADJUSTMENT

This chapter also aims at demonstrating how LSA, which is a general apparatus applied in many fields of engineering and science, can be used for practical applications in the field of pointcloudmetry. In many problems, unknown parameters have to be estimated from observations. LSA is a well-suited method for obtaining an optimal solution if the number of observations is larger than the number of unknowns and the relationship can be expressed mathematically. It is said that the number of observations is redundant so that the relationship results in an overdetermined system of equations. The basic thinking underlying LSA is that the observations are captured in a physical environment using man-made physical devices, for example, optical–mechanical–electronic constructions, that contaminate the observations with noise (random errors). The observations are not the exact values – they are estimates. When calculating the unknown values from the observations, we should allow the observations to change a little bit. That is, they are adjusted to the mathematical model encapsulated in the design matrix **A**. It is supposed that the mathematical model is valid for the task at hand. It can be shown that the solution is the best if the noise on the observation is additive, zero-mean and Gaussian distributed. A non-exhaustive list of applications within the field of pointcloudmetry includes the following:

- Aero-triangulation in photogrammetry
- Georeferencing laser scans
- Least-squares image matching
- Fitting a polynomial through a set of neighbouring points as demonstrated by Equation 4.6
- Determination of the transformation parameters of, for example, a 3D similarity transformation using GCPs
- Fitting a geometric primitive – plane, sphere, cylinder – through a neighbourhood of points.

Linear LSA is considered in Section 11.3 in Chapter 11. Here, we will focus on non-linear LSA to prepare us for least-squares image matching discussed in Section 4.6. As discussed in Section 11.3, the observations are stored in a vector, usually indicated by $\boldsymbol{x}$, and the values of the unknown parameters in a vector are denoted by $\boldsymbol{y}$. Don't confuse $\boldsymbol{x}$ and $\boldsymbol{y}$ with the coordinates in a 2D reference system, such as an image coordinate system, discussed earlier. The mathematical relationship is stored in a matrix $\boldsymbol{A}$, often called a design matrix. When the relationship is linear, it is written in a symbolic way as $\boldsymbol{x} = \boldsymbol{A}\boldsymbol{y}$. The terms are in bold to indicate that they are vectors and matrix; the vectors are in lowercase, and the matrix is in capital letters. If the relationship is non-linear, it is written as:

$$\boldsymbol{x} = \boldsymbol{A}(\boldsymbol{y}) \tag{A.1}$$

It is quite reasonable to give the observations of higher quality a higher weight than the observations of lower quality. This can be incorporated by introducing the weight matrix $\boldsymbol{W}$, as done in Chapter 11 (see Equation 11.14). Often it is possible to express the contamination of the observations with noise quantitatively in the form of variance values. The variance, which is the square of the standard deviation, provides a measure for the weight – the larger the variance, the lower the weight should be. An appropriate weight is the inverse of the variance, which means one divided by the variance. It is convenient to arrange the variances (and covariances) of a set of observations in a so-called covariance matrix $\boldsymbol{C}_{xx}$. If the observations are uncorrelated, only diagonal elements of $\boldsymbol{C}_{xx}$ contain values: The off-diagonal elements all are zero. In some cases, all observations will have the same variance measures. When the observations are uncorrelated and do have the same variance, $\boldsymbol{C}_{xx}$ changes into the identity matrix, which is multiplied by one variance factor (σ^2), which is valid for all observations. The identity matrix contains the value one on the diagonal elements and zeros on off-diagonal elements. The identity matrix is usually indicated by the symbol $\boldsymbol{I}$. In this particular case, the covariance matrix becomes:

$$\boldsymbol{C}_{xx} = \sigma^2 \boldsymbol{I} \tag{A.2}$$

Observations may also be correlated, which means that the value of one observation has a certain relationship with those of one or more other observations. In these cases, all or a part of the off-diagonal elements of $\boldsymbol{C}_{xx}$ will be non-zero. In order to allow the observations to change a little bit, we have introduced an error vector, also called vector of residuals, symbolically indicated by $\boldsymbol{v}$, in Section 11.3. This means, in estimating the values of the unknown parameters, we should allow the observations to deviate from the measured values. The residuals are initially unknown and result from the LSA process. Using the inverse of the covariance matrix as the weight matrix, the least-squares solution becomes (compare with Equation 11.14):

$$\boldsymbol{y} = \left(\boldsymbol{A}^T \boldsymbol{C}^{-1} \boldsymbol{A}\right)^{-1} \boldsymbol{A}^T \boldsymbol{C}^{-1} \boldsymbol{x} \tag{A.3}$$

This solution is valid when the relationship is linear. However, often the relationship will be non-linear, which requires a different approach although the basic principles remain the same:

1. The observations are recorded in a physical environment with man-made devices, resulting in the contamination of the observations with noise; hence, we should allow the observations to change a little bit by introducing an error vector.
2. By introducing a constraint on the error vector in the form stating that the sum of the squared residuals should be as least as possible, we are able to find a unique solution.

If the mathematical relationship between observations and unknowns is non-linear, the relationship has to be expanded in a Taylor series around the unknown parameters. Usually, a first-order expansion suffices. This approach requires (rough) estimates of the values of the unknowns. Storing the approximate values of the unknowns in a vector indicated by $\boldsymbol{y}_0$, the observation $\boldsymbol{x}_0$ can be computed from the linearised form of Equation A.1:

$$\boldsymbol{x}_0 + \Delta \boldsymbol{x} + \boldsymbol{v} = \boldsymbol{A}\left(\boldsymbol{y}_0\right) + \frac{\delta \boldsymbol{A}}{\delta \boldsymbol{y}} \Delta \boldsymbol{y} \tag{A.4}$$

It looks somewhat weird in non-linear LSA that one starts with approximate values of the unknowns and next computes the estimates of the observations from these approximate values. Nevertheless, this apparatus provides an effective means to accurately determine the estimates of the unknowns and even to detect outliers in the observations. After computing the estimates of the observations ($\boldsymbol{x}_0$), $\Delta\boldsymbol{x}$ can be computed:

$$\Delta \boldsymbol{x} = \boldsymbol{x} - \boldsymbol{x}_0 \tag{A.5}$$

where $\boldsymbol{x}$ contains the original observations. Next $\Delta\mathbf{x}$ is used in the linearised part of Equation A.4. Denoting the first derivative of the design matrix $\boldsymbol{A}$, which means $\delta\boldsymbol{A}/\delta\boldsymbol{y}$, as $\Delta\boldsymbol{A}$, the linearised equation becomes:

$$\Delta \boldsymbol{x} + \boldsymbol{v} = \Delta \boldsymbol{A}\, \Delta \boldsymbol{y} \tag{A.6}$$

With least-squares solution:

$$\Delta \boldsymbol{y} = \left(\Delta \boldsymbol{A}^T \boldsymbol{C}^{-1} \Delta \boldsymbol{A}\right)^{-1} \Delta \boldsymbol{A}^T \boldsymbol{C}^{-1} \Delta \boldsymbol{x} \tag{A.7}$$

The initial approximate values of the unknowns ($\boldsymbol{y}_0$) are readjusted according to:

$$\boldsymbol{y}_1 = \boldsymbol{y}_0 + \Delta \boldsymbol{y} \tag{A.8}$$

The readjusted values of the unknowns ($\boldsymbol{y}_1$) are inserted in Equation A.3 to compute the readjusted estimates of the observations ($\boldsymbol{x}_1$). Next, a new $\Delta\boldsymbol{x}$ is computed along the lines indicated by Equation A.5, where $\boldsymbol{x}$ should be replaced by $\boldsymbol{x}_1$. Before doing so, the values in $\Delta\boldsymbol{y}$ are checked on their size. When they are small enough, continuing the iteration process doesn't make sense. To carry out this check, the values in $\Delta\boldsymbol{y}$ have to be compared with predefined convergence measures, one for each unknown or type of unknown. As long as the values in $\Delta\boldsymbol{y}$ are larger than the convergence measures, the iteration process continues. However, sometimes no convergence will occur at all, and so the iterations are inclined to continue endlessly. This may occur when the approximate values of the unknowns are not close enough to the actual values or when the relationship is heavily non-linear. Therefore, it is wise to introduce a threshold on the maximum number of iterations.

Once the final estimates for the unknown parameters have been computed, estimates of the observations can be computed using Equation A.1. In general, these estimates of the observations will differ from the actual observations. When the difference between an actual observation and an adjusted observation – residual – is large, this may be an indication that the observation is an outlier. The residual value can be compared with a threshold: If the value is larger than the prespecified threshold, the observation is removed from the observation vector. When this is the case, the non-linear LSA has to be redone. This is necessary because the outlier has affected the computed values of the unknown parameters. When the standard deviations of the observations are known, outliers can be detected on a sound statistical basis by introducing confidence levels. For example, when we know that all observations have a standard deviation of 15 cm and are Gaussian distributed, a confidence level can be introduced. When the confidence level is set to 95%, the threshold becomes 1.96×15 cm = 29.4 cm. In the course of time, an appropriate method for detecting outliers one by one – called data snooping – has been developed based on the theoretical work of Baarda (1968). Data snooping starts with selecting the largest residual and comparing its value with the prespecified threshold. If the residual doesn't pass the comparison, the concerning observation is removed from the vector of observations. Next, the iterative LSA starts again. The residuals are computed – the largest one is selected and compared with the threshold. This process is repeated until no outliers are detected anymore. This method has been applied to model tunnel profiles described in Section 13.1.1.3 in Chapter 13.

In conclusion, the pipeline of the non-linear LSA involves:

- Linearising the mathematical relationship between observations and unknowns according to a first-order Taylor series expansion
- Identifying the vector of observations
- Selecting proper approximate values for the unknown parameters
- Defining the maximum number of iterations
- Defining a vector of convergence measures, one for each unknown or type of unknowns
- Starting the iteration process

Next, the unknowns are adjusted according to the iterative process described earlier. Now we will turn to the application of the LSA apparatus for the detection and subpixel location of corresponding points in overlapping images.

4.6 Least-Squares Image Matching

Correlation techniques can handle only a linear transformation in intensity and a shift between the two images, a shortcoming already recognised by Ullman (1979). However, due to differences in perspective, there will be more geometric differences between overlapping images than just a shift (Figure 4.20). Photogrammetrists found a solution to this problem by modelling the differences in intensity and geometry explicitly (Ackermann, 1983, 1984; Gruen, 1985). As is the case for correlation techniques, LSA uses the grey values in overlapping image patches and may be categorised as an area-based matching technique. The least-squares solution of the non-linear relationship has to be established by an iterative procedure and requires approximate values of the unknown parameters including the shift.

Despite having computational demands, least-squares image matching (LSM) results in high subpixel accuracy, even up to 1% of a pixel. The shift between the two

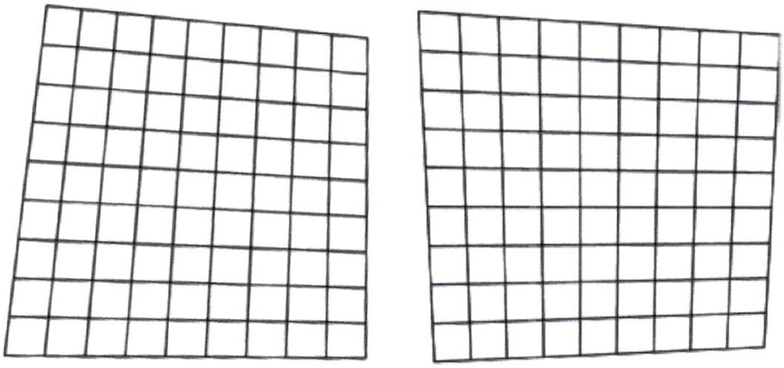

Figure 4.20 Demonstration of geometric differences between overlapping images

corresponding points should be very small, in the order of a few pixels. Therefore, LSM may be better regarded as a method for finding the subpixel location of the corresponding point once a (rough) match has been found in an earlier stage. Hence, LSM is particularly suited as the final matching step for tie point selection in high-definition photogrammetric applications.

LSA is not based on computing similarity measures but on explicitly modelling the radiometric and geometric differences between a pixel in the target patch g with image coordinates (x_t,y_t) and the (hypothesised) corresponding pixel in the search patch h with image coordinates (x_s,y_s). The non-linear relationship can be denoted as:

$$\boldsymbol{g}(x_t, y_t) = \boldsymbol{A}\left(\boldsymbol{h}\left(x_s, y_s\right)\right) \tag{4.12}$$

where $\boldsymbol{A}(\cdot)$ denotes the non-linear relationship. If the time gap between recording the overlapping images is small, for example, in the order of seconds or minutes, the radiometric differences can be modelled by a linear relationship. Accordingly, Equation 4.12 becomes:

$$g(x_t, y_t) = r_1 h(x_s, y_s) + r_0 \tag{4.13}$$

Often the geometric differences between the two images can be modelled as an affine transformation. Compared to a 2D similarity transformation, which models a shift or translation (two additive parameters), rotation (sine and cosine functions of the rotation angle) and scale factor (one multiplicative parameter), the affine transformation introduces two additional parameters describing the geometric differences between two images: non-orthogonality of the coordinate axes (shearing) and a scale difference between the two axes. The affine geometric transformation model can be described in parametric form as a first-order polynomial:

$$\begin{aligned} x_s &= p_1 + p_3 x_t + p_4 y_t \\ y_s &= p_2 + p_5 x_t + p_6 y_t \end{aligned} \tag{4.14}$$

Inserting Equation 4.14 into Equation 4.13 results in:

$$g(x_t, y_t) = r_1 h(p_1 + p_3 x_t + p_4 y_t, p_2 + p_5 x_t + p_6 y_t) + r_0 \tag{4.15}$$

Since Equation 4.15 is non-linear, the expression has to be expanded in a Taylor series around approximate values of the eight parameters, which model the radiometric and geometric differences. Let's assume that there are no radiometric and geometric differences between the stereopair. This assumption is feasible for aerial images; the image planes of the stereopair will be more or less parallel, and the images have been recorded with little time differences. So we may set the approximate values of the radiometric transformation to $r_0^{\,0} = 0$ and $r_1^{\,0} = 1$ and the approximate values of the affine transformation parameters to $p_1^{\,0} = p_2^{\,0} = p_4^{\,0} = p_5^{\,0} = 0$ and $p_3^{\,0} = p_6^{\,0} = 1$. Neglecting the second- and higher-order terms, the linearised equation, which relates the observed grey value difference $\Delta g = g(x_t, y_t) - h(x_s, y_s)$ between the target patch and the search patch, becomes (denoting $h = h(x_s, y_s)$):

$$\Delta g = h dr_1 + dr_0 + h_x dp_1 + h_y dp_2 + x h_x dp_3 + y h_x dp_4 + x h_y dp_5 + y h_y dp_6 \tag{4.16}$$

with $h_x = \partial h / \partial x$ and $h_y = \partial h / \partial y$

Given n grey values in each patch at the pixels 1, …, n and introducing the error vector $\boldsymbol{v}$ consisting of n terms v_i, $I = 1, \ldots, n$, the linear system becomes:

$$\begin{bmatrix} \Delta g_1 + v_1 \\ \cdots \\ \cdots \\ \cdots \\ \cdots \\ \cdots \\ \cdots \\ \cdots \\ \cdots \\ \Delta g_n + v_n \end{bmatrix} = \begin{bmatrix} 1 & h_1 & h_{x1} & h_{y1} & xh_{x1} & yh_{x1} & xh_{y1} & yh_{y1} \\ & & & & \cdots & & & \\ & & & & \cdots & & & \\ & & & & \cdots & & & \\ & & & & \cdots & & & \\ & & & & \cdots & & & \\ & & & & \cdots & & & \\ 1 & h_n & h_{xn} & h_{yn} & xh_{xn} & yh_{xn} & xh_{yn} & yh_{yn} \end{bmatrix} \begin{bmatrix} dr_0 \\ dr_1 \\ dp_1 \\ dp_2 \\ dp_3 \\ dp_4 \\ dp_5 \\ dp_6 \end{bmatrix} \tag{4.17}$$

This system of equations can be solved by the non-linear LSA process, described in the side bar.

4.7 Feature-Based Image Matching

Feature-based Image Matching (FBM) algorithms first extract distinctive points just as any correlation-based matching algorithms do. However, FBM detects distinctive points not only in the target image but also in the search image. Furthermore, FBM does not compare the grey values of supposed corresponding image patches but assigns features to the distinctive points derived from the grey values within the neighbourhood. Next, each distinctive point in the target image is linked to the point with the most similar features in the search image. The method requires the definition of the size of the image patch to calculate the features from. A variety of features may be used, including grey value variance within the image patch or texture measures. Features used in image matching are treated in the general chapter on feature extraction from images and point clouds (Chapter 11). FBM is often used to establish matches at pixel-level precision. Thereafter, the match can be refined to subpixel precision using least-squares matching.

There are six properties of overlapping image pairs which can strongly influence the reliability of FBM (Barnard and Thompson, 1980; Förstner, 1986):

1. Distinctiveness or discreteness – how distinct the individual interest point is from its neighbourhood.
2. Similarity – this indicates the degree of resemblance between two candidate corresponding points.
3. Global consistency – a measure for the conformity of a particular match between two points with matches in their vicinity.
4. Surface fluctuations – surfaces should vary smoothly, and the number of surface discontinuities should be limited, which means the differences in orientation and scale should be low or absent.
5. Stability – the object space point must have a high probability to appear in the image overlap.
6. Seldomness – the selection of interest points on nearby identical objects (repetitive structures) should be avoided.

These properties or requirements implicitly induce the limitation that feature-based matching, as is also the case for area-based matching methods, is not suited for creating dense point clouds. The methods developed in the 1970s and 1980s were inspired by different geometrical and statistical approaches, but all included the distinctiveness criterion. Interest points were characterised by a steep autocorrelation function (Hannah, 1974), high variances (Moravec, 1977; Förstner, 1986) and steep gradients in all directions (Dreschler and Nagel, 1982). Global consistency is checked after the local matches have been established. The check aims at discovering discrepancies between the local matches and the trend of a proposed surface, which is usually assumed to vary smoothly. A local match is the assignment of corresponding points based on similarities between their respective image patches.

An interesting interest operator often applied is the SIFt operator. This operator, developed within the realm of computer vision, has the advantage that it is invariant to shift, scale and rotation. The SIFT image-matching algorithm, discussed in Chapter 11 in the framework of feature extraction, is probably the most preferred method in computer vision. It can adequately handle differences in illumination, orientation and scale. Since the implementation appeared to be prohibitively slow, alternatives were sought. One of them was based on a further simplification of the Laplacian of Gaussian by replacing the Difference of Gaussian implemented in SIFT with a box filter. It was called speeded up robust features (SURF) (Bay et al., 2008). Another algorithm to avoid excessive computer resources and aimed at implementation in mobile robots, which have limited processing capacity, was the features from accelerated segment test (FAST) (Rosten and Drummond, 2006). Babri et al. (2016) present a comparative study on the performance of these three matching algorithms, which are prevalent in computer vision. They found the following: Both SIFT and SURF detect many interest points. Often the image patches covered by the interest points detected by SURF do not contain sufficient information to establish matches, even at places where texture is present. Nevertheless, the many interest points detected result in a rather dense point cloud. The number of interest points detected by SURF is not so high as that generated using SIFT, but it rarely occurs that the image patches do not contain sufficient information for establishing matching. The number of interest points detected by FAST is lower than the number of interest points that SIFT and SURF are able to detect, since the operator wields a high distinctiveness criterion, which means the interest points should have high contrast compared to their vicinity. The result is that nearly all interest points detected are matched. This means that FAST does the job it has been designed for: guiding robots.

Wouldn't it be possible to release the distinctiveness criterion and incorporate the smoothness constraint in the matching step itself? In this case, the assignment of corresponding points would be simultaneously guided by the similarity of their image patches and the distance of the match to the trend of the surface; the smaller the distance to this surface, the higher the score should be. Implementation of this idea, called global matching, has been retained for a long time because a rigorous implementation would consume vast amounts of computation time.

4.8 Dense Image Matching

The point density of airborne Lidar rapidly increased from 1 point/20 m^2 to 20 points/m^2, in a time span of just 15 years from 2000 onwards. This made airborne Lidar a preferred geodata acquisition technology for many applications, including the generation of DSMs and DTMs. Compared to photogrammetry, Lidar surveys are expensive, and this induces the question of whether photogrammetry could produce point clouds with a density and accuracy similar to those of Lidar as an alternative. It has been well known for a long time that photogrammetric image matching could deliver dense point clouds, but the further development of methods and their implementation were withheld due to the required exorbitantly huge requirements on processing time and memory consumption. The approach was regarded as impractical, even unworkable, for a long time. With the rapid expansion of computer power and storage facilities, and the possibility to programme graphical processing units (GPU), which are much faster than CPUs, DIM, exploiting global matching principles, increasingly gained the attention of researchers, resulting in a solution called SGM. Before discussing SGM in greater detail, we will first have a look at two geometric properties of stereo images: epipolar geometry and disparity. Their exploration is essential for understanding SGM.

4.8.1 Epipolar Geometry and Disparity

The collinearity condition states that a point P in the object space, its projection p in the image and the projection centre O of the camera lie on the same straight line. Does similar systematics exist for a stereopair of images? Also, for a stereopair, more points are part of a geometric figure than are strictly necessary to construct that figure. For a stereopair the geometric figure is not a line but a plane, called epipolar plane. A point in the object space, its projection in the two images and the two projection centres lie in a common plane (Figure 4.21). This property is called coplanarity. In total, five points span the plane, but only three points are necessary to define a plane. Coplanarity can be used for finding the corresponding point in the other image, if the position of the two projection centres and the orientations of both optical axes are known. It reduces the search space from two dimensions to one dimension and is often exploited in photogrammetric image matching to reduce computation time.

Disparity in stereo images refers to the differences in the location in the images of the same point in the object space. Height discontinuities cause different disparities. When a surface is smooth (red points in Figure 4.22), the difference in disparity will be

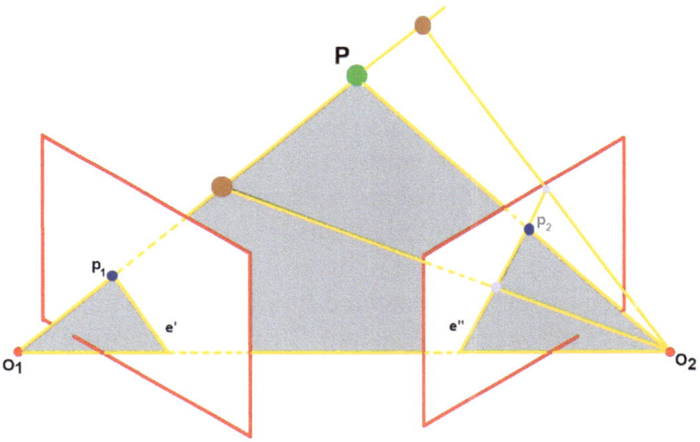

Figure 4.21 Epipolar geometry: an epipolar plane is defined by projection centres O_1 and O_2 and an object point P. Epipolar lines e′ and e″ are intersections of the epipolar plane with the image planes

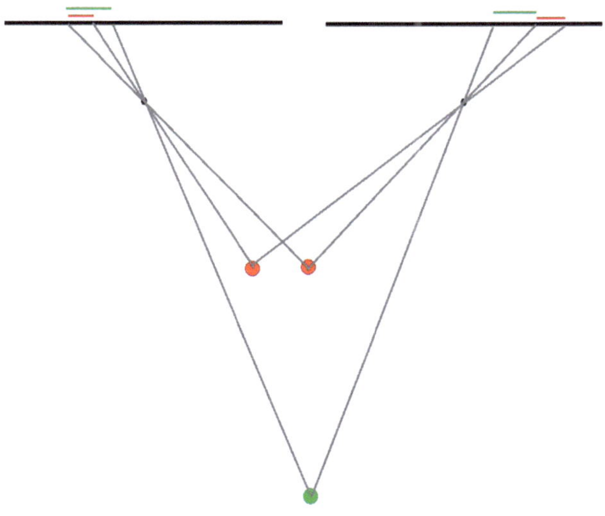

Figure 4.22 For a smooth surface (red points), the difference in disparity will be (close to) zero, while large height discontinuities cause large differences in disparity

(close to) zero. Large height discontinuities cause large differences in disparity. In SGM, the difference in disparity between adjacent pixels is used as a smoothness constraint. Disparity is defined as the coordinate difference of the point in the right and left images measured along the epipolar line.

4.8.2 Global Matching

The area-based and feature-based matching methods, outlined earlier, evaluate the similarities of image patches with high contrast/texture without considering whether the matches fit within those found in the vicinity. Therefore, they are called local methods. The local approach is quick and accurate but does not find corresponding points, or if it finds the match, it will be unreliable in areas with low contrast/texture and will thus fail if the surface consists of desserts, road pavement, dunes or other low-contrast areas with few distinctive points and little texture. In addition, in the presence of neighbouring repetitive patterns, wrong correspondences are likely to be found (Figure 4.23). Furthermore, the local approach will also fail in the presence of height discontinuities and occlusions. In areas with little contrast or texture, the point density will be low, resulting in an inhomogeneous distribution of points, which is unfavourable for many applications. This is in contrast to Lidar, which is an active system emitting electromagnetic (EM) beams itself. To get rid of mismatches in local matching methods, they are regarded as outliers, which are detected in a postprocessing step by introducing a mathematical model of the surface behaviour. This postprocessing step is called a global consistency check. Note that this check is conducted after the local consistency check as the last step. The check aims at discovering discrepancies between the local matches and the trend of a proposed surface model. Usually, the surface is assumed to vary smoothly.

If the surface smoothness constraint is implemented in the matching step itself, the assignment of corresponding points will be simultaneously guided by the similarity of their image patches, as is done in local matching, and by a built-in constraint on how well the location of the match fits the global trend surface. This notion expresses the basic idea of global matching. This means global matching simultaneously balances the similarity values between candidate corresponding points and the smoothness of the surface on which these points lie. As said, the implementation of this idea has been retained for a long time, because a rigorous implementation would consume vast amounts of

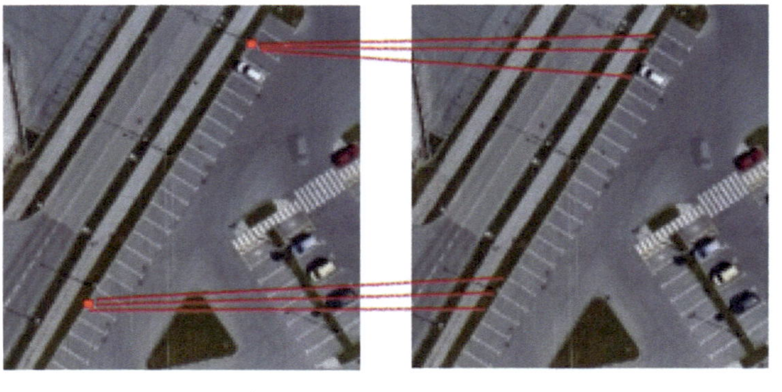

Figure 4.23 Repetitive patterns such as corners of parking lot outlines cause match ambiguities

computation resources beyond availability. Particularly, in aerial photogrammetry in which a scene may be captured by hundreds, thousands or even tens of thousands of images while each of the images consists of millions of pixels, implementation, which would be feasible in practice, was far beyond realisation for a long time.

4.8.3 Semi-Global Image Matching

The first solution to reduce the exorbitant computational workload was by shrinking global from a 2D problem to a 1D problem by limiting the search space from the entire overlap to epipolar lines – this requires an accurate relative orientation carried out in advance. Next, the epipolar lines are resampled to correspond with rows in the image. This approach considerably reduces computation time but has the disadvantage of disconnected features since the optimal matching along one row may result in disparities which deviate from those of the adjacent rows. This effect is visible as disconnected straight boundaries or straight boundaries appearing in a zigzag pattern with stripes. Therefore, this effect is called streaking. Hirschmüller (2005, 2008) found a solution to the problem that matches found in one epipolar line may be inconsistent with those found along adjacent epipolar lines. He extended the idea of matching along epipolar lines through computation along several paths, symmetrically from all directions through the overlap, and found that the number of paths should be at least 8 and preferably 16 in order to approximate global coverage as much as possible. In the case of eight paths, the directions correspond with rows, columns and the two diagonals of the image raster (Figure 4.24). So, instead of involving all pixels in the overlap, only samples along regularly spaced paths are used, reducing global matching to SGM.

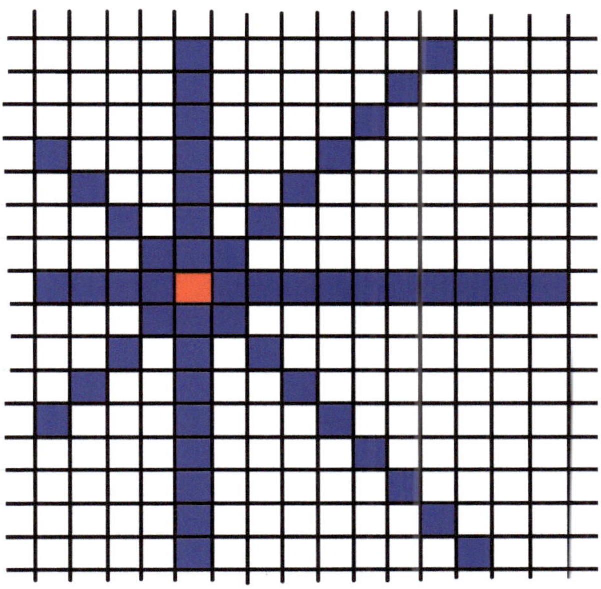

Figure 4.24 Semi-global matching (SGM) based on defining eight paths symmetrically from all directions through the overlap

Next, Hirschmüller faced the problem of balancing the degree of similarity and global consistency in a mathematical way. Intuitively a balance can be achieved by summing up the similarity values within the SGM space and subtracting from this the sum of the differences in disparities between adjacent pixels according to a balancing factor. The balancing factor provides a means for defining how much value is attached to similarity and how much to global consistency. For example, if discontinuities are allowed, the weight of the consistency part may be weakened. By SGM space, we mean the pixels involved in the matching process. Usually, the combination is done by introducing an additive cost function:

$$E(D)=\sum_{p}\left[C\left(p,D_{p}\right)+\sum_{q\in N_{p}}PT\left(\left|D_{p}-D_{q}\right|\geq 1\right)\right] \tag{4.18}$$

The first term of the two additive terms sums up all pixel-wise matching costs over the paths, and the second term defines a penalty for all pixels with neighbours that have a different disparity. In case of occlusion, SGM provides – as all image-matching techniques – unreliable results. These can be partially detected by conducting matching between the right image and the left image, in addition to the matching between the left and the right image.

SGM enables a pixel-by-pixel matching, which means a height value can be obtained for each and every pixel, resulting in several favourable advantages compared to conventional local matching methods. SGM requires only a minority of image patches showing sufficient variation in grey value. Therefore, areas with little texture and low contrast, such as dunes or buildings, can also be matched. They produce dense point clouds with a GSD similar to the imagery from which they are derived. SGM can detect fine structures and discontinuities. Hence, SGM is suited for the reconstruction of buildings and it supports the creation of 3D city maps.

SGM has been extensively investigated for a variety of geomatics applications and image types including aerial images (Hirschmüller et al., 2005; Hirschmüller, 2008), terrestrial and extraterrestrial satellite data (Krauss et al., 2008; Alobeid et al., 2010) or video sequences (Heinrichs et al., 2007; Gerke, 2008). Figure 4.25 shows a DSM of Mt.

Figure 4.25 Digital surface of Mt. Everest obtained from overlapping WorldView satellite images (Digital Globe), GSD 50 cm, using SGM (source: Hirschmüller, 2011)

Everest obtained from overlapping WorldView satellite images using SGM. Furthermore, SGM is being recognised and deployed in the industry. SGM has been quickly picked up by photogrammetric researchers, and its suitability has been tested for photogrammetric applications. The potential of SGM for creating dense DEMs from aerial images has been convincingly demonstrated by Rothermel and Haala (2011) from the University of Stuttgart, Germany. The results appeared to be impressively good, especially for images with little difference in perspective (short baselines) and sufficient image contrast. For areas with similar intensities, the matching quality decreased. For 60% of the pixels, matches were found, which proves that SGM has the capability to create DSMs from aerial images with high point densities. The research at Stuttgart resulted in a software package – called SURE – based on SGM (Rothermel et al., 2012). This software package has been brought to the marketplace by nFrames. Applying SGM to Ikonos satellite images, Alobeid et al. (2010) found good results for large buildings, but in urban areas with small buildings and narrow streets, such as in Istanbul, no matches were generated in occluded image parts. Furthermore, the amount of streaking was very limited.

4.8.4 Software and Examples

SGM has become a regular part of many commercial photogrammetric mapping software solutions, including Agisoft, Bentley's ContextCapture, PIX4Dmapper, SimActive's Correlator3D, Racurs' PHOTOMOD, SURE offered by nFrames and Trimble's Inpho. The abilities of SGM have been demonstrated in a research project of MSc geomatics students from Delft University of Technology (TU Delft), the Netherlands (Arapakis et al., 2018). The project group – supervised by Dr Liangliang Nan and Dr Ravi Peters and coordinated by Dr Mathias Lemmens – compared SIFT-based matching, an FBM image-matching approach, with PIX4Dmapper using UAS images of a white cottage on the TU Delft campus (Figure 4.26). The roof and facades were coated white, and PIX4Dmapper dealt well with them. The SIFT-based FBM approach extracted points on corners, ribbings and other locations with sufficient contrast. Figures 4.27–4.29 show examples of dense point clouds generated with SGM using images captured by an oblique camera system and a UAS.

In conclusion, SGM shows a good balance between computational requisites, accuracy and detail and is thus well suited for creating dense point clouds for many practical applications including the creation of DEMs and 3D city models. Each type of image and application will require tailor-made postprocessing in order to get rid of imperfections

Figure 4.26 Building with little texture (left), point cloud generated by a SIFT-based FBM approach (middle) and point cloud generated by SGM as implemented in PIX4Dmapper (courtesy: Delft University of Technology, the Netherlands)

Figure 4.27 Point cloud of the centre of Graz generated from images captured by Vexcel's oblique camera system UltraCam Osprey using the SGM implementation in PHOTOMOD (courtesy: Racurs, Moscow, Russia)

Figure 4.28 SGM point cloud of the Cathedral of Rostov-on-Don generated from UAS images (courtesy: Racurs, Moscow, Russia)

Figure 4.29 SGM point cloud of a residential area generated from UAS images (courtesy: Racurs, Moscow, Russia)

and to increase quality. Manufacturers of photogrammetric software are continuously extending and enhancing their SGM software.

4.9 Concluding Remarks

The development and improvement of image-matching techniques has been one of the most investigated research topics in photogrammetry and computer vision over the past decades. Until 2010, DEM generation of aerial imagery relied on local matching techniques. The resulting sparse point clouds, together with the steadily increasing point density of Lidar, triggered the development of semi-global image-matching techniques. Determination of tie points for high-precision applications requires a method which provides high subpixel accuracy, while DIM for creating DEMs requires a method which incorporates constraints on the smoothness of the surface, provides matches in low-texture areas and handles repetitive structures in the object space. The creation of point clouds from overlapping images consists of a sequence of consecutive steps which are implemented in commercially available software. The setting of the values of tuning parameters, such as window size and thresholds, has a significant influence on the quality of the resulting point cloud. Usually, commercial software appears to be a black box, and reading the manual alone is not enough. The exploration of image-matching software requires a basic knowledge of the principles of photogrammetry.

References

Ackermann, F. (1983) High Precision Digital Image Correlation. In *Proceedings 39th Photogrammetric Week*, Institut für Photogrammetrie, Stuttgart, Heft 9, pp. 231–243.

Ackermann, F. (1984) Digital Image Correlation: Performance and Potential Application in Photogrammetry. *Photogrammetric Record*, 11(64), pp. 429–439.

Alobeid, A., Jacobsen, K., Heipke, C. (2010) Comparison of Matching Algorithms for DSM Generation in Urban Areas from Ikonos Imagery. *Photogrammetric Engineering & Remote Sensing*, 76(9), pp. 1041–1050.

Arapakis, T., van Heerden, N., Rodriguez-Mon Barrera, G., Wang, Q., Wang, X. (2018) Digitizing Real-World Scenes from Images. *Geomatics Synthesis Project*, Delft University of Technology.

Atkinson, K.B. (Ed.) (2003) *Close Range Photogrammetry and Machine Vision*. Whittless Pub., Dunbeath, Caithness, Scotland, UK.

Baarda, W. (1968) *A Testing Procedure for Use in Geodetic Networks*. Publications on Geodesy 9, Vol. 2, no. 5, Netherlands Geodetic Commission, Delft.

Babri, U.M., Tanvir, M., Khurshid, K. (2016) Feature Based Correspondence: A Comparative Study on Image Matching Algorithms. *International Journal of Advanced Computer Science and Applications*, 7(3), pp. 206–210.

Ballard, D.H., Brown, Ch.M. (1982) *Computer Vision*. Prentice Hall, Inc., Englewood Cliffs, NJ, US.

Banks, J., Corke, P. (2001) Quantitative Evaluation of Matching Methods and Validity Measures for Stereo Vision. *The International Journal of Robotics Research*, 20(7), pp. 512–532.

Barnard, S.T., Thompson, W.B. (1980) Disparity Analysis of Images. *IEEE Transactions on Pattern Analysis and Machine Intelligence*, PAMI-2(4), pp. 333–340.

Bay, H., Ess, A., Tuytelaars, T., Van Gool, L. (2008) Speeded-Up Robust Features (SURF). *Computer Vision and Image Understanding*, 110(3), pp. 346–359.

Dreschler, L., Nagel, H.-H. (1982) Volumetric Model and 3D Trajectory of a Moving Car Derived from Monocular TV Frame Sequences of a Street Scene. *Computer Graphics and Image Processing*, 20(3), pp. 199–228.

Förstner, W. (1986) A Feature Based Correspondence Algorithm for Image Matching. *The International Archives of the Photogrammetry*, 26(3), pp. 1–17.

Fraser, C.S. (1997) Digital Camera Self-Calibration. *ISPRS Journal of Photogrammetry & Remote Sensing*, 52(4), pp. 149–159.

Gerke, M. (2008) Dense Image Matching in Airborne Video Sequences. In *The International Archives of the Photogrammetry, Remote Sensing and Spatial Information Sciences*, Beijing, China, Vol. XXXVII, Part B3b, pp. 639–644.

Graham, R., Koh, A. (2013) *Digital Aerial Survey: Theory and Practice*. Whittles Publishing, Dunbeath, Caithness, Scotland, UK.

Gruen, A.W. (1985) Adaptive Least Squares Correlation: A Powerful Image Matching Technique. *South African Journal of Photogrammetry, Remote Sensing and Cartography*, 14(3), pp. 175–187.

Gruen, A., Huang, T.S. (Eds.) (2001) *Calibration and Orientation of Cameras in Computer Vision*. Springer, Berlin.

Hannah, M. (1974) Computer Matching of Areas in Stereo Images. Ph.D. Thesis, University of Stanford, US.

Hartley, R., Zisserman, A. (2004) *Multiple View Geometry in Computer Vision*. 2nd Edition, Cambridge University Press, Cambridge.

Heinrichs, M., Rohdehorst, V., Hellwich, O. (2007) Efficient Semi-Global Matching for Trinocular Stereo. In *The International Archives of the Photogrammetry, Remote Sensing and Spatial Information Sciences*, Munich, Germany, Vol. XXXVI, Part 3/W49A, pp. 185–190.

Hirschmüller, H. (2005) Accurate and Efficient Stereo Processing by Semi-Global Matching and Mutual Information. In *IEEE Conference on Computer Vision and Pattern Recognition*, Vol. 2, pp. 807–814.

Hirschmüller, H. (2008) Stereo Processing by Semiglobal Matching and Mutual Information. *IEEE Transactions on Pattern Analysis and Machine Intelligence*, 30(2), pp. 328–341.

Hirschmüller, H. (2011) Semi-Global Matching – Motivation, Developments and Applications. In *Photogrammetric Week '11*, pp. 173–184.

Hirschmüller, H., Scholten, F., Hirzinger, G. (2005) Stereo Vision Based Reconstruction of Huge Urban Areas from an Airborne Pushbroom Camera (HRSC). In *Lecture Notes in Computer Science: Pattern Recognition, Proceedings of the 27th DAGM Symposium*, Vol. 3663, Springer, pp. 58–66.

Kraus, K. (2007) *Photogrammetry: Geometry from Images and Laser Scans*. 2nd Edition, Walter de Gruyter, Berlin.

Krauss, T., Lehner, M., Reinartz, P. (2008) Generation of Coarse 3D Models of Urban Areas from High Resolution Stereo Satellite Images. In *International Archives of the Photogrammetry and Remote Sensing*, Vol. XXXVII, Part B1, pp. 1091–1098.

Lemmens, M.J.P.M. (1988) A Survey on Stereo Matching Techniques. *International Archives of Photogrammetry and Remote Sensing*, 27(B8), pp. 11–23.

Lemmens, M. (2014a) Digital Oblique Aerial Cameras (1) – A Survey of Features and Systems. *GIM International*, 28(4), pp. 20–25.

Lemmens, M. (2014b) Digital Oblique Aerial Cameras (2) – A Survey of Features and Systems. *GIM International*, 28(5), pp. 20–25.

Lemmens, M., Lemmen, C., Wubbe, M. (2007) Pictometry: Potentials for Land Administration. In *6th FIG Regional Conference*, San José, Costa Rica.

Levine, M.D., O'Handley, D.A., Yagi, G.M. (1973) Computer Determination of Depth Maps. *Computer Graphics & Image Processing*, 2, pp. 131–150.

Linder, W. (2016) *Digital Photogrammetry: A Practical Course*. 4th Edition, Springer.

Luhmann, T., Robson, S., Kyle, S., Boehm, J. (2019) *Close-Range Photogrammetry and 3D Imaging*. 3rd Edition, De Gruyter, Berlin, Boston.

Marr, D. (1979) Vision, Freeman and Company, San Francisco, US. Reprinted: in 2010, The MIT Press.

Marr, D., Poggio, T. (1979) A Computational Theory of Human Stereo Vision. *Proceedings of the Royal Society of London*. Series B, 204, pp. 301–328.

McGlone, J. C. (ed.) (2013) *Manual of Photogrammetry*. 6th Edition, American Society for Photogrammetry and Remote Sensing, Bethesda, MD.

Mikhail, E.M., Bethel, J.S., McGlone, J.C. (2001) *Introduction to Modern Photogrammetry*. Wiley, New York, Chichester.

Moravec, H.P. (1977) Towards Automatic Visual Obstacle Avoidance. In *Proceedings of the International Joint Conference on Artificial Intelligence*, p. 584. Cambridge, Massachusetts, USA.

Mori, K., Kidode, M., Asada, H. (1973) An Iterative Prediction and Correction Method for Automatic Stereocomparison. *Computer Graphics & Image Processing*, 2, pp. 393–401.

Op't Eyndt, T., Volkmann, W. (2013) UAS as a Tool for Surveyors – Virtual Surveying instead of Tripods and Trucks. *GIM International, UAS Special*, pp. 26–29. https://www.gim-international.com/content/article/uas-as-a-tool-for-surveyors.

Rosnell, T., Honkavaara. E. (2012) Point Cloud Generation from Aerial Image Data Acquired by a Quadrocopter Type Micro Unmanned Aerial Vehicle and a Digital Still Camera. *Sensors*, 12, pp. 453–480.

Rosten, E., Drummond, T. (2006) Machine Learning for High Speed Corner Detection. In: Leonardis, A., Bischof, H., Pinz, A. (Eds.) *9th European Conference on Computer Vision*, vol. 1, pp. 430–443. Springer, Berlin, Heidelberg.

Rothermel, M., Haala, N. (2011) Potential of Dense Image Matching for the Generation of High Quality Digital Elevation Models. In *International Archives of the Photogrammetry, Remote Sensing and Spatial Information Sciences*, Vol. XXXVIII-4/W19, pp. 271–276. Hannover, Germany

Rothermel, M., Wenzel, K., Fritsch, D., Haala, N. (2012) SURE: Photogrammetric Surface Reconstruction from Imagery. In *Proceedings of LC3D Workshop*, Berlin.

Schenk, T. (1999) *Digital Photogrammetry, Vol 1: Background, Fundamentals, Automatic Orientation Procedures*, Terra Science, Laurelville, OH.

Shapiro, L.G., Stockman, G.C. (2001) *Computer Vision*. Prentice Hall.

Ullman, S. (1979) *The Interpretation of Visual Motion*. MIT Press, Cambridge, MA.

Wolf, P.R., Dewitt, B.A. (2000) *Elements of Photogrammetry with Applications in GIS*. 3rd Edition, McGraw-Hill, New York, NY.

Zabih, R., Woodfill, J. (1994) Non-Parametric Local Transforms for Computing Visual Correspondence. In Eklundh, J-O, *Proceedings of Third European Conference of Computer Vision*, Stockholm Sweden, pp. 151–158. Springer, Berlin, Heidelberg.

Airborne Lidar

CHAPTER

5

Since the mid-1990s, airborne Light Detection and Ranging (Lidar), also called airborne laser scanning (ALS), has matured into a mapping technology routinely used for 3D modelling of urban areas, capturing boreal forests, seabed mapping and many other applications all over the world. The speed with which the laser pulses are fired – pulse rate frequency – continues to soar, and for a number of commercial systems the pulse frequency has reached the impressive number of one million pulses per second and it may be expected that the pulse frequency will gradually rise over time. Since the mid-1990s, airborne Lidar has made big progress and has gone mainstream, joining aerial photogrammetry as an important geodata acquisition technology for creating digital surface models (DSMs), digital terrain models (DTMs), orthophotos and derived products, 3D models of cities and 3D digital landscape models. Indeed, airborne Lidar has developed into a most successful geodata acquisition technology used for multiple types of applications, including forest stand monitoring, urban planning and monitoring, natural resource management, disaster management, corridor mapping, hydrological modelling and forest carbon monitoring.

The principles of using laser light for measuring ranges have been outlined in Chapter 3. In this chapter, we look at how laser light can be turned into a surveying technology. Section 5.1 considers how to put laser light at work for capturing point clouds, particularly from the perspective of an airborne platform. A laser device mounted in an aircraft captures ranges by measuring the travel time of laser pulses forth and back. The transformation of these ranges to 3D coordinates of points in a unified reference frame requires additional simultaneous measurements which come together with the laser range in a storage medium. In particular, the equipment onboard the aircraft should be able to measure the 3D coordinates of the position of the laser sensor at the moment of firing the pulse. The equipment should also be able to measure the orientation in space of the laser sensor, defined by three rotation angles, at the moment of firing the pulse. The six parameters of position and orientation – exterior orientation parameters – should be captured or interpolated from nearby measurements for each and every emitted pulse. The system components, which measure the parameters essential for transferring ranges to 3D coordinates, are outlined in Section 5.2. An important facility to capture a strip across both sides of the track of the aircraft is a scanning mirror which distributes the pulses across the flight direction. Most systems use an oscillating mirror or a mirror which rotates 360 degrees. The type of rotation determines the distribution of the points on the ground. The distribution results in a certain scan pattern or scan configuration, which is the topic of Section 5.3. In addition to the scan pattern, the point density is an important parameter, which determines the usability of the point cloud for a particular task at hand. Point density and its counterpart point spacing are the subject of Section 5.4.

Section 5.5 considers georeferencing and briefly touches co-registration. These topics have been extensively treated in several books. A recent one is Shan and Toth (2018). This book is for the major part devoted to airborne and terrestrial laser scanning. In course of time, the design and engineering of Lidar systems mainly focused on the improvement of the point density and the amount of attributes recorded for individual points. The main solutions for achieving higher point densities are mounting two or more laser scanners in one of the devices, multiple pulses in air and single-photon Lidar (SPL). These topics are considered in Section 5.6. Increasing the number of attributes or features per point supports further processing and analysis of the airborne Lidar point cloud. Particularly, recording multiple returns, full-waveform digitisation and multi-spectral Lidar are technological advances, which increase the number of attributes per point. These topics are outlined in Section 5.7. Since the mid-2010s, unmanned airborne systems (UASs), by the general public usually named drones, have become so powerful that they can lift lightweight laser scanners into the air. UAS Lidar is outlined in Section 5.8. This chapter concludes with Section 5.9, in which the issue of multipath of Lidar pulses is briefly touched. Parts of this chapter are based on Lemmens (2011, 2017).

5.1 Laser as a 3D Mapping Technology

As an acronym of Light Detection and Ranging – some prefer to read Lidar as Laser Imaging Detection and Ranging – the term has become a "proper name" with the initial letter the only capital. Although the rise of Lidar as a widely used operational system began just at the end of the 1990s, its history dates back to the 1960s when it was first tried out (Ackermann, 1999). In the 1970s, experimental systems were developed. Accurate positioning remained a troublesome issue until, in the early 1990s, the Global Positioning System (GPS) became a reliable, stable and precise positioning technology. Lidar features high accuracy and precision and a high level of detail.

Today, many professionals working in the geo-domain have recognised airborne Lidar as an advanced technology with many beneficial characteristics for mapping the earth's surface and objects present on that surface. Areas, which extend over thousands of square kilometres, can be captured in a few hours. The level of automated data processing is high. 3D maps of cities at a course level of detail (LoD1) can be created with little intervention from a human operator, especially when digital maps containing the outlining of building footprints are available. The coordinates of airborne Lidar point clouds can be determined with precision at the sub-decimetre level. The achievable point density lies in the order of hundreds of points per square metre. As a result, airborne Lidar has gained widespread recognition as a major source of point clouds from which many valuable products can be derived, such as accurate DTMs and 3D-city maps. The resulting products offer many benefits for planning, design, inspection and maintenance of infrastructure. Dunes, beaches, dykes and floodplains can be captured accurately and with a high level of detail, enabling flood-risk management, estimating coastal erosion, and designing maintenance and renovation works of dykes, dunes and levees. Compared to aerial photogrammetry, which is based on creating point clouds from overlapping images exploiting automated image matching techniques applied to digital stereo imagery, airborne Lidar has the following advantages (Baltsavias, 1999a; Wehr and Lohr, 1999; Romano, 2004; Meng et al., 2010):

- Lidar facilitates the recording of multiple returns and full-waveform digitisation of the backscattered pulse enabling mapping of the bare ground surface in forests and other regions with dense vegetation.
- In contrast to overlapping images, no texture is needed to create an accurate and reliable point cloud.
- Since Lidar is an active system, visibility and weather conditions only slightly affect the operation of an airborne Lidar survey, which makes the technology fairly independent of season and daytime.

The maximum distance a Lidar pulse can bridge from sensor to ground and back depends on the power used, the pulse frequency and the reflectivity of the part of the object hit by the pulse.

For example, when the reflectivity is at least 20% for all terrain objects, the Optech Galaxy, which belongs to the major workhorses for capturing airborne Lidar point clouds of extent areas, can capture up to one million returns per second for surveys up to 1.5 km above ground level (AGL). Flying at an altitude of 2.5 km, the number of returns reduces to 750 kHz.

5.2 System Components

Airborne Lidar devices are multi-sensor systems, composed of a laser sensor, which is the principal device, a positioning system based on the Global Navigation Satellite System (GNSS) and an inertial navigation system (INS), usually called inertial measurement unit (IMU), which enables the determination of the orientation of the sensor with respect to a reference system. The laser sensor consists of a laser pulse emitter, scanning optics to direct and distribute the laser pulses across the flight path and a receiver. Added to these devices, the aircraft is usually further equipped with a digital camera or video recorder. Basically, airborne Lidar works similar to terrestrial laser scanning (TLS), which will be outlined in Chapter 6. However, airborne Lidar is more demanding as the six parameters of the exterior orientation (three positions and three orientations) have to be determined in a dynamic environment, that is, while the system is moving through the air. The positioning component of the exterior orientation is determined by a GNSS receiver installed onboard and one or more receivers installed on base stations. GNSS base stations are installed on points of which the coordinates are known with high precision. The use of one or more GNSS base stations enables obtaining high precision measurements using differential GNSS (DGNSS) (see Section 5.3). The IMU measures the orientation in space of the sensor. An IMU consists of a combination of accelerometers and gyroscopes to detect changes in acceleration and orientation. The

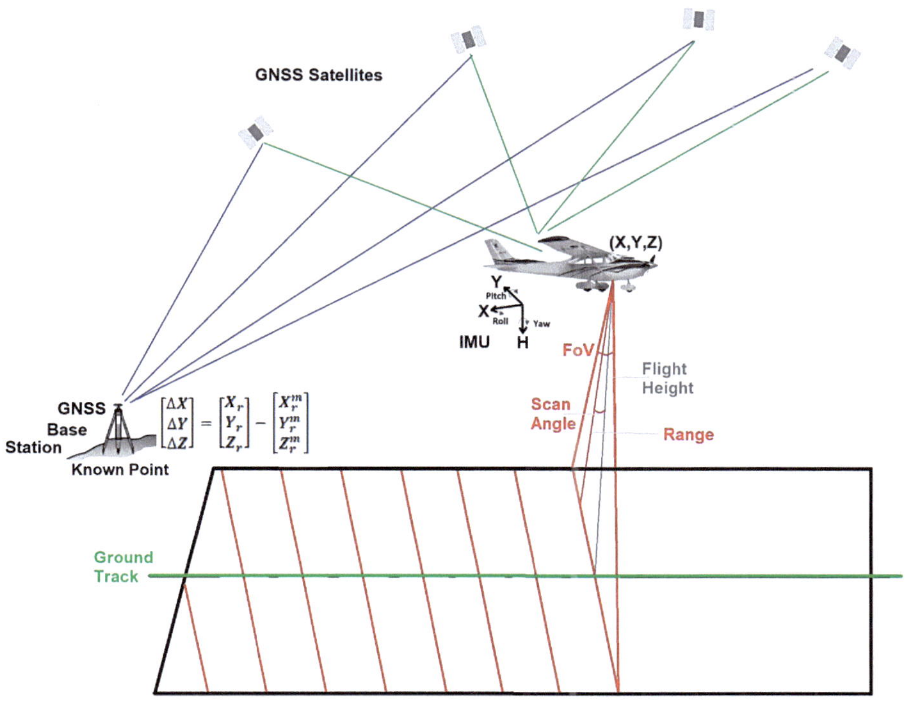

Figure 5.1 During an airborne Lidar survey, GNSS and IMU continuously measure the six parameters of the exterior orientation enabling automatic calculation of the 3D coordinates of ground points from range and scan angle; the scan pattern of parallel lines is typical for a rotating mirror

orientation is usually expressed in terms of pitch, roll and yaw or, from the photogrammetric point of view, ω, φ and κ (omega, phi and kappa). Figure 5.1 provides a schematic overview of the components, which enable the computation of 3D coordinates of points from measuring laser ranges.

5.3 Scan Pattern

The scan pattern on the ground is mainly determined by the scanning mechanism, flight speed and elevation differences of the terrain. This section focuses on the relation between the scanning mechanism and scan pattern. Manufacturers of airborne Lidar systems have developed four types of mechanisms for the distribution of the laser pulses across the flight direction. All mechanisms are based on a mirror, which moves around a fixed axis, mounted in front of the laser pulse emitter. The movement of the mirror results in a sideward spreading of the pulses. According to their type of movement, the mirrors can be distinguished into three types (Figure 5.2):

1. Oscillating
2. Rotating
3. Nutating

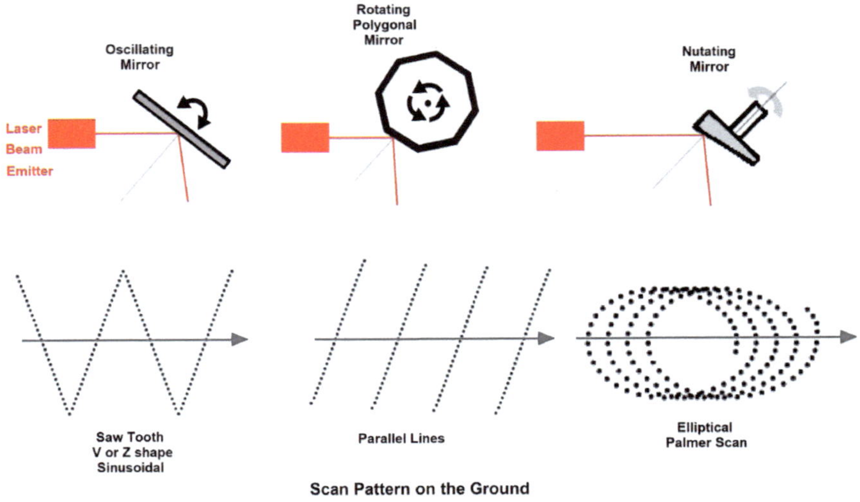

Figure 5.2 Scanning mechanisms (top) and resulting scan patterns on the ground

In addition to a rotating mirror, an array of glass fibre can be mounted in front of the pulse emitter and pulse receiver. This is the fourth type of scanning mechanism resulting in a push broom fibre scanning pattern.

Oscillating and rotating mirrors produce a scan in which the pulses are distributed along straight lines across the flight path more or less perpendicular to the flight direction. These straight lines are not exactly perpendicular to the flight path. Due to the movement of the aircraft while scanning across the flight path, the pulse emitted at the one side of the strip or field of view (FoV) will lie behind the pulse which reaches the ground on the other side of the strip, a fraction of a second later. Oscillating mirrors generate a saw tooth (V-shaped, also called Z-shaped) scan pattern. An oscillating mirror must be slowed down and brought to a complete standstill before the scanning can start in the opposite direction. As a result, the pulse may suffer from mispointing at both ends of the swath: The direction of emission may slightly differ from the measured scan angle causing an error in the computed 3D coordinates. Because of the mispointing characteristics, some remove the points at both ends of the strip. At the same aircraft speed, nutating mirrors acquire two times as many points as rotating mirrors, which generates a two times higher point density in the along-track direction. Added to this, the forth and back movements of the nutating mirror yield a scanning in which the terrain is scanned from two different directions. As a result, objects which are not captured in the one scan because of occlusion may be captured in the other scan.

Rotating mirrors produce parallel scan lines by making complete rotations of 360 degrees. Only during a part of the rotation of the polygon, on which the mirrors are mounted, the pulses are emitted in the direction of the terrain and only that part is useful for the survey. However, the uniform movement of the mirror results in a constant rotation speed so that the measurement precision of this scanning mechanism is less affected by mispointing than is the case for oscillating scanning mechanisms. The nutating mirror generates an elliptical scan pattern. This type of scanning is also called Palmer scanning.

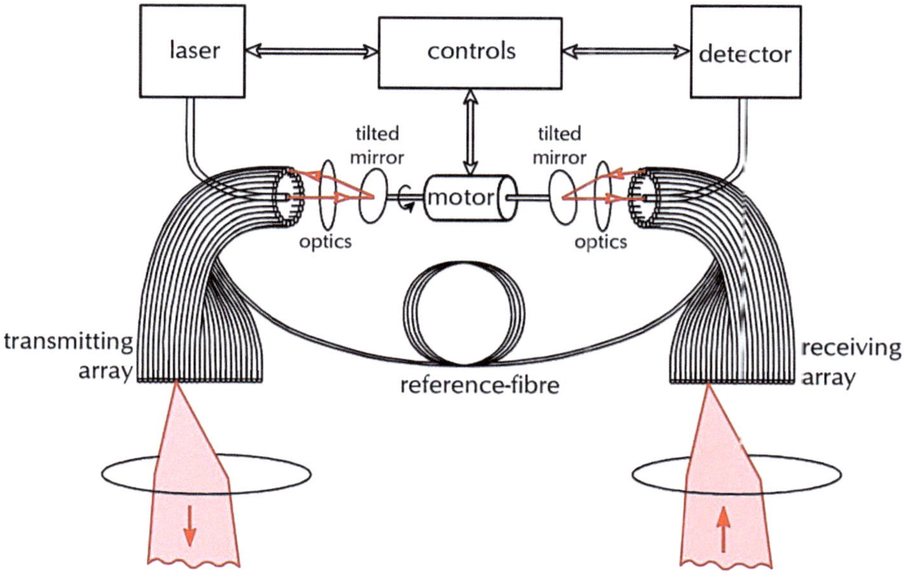

Figure 5.3 The aim of the reference fibre is to compensate for variations in the measurement electronics and mechanics including pointing jitter (source: Schnadt and Katzenbeißer, 2004)

The mirrors rotate with high speed and make hundredth or more rotations or back and forth movements. The movements of the mirror are driven by mechanics and thus may be affected by small fluctuations in rotation speed, for example due to sudden aircraft shocks caused by air turbulence. Small fluctuations in the nominal rotation speed cause deflection errors: The measured scan angle differs from the actual angle under which the pulse is fired into the air. The fluctuations in mirror speed are also called pointing jitter. The resulting pointing errors can be compensated for by mounting two arrays of glass fibres in the Lidar system, one in front of the emitter and the other in front of the receiver (Figure 5.3). The number of pulses emitted across the flight direction is equal to the number of glass fibres. The glass fibres are assembled such that each fibre forces the incident laser pulse to move in a predefined, fixed direction. The effect of the glass fibre provision is that the direction, in which the pulse is emitted, is stabilised. Pointing jitter may cause the energy of a pulse to fall only partly on the corresponding glass fibre. This generates a loss of energy but will not affect deflection and thus not the measurement precision. The fibres can also be arranged in a circle resulting in a nutating scan pattern.

5.4 Point Spacing and Point Density

The point spacing and thus average point density of an airborne Lidar system is determined by the following factors (Lemmens and Fortuin, 1997; Baltsavias, 1999b):

- Pulse frequency, also called pulse repetition rate; this is the number of pulses emitted per second – the dimension is Hertz (Hz).

- Mirror frequency, also called scan frequency; this is the number of complete rotations or forth and back movements of the mirror (Hz).
- Speed of the aircraft (v_a [m/s]).
- Flight height (H [m]).
- FoV, which determines the width of the strip and is equal to two times the maximum scan angle (φ_{max}).

The effects of aircraft speed, pulse frequency and mirror frequency on the pulse distribution on the ground of a nutating mirror are shown in Figure 5.4. The point spacing across the flight direction – here called intra-line point spacing – differs from the point space in along-track direction – here called inter-line point spacing. The inter-line point spacing corresponds to the distances between the scan lines and is defined by the mirror frequency and the speed of the aircraft and can be computed by dividing the aircraft speed (v_a) by the mirror frequency (f_m):

$$PS_{\text{inter}} = \frac{v_a}{f_m} \tag{5.1}$$

This division results in a parameter with the dimension metre if the flight height is expressed in metres. Suppose the mirror frequency is 200 Hz, which means the mirror rotates with 200 rotations per second, and the speed of the aircraft is 250 km/h (60 m per second); then the inter-line point spacing becomes 60/200 m = 30 cm. For systems with a rotating mirror, the inter-line point spacing is constant along the width of the strip, but this is not the case for oscillating mirrors. The inter-line point spacing between two lines will be small at the turn of the mirror to increase linearly till the mirror covers the other end of the strip. The average intra-line point spacing (PS_{intra})

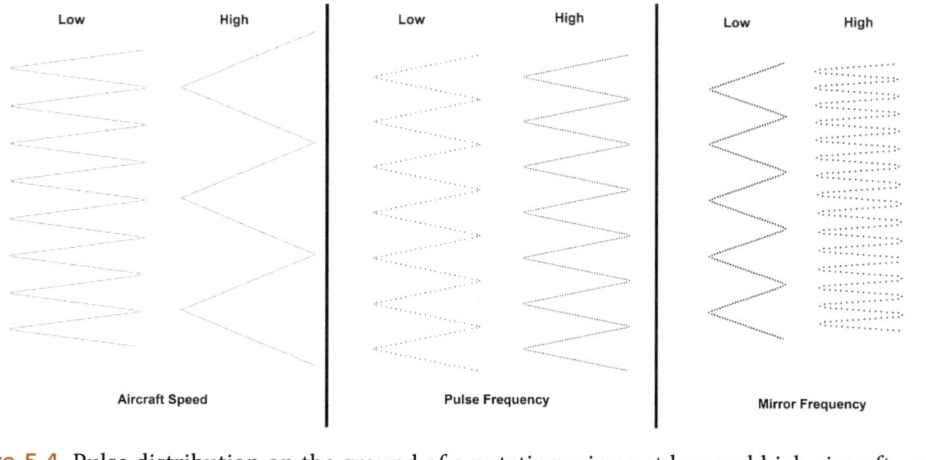

Figure 5.4 Pulse distribution on the ground of a nutating mirror at low and high aircraft speed, low and high pulse frequency and low and high mirror frequency

is determined by the pulse frequency (f_p), mirror frequency (f_m) and strip width (Sw), according to:

$$PS_{\text{intra}} = S_w \cdot \frac{f_m}{f_p} \tag{5.2}$$

The strip width (S_w) is determined by the flight height (H) and FoV, which is two times the maximum scan angle ($\varphi_{\max}$) according to:

$$S_w = 2H \cdot \tan\varphi_{\max} \tag{5.3}$$

Suppose an airborne Lidar system has a pulse frequency of 1,000 kHz, a maximum scan angle of 37.5° (i.e. an FoV of 75°) and a mirror frequency of 200 Hz, resulting in 200 lines per second. The system is operating at a flight height of 1,000 m. The resulting intra-line point spacing becomes $2 \times 1{,}000 \times 0.77 \times (200/1000{,}000) = 31$ cm. The intra-line point spacing halves to 15 cm by reducing the FoV to 40°.

The average point density (PD) is given by:

$$PD = \frac{1}{PS_{\text{inter}} \cdot PS_{\text{intra}}} = \frac{f_p}{2 \cdot v_a \cdot H \cdot \tan\varphi_{\max}} \tag{5.4}$$

5.5 Georeferencing

Georeferencing relies on the accurate recording of the six exterior orientation parameters of the Lidar system during the survey using GNSS and IMU onboard the aircraft. Measuring instruments are afflicted with random and systematic errors, due to optical, electronic and mechanical causes as well as their operation in a physical, dynamical environment, which is sometimes hostile. These errors can be compensated for by a proper survey design, particularly by taking care that there are enough along-track and across-track overlaps between the strips. Nevertheless, discrepancies between adjacent strips will frequently occur, although their extent has been substantially reduced compared to the mid-1990s, when commercial airborne Lidar systems became operational (Toth, 2009). One of the methods – and also the most popular one – to get rid of these discrepancies is to carry out a block adjustment much in the same way as is done in photogrammetry for the co-registration of overlapping images. In photogrammetry, co-registration is usually called relative orientation. Block adjustment aimed at stitching together overlapping strips to eliminate systematic errors requires the identification of corresponding points in the overlaps, which act as tie points (Lichti and Skaloud, 2010). The tie points may be individual points or patches of points. The similarity measures may be based on several feature types, such as reflectance value, surface normal or the three eigenvalues of the covariance matrix of the 3D coordinates of a small group of adjacent points. As is the case for photogrammetry, the strips should have sufficient overlap.

To improve point accuracy, differential GNSS is applied which requires a base station. Deviations may have several causes. One of them is that GNSS signals travel through the atmosphere, which is full of small particles which affect the speed of electromagnetic (EM) beams. Particularly, the delays due to the particles in the ionosphere and the troposphere affect positioning precision (Klobuchar, 1996). In course of time, several remedies have been developed to decrease the degrading effects of the atmosphere and to improve the precision of positioning, including the following:

- Omitting measurements to satellites which are less than 15° (cut-off angle) above the horizon. This remedy is underpinned by the observation that the accuracy decrease is the smallest when the satellite is in the zenith, that is, right above the receiver. The accuracy decreases gradually when the satellite moves from the zenith position to the horizon, since the closer to the horizon, the longer the GNSS signal will travel through the atmosphere.
- Removal of the ionospheric effects by using not one signal transmitted by the satellite (single frequency) but using several frequencies of the GNSS satellites; the delay of microwave signals in the ionosphere depends on frequency. The signal delays caused by the troposphere cannot be eliminated.
- Use of differential GNSS (DGNSS); the simultaneous use of two receivers, one positioned on a point at which the 3D coordinates (planimetric X and Y coordinates and height) are known in a geodetic reference frame. Such a point is called a base point, reference point or known point.

Particularly, DGNSS allows arriving at computations of coordinates in a National Geodetic Reference System suitable for surveying purposes and merging the point cloud with other geodata in a GIS package. The use of DGNSS enables the correction not only for measurement errors caused by ionospheric and tropospheric delays, but also for errors due to receiver clock errors and orbital errors. The principle of DGNSS is usually implemented by accompanying the user's receiver, called rover, with a second GNSS receiver positioned at a location of which the coordinates are known in a geodetic reference system (Figure 5.5). The latter GNSS receiver is called the base station. Since the coordinates of the base stations (X_r, Y_r, Z_r) are known, they can be compared with the coordinates measured with the GNSS receiver located on the reference point (X_r^m, Y_r^m, Z_r^m). The differences are computed according to:

$$\begin{bmatrix} \Delta X \\ \Delta Y \\ \Delta Z \end{bmatrix} = \begin{bmatrix} X_r \\ Y_r \\ Z_r \end{bmatrix} - \begin{bmatrix} X_r^m \\ Y_r^m \\ Z_r^m \end{bmatrix} \tag{5.5}$$

When the assumption is feasible that the atmospheric distortions and other error sources are more or less the same for the base station position and the position of the GNSS receiver at an unknown position, the corrections computed from measurements at the base station are also applicable to the measurements taken at the unknown position. The shorter the distance between the base station and the rover, also called baseline, the more

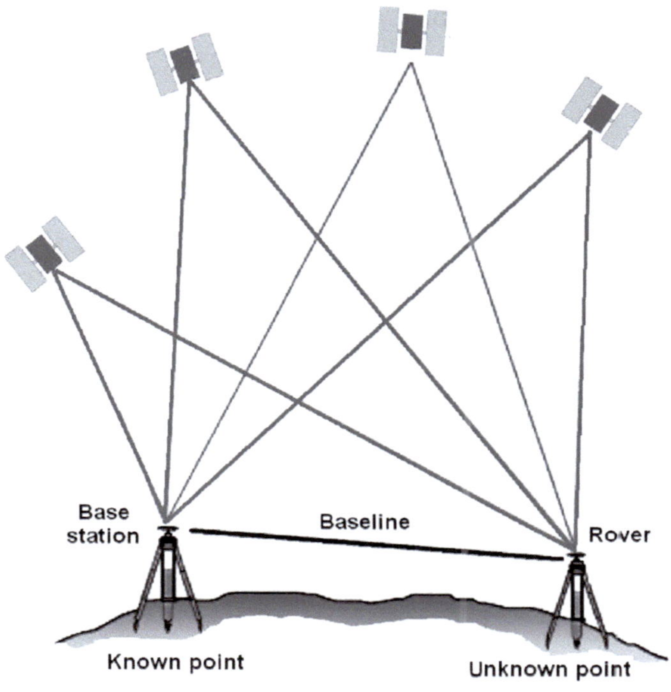

Figure 5.5 Principle of DGNSS: the position of a GNSS receiver, called rover, can be determined relative to the position of the base station (source: Lemmens, 2011)

resemblance between the atmospheric conditions – and hence distortions – the better the corrections will fit. Differential corrections may be transmitted in real time by radio link or mobile phone to the rover, enabling the user instantaneous capture of position.

The use of DGNSS enables the determination of the 3D coordinates of the points directly in national, or other preferred, reference systems. When combining the point cloud with other geodata referenced with respect to another coordinates system, a transformation from the one reference frame to the other reference frame is necessary. Assuming that all systematic errors have been removed, the geometric transformation model can be limited to including the following parameters: one scale factor, three translations and three rotations. This geometric transformation model is also called 3D similarity transformation, Helmert transformation or seven-parameter transformation (see Section 6.2.2).

5.6 Increasing Point Density

Increasing the PD of an airborne Lidar point cloud can be achieved by a proper survey design, including flying at low attitudes, flying at low speed and capturing the same area two or more times through high overlaps between adjacent strips or capturing the area a second time by carrying out across flights. Figure 5.6 shows the relationship between PD, distance to object (range) and aircraft speed of the Phoenix Alpha AL3-32 Lidar system. These solutions for increasing the PD are based on adjusting the survey and

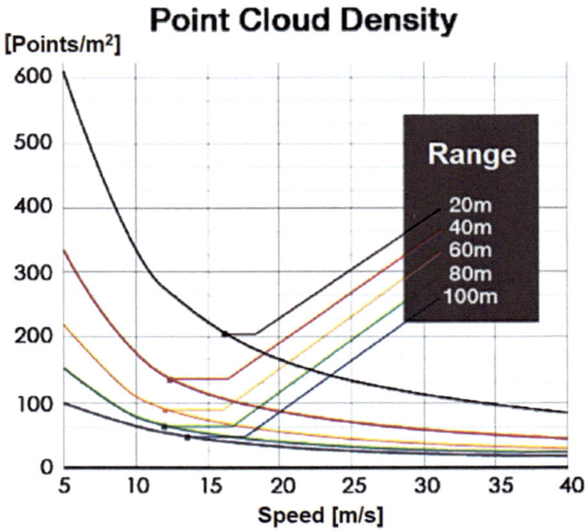

Figure 5.6 Relationship between point density, aircraft speed and flight height of the Phoenix Alpha AL3-32 Lidar system, weighing 3.2 kg (courtesy: Phoenix LiDAR Systems, USA)

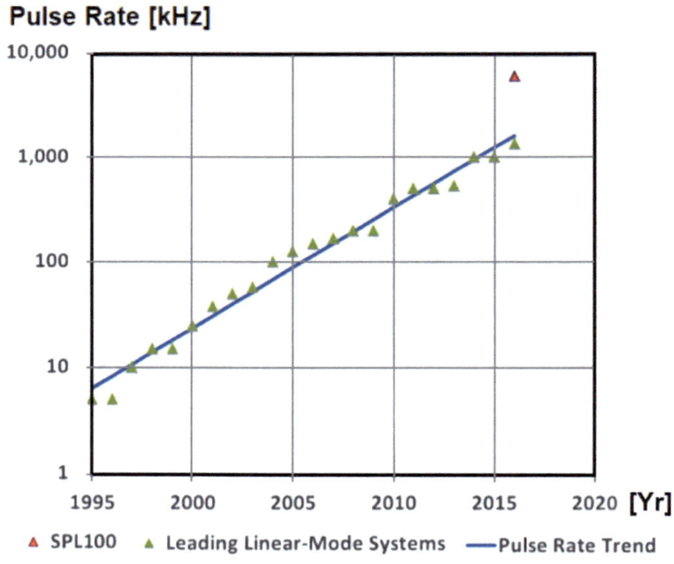

Figure 5.7 Increase of pulse frequency of airborne Lidar systems since the mid-1990s; the red triangle in the upper-right corner shows the quantum leap resulting from single photon Lidar; note the logarithmic scale of the vertical axis (source: Sirota and Roth, 2017)

considerably increase survey time. Another solution is to increase the pulse frequency of the Lidar system. Currently, the pulse frequency of airborne Lidar systems of the major manufacturers lies around one million pulses per second. Although the pulse frequency increased by around 25% per year since the mid-1990s and still continues to rise (see Figure 5.7), realising higher pulse frequency in a single Lidar system is limited by the power consumption necessary to arouse the laser pulses. In the remaining part of this section, we outline three solutions, which are sensor related, for increasing the PD: mounting two or more laser sensors in one system, multiple pulses in air and SPL.

5.6.1 Dual Channel

One of the solutions to increase PD is by mounting two or more Lidar sensors in one device. In March 2014, Swedish firm Airborne Hydrography AB (AHAB), which has been part of Leica Geosystems since October 2013, launched the dual head consisting of two scanners, each emitting up to 500,000 pulses per second, totalling a pulse frequency of 1 MHz. To date, this airborne Lidar system is called DragonEye (Figure 5.8). The laser wavelength is 1.064 μm, and the average PD is higher than 16 points per square metre at a flight height of 1 km AGL and an aircraft speed of 60 m/s. In order to increase PD, RIEGL introduced a dual-channel system, which was released in 2016: the VQ-1560i (Figure 5.9). One channel emits green pulses (0.532 μm) and the other channel near-infrared (NIR) pulses (1.064 μm). The amount of EM energy which is reflected by objects depends on the wavelength. So the use of these two wavelengths yields independent reflectance values, which is beneficial for further processing and analysis. Each channel of this full-waveform processing system is able to emit up to one million pulses per second. This means that in total two million pulses can be sent in the direction of the earth's surface, resulting in a pulse frequency of up to 2 MHz. In this way, on the ground, in total, more than 1.3 million measurements per second may be obtained, according to the product specifications of the manufacturer. The two channels create straight parallel scan lines, which make an angle of 28° with each other to provide good ground coverage of the pulses (Figure 5.10). The operation altitude is up to 5 km, making the system suitable for multiple tasks including ultrawide area mapping, mapping of complex urban environments, city modelling, corridor mapping, agriculture and forestry.

Figure 5.8 Leica DragonEye consisting of two oblique laser scanners and accompanied by one camera, which captures four spectral bands: RGB and NIR (courtesy: Leica Geosystems)

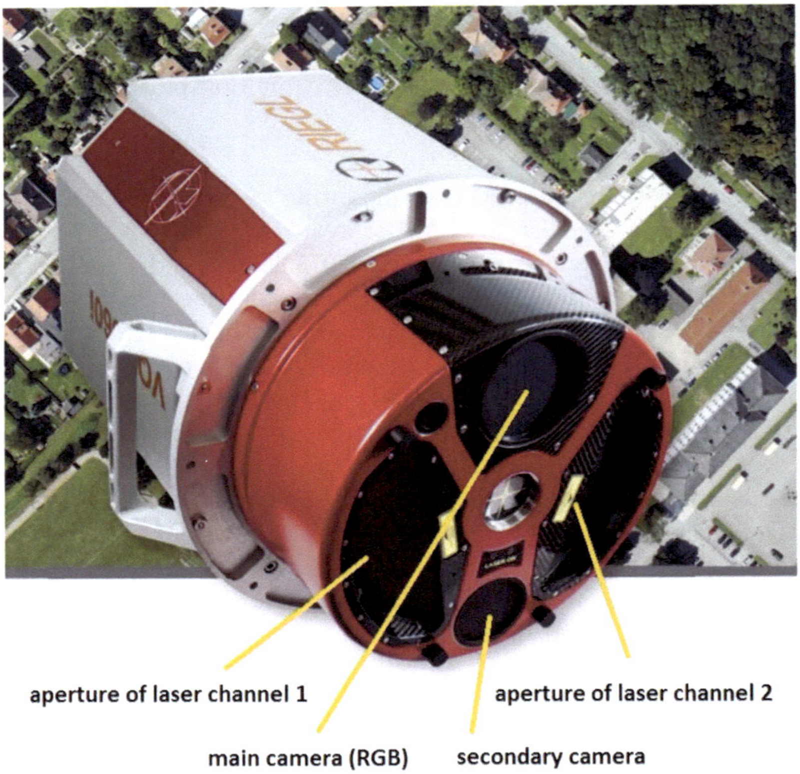

Figure 5.9 RIEGL VQ-1560i, consisting of two laser scanners and two cameras (source: Riegl, Austria; modified)

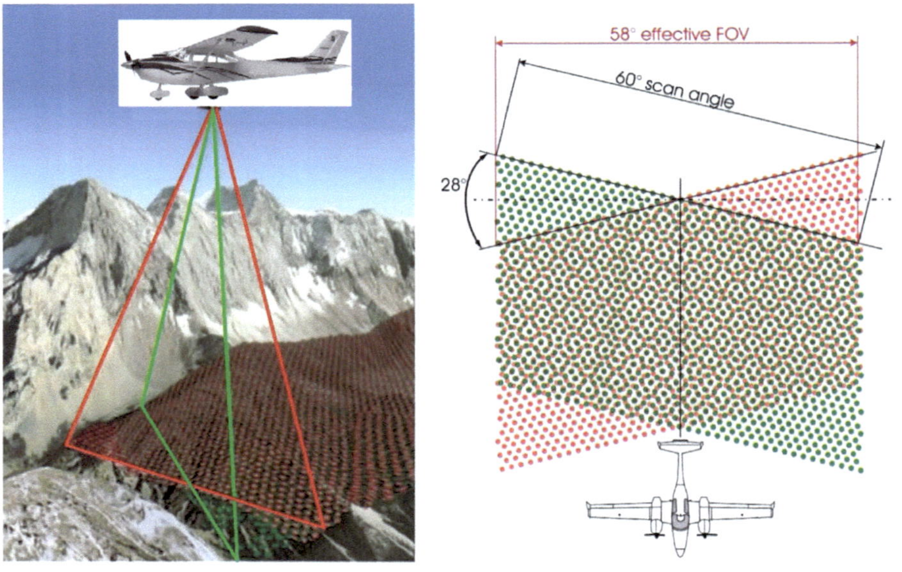

Figure 5.10 Scan pattern of the dual-channel system of Riegl (courtesy: Riegl, Austria; modified)

5.6.2 Multiple Pulses in Air

PD depends on a number of system parameters including pulse frequency, rotating speed of the scanning mirror, FoV, the speed of the aircraft and the flight height. One of the limitations of the early Lidar systems to increase PD was that the return of the previous pulse had to be captured by the sensor before the next pulse could be emitted. This was necessary to avoid confusion between returns of subsequent pulses. At a flight height of 1 km and given that the speed of light through the atmosphere is approximately 300,000 km/s, the pulse frequency, also called pulse repetition rate, could thus not exceed 150 kHz. However, using smart solutions embedded in the system, the phenomenon of multiple pulses in the air was introduced, largely eliminating the limitation discussed earlier. Indeed, a part of the centupling of PD since the mid-1990s has been established by the multiple pulses in air technology, which has become commercially available by 2007. This technology allows the firing of the next laser pulse before the reflection of the previous pulse has been received. The basic advantage of the multiple pulses in air technology becomes clear when surveying at higher altitudes. It reduces acquisition time (and hence costs) and also reduces occlusion by relief variations; the latter enables the reduction of the width of across overlap. Furthermore, air turbulence at higher altitudes is less severe than closer to the ground, so the flight is more comfortable and the crew can stay in the air longer.

5.6.3 Single Photon Lidar

Around 2015, an innovative solution came into the commercial market which enables the capture of multiple adjacent points from a single laser pulse. The solution exploits an array of Geiger-mode Avalanche Photodiode (GmAPD) detectors, which are able to detect returns just consisting of a few individual photons or even a single photon (Clifton et al., 2015). The very sensitive detectors not only allow high PD but also flights at high altitudes, enabling the capture of large areas within short time spans. In conventional Lidar systems, one pulse provides data on the reflectivity, the range and, when using (full) waveform digitisation (see Section 5.7.1), the surface structure of the footprint of the single pulse on the object – one pulse results in an information nucleus for one object point. The return signal contains thousands of photons. In contrast, the SPL approach creates an array of points from a single pulse fired by the system – one pulse is subdivided into dozens or hundreds of sub-pulses. The partition of one pulse into many is enabled through the use of diffractive optics which split the outgoing pulse into an array of sub-pulses. The optics can be tailored to the needs of the user; the size of the array may be, for example, such that a quadrangle of 10 × 10 sub-pulses is generated from one pulse emitted by the sensor. The partition of the pulse in a 10 × 10 array enables one pulse to capture a point cloud of up to 100 points. The returns from the individual sub-pulses are captured by a receiver also consisting of a 10 × 10 array. So one pulse does not cover one footprint, as conventional airborne Lidar does but rather captures multiple individual adjacent points, resulting in a high PD. The sensitivity of the sensor is so high that the range to the surface of an object can already be determined even if just one photon is present in the return signal (Stoker et al., 2016). As a result, the distance from the sensor to the object may be much larger than is the case for conventional

Lidar. Similarly, the swath width may be larger, which reduces the number of flight lines and hence data acquisition time, without affecting PD. A typical conventional airborne Lidar survey may be flown at 1,000–1,500 m, while a SPL survey may achieve equivalent point densities at a flight height of 4,000–5,000 m and the number of flight lines may be reduced by a factor of three.

Photon Lidar counts only the individual photons and registers the time of flight but not the strength of the return signal and thus no waveform digitisation is possible. However, multiple returns can be detected by resetting the photon counter at small time intervals corresponding to a travel distance at the level of a few decimetres. The photons counted by the same detector after the reset are regarded as stemming from a new return (Sirota and Roth, 2017). Since the length of each interval is only a few decimetres, multiple returns can be detected. In forest areas this means that returns from the canopy and the ground surface can be separated and measured. A substitute for the signal strength of the return can be introduced by implementing a red, green and blue (RGB) or NIR camera alongside the SPL system. The reconstruction of the surface structure, which is the main asset of full-waveform digitisation, is possible through the reset facility of the photon contour discussed earlier. The ability to operate at low power levels is an advantage but requires, on the other hand, highly sensitive sensors, which may wrongly detect solar photons as return signals. The effects of this type of noise may be diminished by a careful design of beam divergence, spectral width, filters and other system parameters.

In early 2015, Harris Corporation, based in the United States, brought its IntelliEarth Geiger-mode Lidar system to the marketplace. A 32 × 128 pulse splitter array divides each pulse into 4,096 sub-pulses. In early 2017, Leica brought an SPL system to the marketplace. This system, baptised SPL100, proves that innovative solutions to increase the pulse return rates do not always walk in step with miniaturisation. Leica's SPL100 Lidar system weighs 83.8 kg and is table high. Initially, the system has been developed by Sigma Space, based in the United States. A 10 × 10 pulse splitter array divides each pulse into 100 sub-pulses. The system generates 60,000 pulses per second resulting in a hundredfold sub-pulses, which means 6 million sub-pulses are emitted. Figure 5.11

Figure 5.11 Projection of the sub-pulses generated by the single-photon Lidar system SPL100 (left) on a wall (courtesy: Leica Geosystems; source: Sirota and Roth, 2017)

shows the projection of the 100 sub-pulses on a wall. The figure also shows that this SPL system emits green light: The wavelength of the laser is 0.532 µm. The operational flight height is 2–4.5 km AGL. The laser sensor is accompanied by a multispectral camera. According to the manufacturers, SPL systems have great potential for mapping large areas. Operating the SPL100 system from an altitude of 3,800 m over a productive forest site of 1,300 ha in southern Sweden, Wästlund et al. (2018) found the following:

- PD is similar to a conventional Lidar flying at an altitude of 400 m AGL.
- Dense canopies can be better penetrated compared to conventional systems.
- High flight altitude and aircraft speed without compromising high PD enable high temporal coverage of large areas.
- Atmospheric conditions affect the extraction of useful information.
- Solar noise requires advanced noise filtering.

Although operational SPL systems are available for the commercial collection of point clouds, the benefits, which SPL can bring for applications such as forest management, are still in the research stage. Before SPL will become a proven technology, a lot of further in-depth research has to be carried out.

5.7 Increasing Number of Attributes

Initially, most airborne Lidar systems detected the first return reflected from objects on the earth's surface hit by the laser pulse. Forestry became an important application area in the early days, and foresters discovered that it would be beneficial when not only the reflection on the canopy (first return) but also the ground underneath the trees would be captured to generate a bare ground terrain model (DTM). So manufacturers brought in facilities to capture also the last return. The approach was based on the observation that the foliage cover is usually not completely impermeable but semi-porous: A part of the pulse may reflect on leaves but gaps between the leaves allow the pulse to reach the ground and be scattered back to the Lidar receiver. That is, the first return reflects on the foliage, indicating the canopy and the last return on the bare ground, indicating the bare ground. The difference between the last and first returns provides a measure of the height of the tree and ultimately the forest stands as a whole. The recording of two returns only was soon extended to more returns. For deriving more information from each pulse, one started to record also returns between the first and the last return, finally resulting in full-waveform digitisation (FWD). The principles of multiple return recording and FWD are sketched in Figure 5.12.

5.7.1 Full-Waveform Digitisation

FWD samples the entire return signal from its leading edge to its trailing edge in regular intervals, for example 1 ns. Depending on the sample interval and the height of the object, the digitisation is tuned to take, for example, 64, 128 or 256 samples at regular intervals. FWD increases the demands on data storage capacities tremendously but benefits from the fact that more attribute data for each and every point becomes available, resulting

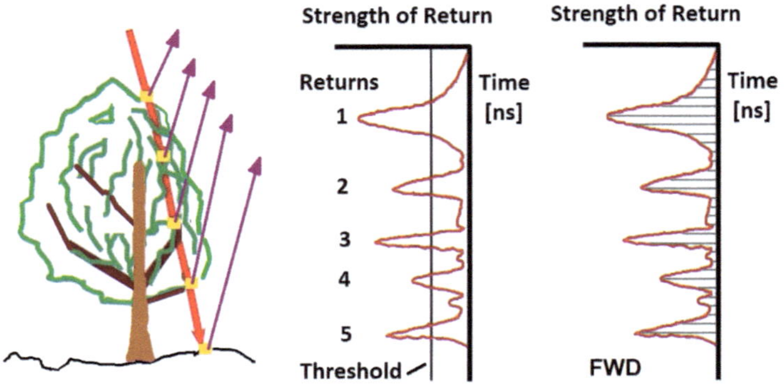

Figure 5.12 Sketch demonstration of the principles of capturing multiple returns and full-waveform digitisation (FWD)

in a more reliable classification of objects and refined classification schemes. Systems using FWD technology became operational in 1999, primarily for forest applications, and by 2004 full-waveform facilities were implemented in commercial systems. FWD of each backscattered pulse enables the determination of the vertical surface structure such as roughness, height and shape of objects, canopy density and height of trees and reflectivity. The digitised samples of the returns provide – so to say – fingerprints of the object, which were hit by the pulses. Capturing the full waveform provides additional and more detailed information on the structure and the physical characteristics of the surfaces of objects (Mallet and Bretar, 2009). For example, scrub and other low vegetation may be differentiated from the ground surface and height measurements of the canopy may be improved (Hug et al., 2004). The capability of airborne Lidar systems on digitising the return signal in parts is especially appreciated by forestry practitioners. Such signals enable foresters to estimate vegetation structural variables (Chasmer et al., 2006). The waveform indicates the top of the canopy, underlying layers and the ground surface, allowing a more accurate determination of ground surface DTM and canopy height. FWD of small-footprint Lidar provides information on the strength, width and cross-section of the return signal which provides means to separate vegetation from the ground surface (Wagner et al., 2008). Full-waveform airborne Lidar is also investigated for the classification of tree species. Using the intensity of the return signal, its width and the number of targets popping up in the signal, Heinzel and Koch (2011) could classify six tree species with an overall accuracy of 57% and four tree species with 78% accuracy. Conifers and broadleaved trees could be separated with 91% accuracy. In urban areas, roofs can be distinguished from trees and cars from garages. The analysis of FWD of small-footprint Lidar pulses provides also information suited for determining the biomass of forests in areas with a subtropical climate (Cao et al., 2014).

5.7.2 Multispectral Lidar

Increasing the number of attributes per ground point can also be established by multispectral Lidar. Monospectral airborne Lidar systems exploit one out of three

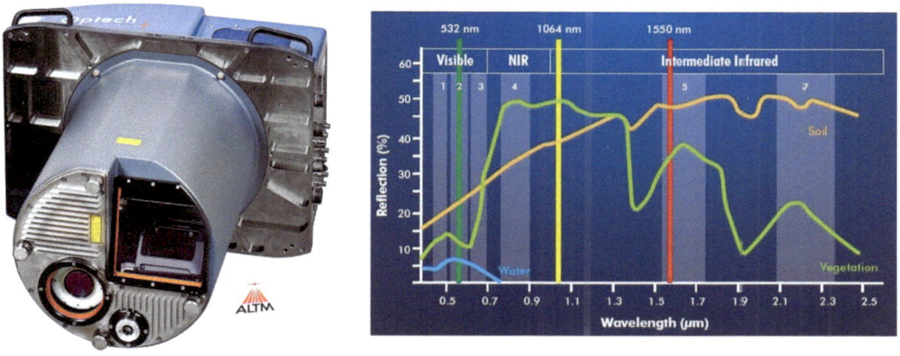

Figure 5.13 The Titan (left) airborne Lidar system emits pulses in three narrow spectral bands: 0.532 µm, 1.064 µm and 1.550 µm (right) (courtesy: Teledyne Optech)

narrow spectral bands with the following wavelengths: 0.532 µm, 1.064 µm or 1.550 µm. Multispectral airborne Lidar point clouds can be obtained in three ways:

1. Combining a monospectral airborne Lidar point cloud with multispectral areal or satellite imagery to enrich each point with the multispectral information of the corresponding pixel.
2. Combining the point clouds captured by different airborne Lidar systems, which exploit different wavelengths, during independent surveys.
3. Combining the three Lidar sensors capturing the above-mentioned wavelengths in the same system (multispectral Lidar).

Simultaneous acquisition of multispectral Lidar point clouds avoids the time gap and thus discrepancies with which the other two methods – combining airborne Lidar points acquired during different surveys or enriching a monospectral Lidar point cloud with spectral data stemming from imagery – are afflicted (Wichmann et al., 2015).

In December 2014, Teledyne Optech introduced the world's first commercial multispectral airborne Lidar system, called the Titan ALTM system. The Titan emits pulses in three wavelengths from a single sensor system. The three independent pulses – wavelengths 0.532 µm, 1.064 µm and 1.550 µm – are emitted, each with a 300 kHz effective sampling rate for a combined ground sample rate of 900 kHz. The signal strength of the returns in three different bands provides an indication of the type of objects the pulse hits, which is beneficial for analysis and classification purposes (Figure 5.13). When the PD is high, the system produces, after gridding, a multispectral image-like picture (Figure 5.14). Applications include topographic surveying, 3D land cover classification, environmental modelling, vegetation mapping and shallow water bathymetry.

5.8 UAS Lidar

UASs have garnered the earnest interest of a broad group of geomatics professionals, specialised in geodata acquisition. UASs have proven to provide reliable technology for

Figure 5.14 Three images created from the three spectral bands of Titan Lidar pulses; they form together a multispectral image (courtesy: Teledyne Optech)

capturing 3D geodata of small areas, individual buildings or complexes of man-made structures using cameras. Up until recently, the use of airborne Lidar was largely confined to manned flights, mainly because Lidar sensors used to be heavy – in the order of tens of kilograms – and consumed a lot of power which constrained their use on small, unmanned aircraft. As a result, until 2015, UAS Lidar surveys were unfeasible for all but the very largest UASs. Overcoming this barrier requires small dimensions as well as low weights and modest power consumption. To demonstrate the recent and ongoing miniaturisation of geodata acquisition systems, a few of the airborne Lidar systems suited for operation from small, lightweight UASs are considered here, ranked according to weight.

RIEGL's VUX-1, introduced in 2016, weighs around 3.5 kg and is compact enough to be mounted on a variety of small fixed-wing and rotary unmanned aircraft (Figure 5.15). The system captures up to 200 parallel scan lines per second while up to 500,000 measurements per second can be recorded. The manufacturer claims an accuracy of 10 mm. The maximum flight height is over 300 m, and the FoV is up to 330°. Next in line is the Alpha AL3-32, manufactured by Phoenix Lidar Systems based in Los Angeles, California, USA. This Lidar device weighs 3.2 kg, the maximum flight height is 100 m, the pulse frequency is 700 kHz, the FoV is 360° and two returns per pulse can be recorded. Smaller than the AL3-32 is the Alpha AL3-16. This system fits in a box measuring less than 25 cm square and weighs 2.2 kg. The maximum flight height is 100 m, the pulse frequency is 300 kHz, the FoV is 360° and two returns per pulse can be recorded. For both the AL3-32 and the Alpha AL3-16, the manufacturer claims an absolute accuracy of 25/35 mm at a flight height of 50 m. As the systems operate fully autonomously, they can be mounted not only on a UAS but also on cars, boats or backpacks.

Figure 5.15 The RIEGL VUX-1 Lidar system is small and lightweight enough to be carried by a UAS (courtesy: Riegl, Austria)

Figure 5.16 YellowScan Mapper (courtesy: YellowScan, France)

At 2.1 kg, the YellowScan Mapper, introduced in 2014, weighs a little bit less than the Alpha AL3-16 (Figure 5.16). The maximum flight height of the Mapper is 100 m, the pulse frequency is 40 kHz, the FoV is up to 100° and three returns per pulse can be recorded. According to the manufacturer, its main applications are archaeology, construction, forestry and corridor mapping. In 2016, YellowScan, based in France, introduced a system called the Surveyor with an even lower weight: just 1.5 kg. The maximum flight height is 50 m, the pulse frequency is 300 kHz, the FoV is up to 360° and a maximum of two returns per pulse can be recorded. The applications are focused on mining, civil engineering and corridor mapping. The Velodyne Puck fits into the palm of a human hand and weighs 830 g (Figure 5.17). The maximum range is 100 m, the pulse frequency is 600 kHz and the system captures two returns. As stated earlier, the maximum distance a Lidar pulse can bridge from sensor to ground and back depends on the power used, the pulse frequency and the reflectivity of the part of the object hit by the pulse. For example, when the reflectivity is 20%, assuming flat terrain and the pulse

Figure 5.17 Miniaturisation enables manufacturers to produce laser scanners which become smaller and more lightweight by the year; shown is the Velodyne Lidar Puck VLP-16 (source: Velodyne LiDAR, USA)

frequency is 50 kHz, the maximum operational flight height of RIEGL's VUX-1 is 350 m. This figure reduces to 55 m when the pulse frequency is increased to 550 kHz and the pulses are emitted with reduced power.

5.9 Concluding Remarks

This chapter outlined the main measurement principles of operational Lidar systems. The 3D coordinates of airborne Lidar point clouds can be determined with a precision at the sub-decimetre level and the achievable PD lies in the order of hundreds of points per square metre. The random errors are small, and systematic errors can be modelled or largely eliminated through stitching adjacent strips through block adjustment. There remains one error source which may introduce outliers of considerable size: multipath. Multipath is caused by multiple reflections of the laser pulse. An electromagnetic beam hitting a surface may interact with it in one of three ways: The signal may be reflected, absorbed or transmitted. Having once been reflected, a laser pulse may be reflected a second time or even a third time from another surface of a nearby object such as a road or a building wall. In airborne Lidar, multiple reflection results in the recording of longer pulse travel times that ultimately appear as dips in the reconstruction of the

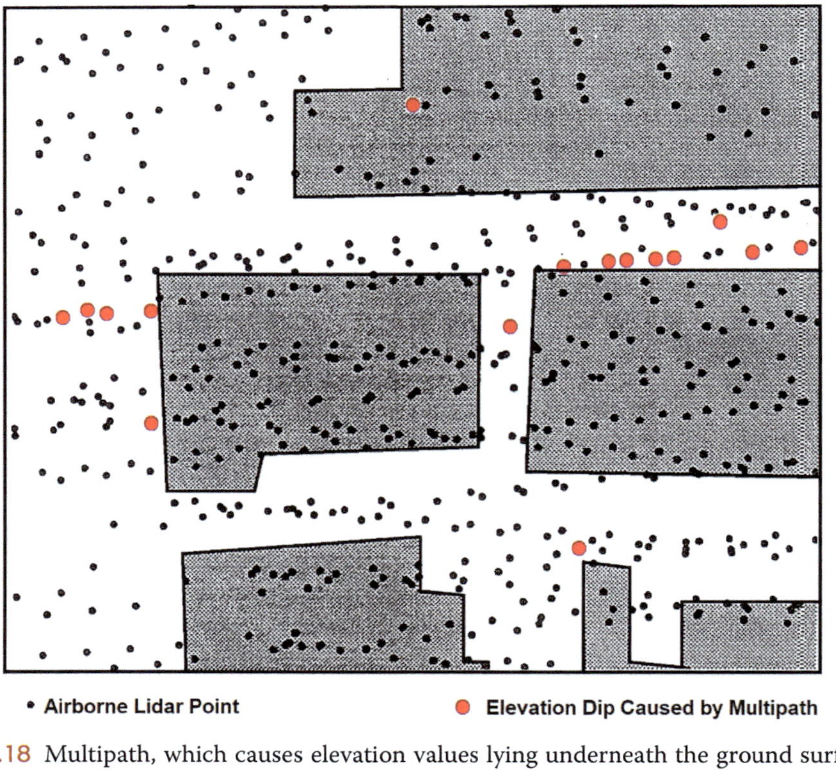

Figure 5.18 Multipath, which causes elevation values lying underneath the ground surface, are isolated occurrences, which can be largely eliminated through applying proper ground filtering techniques (source: Lemmens et al., 1997)

terrain. Since these outliers are usually isolated events (see Figure 5.18), they can be detected by spatial filtering. The topic of multipath has been outlined in Chapter 2.

Airborne Lidar is flourishing as a prevalent geodata acquisition technology and continues to show a fierce rise in terms of advancements and applications. Since the mid-1990s, airborne Lidar has made admirable progress and has gone mainstream, joining aerial photogrammetry as an essential geodata acquisition technology. A promising and advanced development is SPL. An SPL system allows the capture of extensive areas with high detail within reasonable amounts of time.

References

Ackermann., F. (1999) Airborne Laser Scanning Present Status and Future Expectations. *ISPRS Journal of Photogrammetry and Remote Sensing*, 54, pp. 64–67.

Baltsavias, E.P. (1999a) A Comparison between Photogrammetry and Laser Scanning. *ISPRS Journal of Photogrammetry and Remote Sensing*, 54(2–3), pp. 83–94.

Baltsavias, E.P. (1999b) Airborne Laser Scanning: Basic Relations and Formulas. *ISPRS Journal of Photogrammetry and Remote Sensing*, 54(2–3), pp. 199–214.

Cao, L., Coops, N.C., Hermosilla, T., Innes, J, Dai, J., She, G. (2014) Using Small-Footprint Discrete and Full-Waveform Airborne LiDAR Metrics to Estimate Total Biomass and Biomass Components in Subtropical Forests. *Remote Sensing*, (6), pp. 7110–7135.

Chasmer., L., Hopkinson, Ch., Treitz, P. (2006) Investigating Laser Pulse Penetration through a Conifer canopy by Integrating Airborne and Terrestrial Lidar. *Canadian Journal of Remote Sensing*, 32(2), pp. 116–125.

Clifton, W.E., Steele, B., Nelson, G., Truscott, A., Itzler, M., Entwistle, M. (2015) Medium Altitude Airborne Geiger-mode Mapping LIDAR System. *Proceedings of the SPIE*, 2015, p. 9465.

Heinzel, J., Koch, B. (2011) Exploring Full-waveform Lidar Parameters for Tree Species Classification. *International Journal of Applied Earth Observation and Geoinformation*, 13(1), pp. 152–160.

Hug, C., Ullrich, A., Grimm, A. (2004) LiteMapper-5600: A Waveform Digitizing Lidar Terrain and Vegetation Mapping System. *International Archives of the Photogrammetry, Remote Sensing and Spatial Information Sciences*, XXXVI(Part8/W2), pp. 24–29.

Klobuchar, J.A. (1996) Ionospheric Effects on GPS. In Parkinson, B.W., Spilker, J.J. (Eds.), *Global Positioning System: Theory and Applications*, Vol. 1. American Institute of Astronautics and Aeronautics, Washington, DC, pp. 485–515.

Lemmens, M. (2011) *Geo-information: Technologies, Applications and the Environment*. Springer, Science + Media Business, Berlin.

Lemmens, M. (2017) The Fierce Rise of Airborne Lidar: A View on Status, Developments and Trends. *GIM International*, 31(1), pp. 16–19.

Lemmens, M., Fortuin, E. (1997) *Error Budget of Airborne Laser Scanning (Fouten-Analyse van Vliegtuig-Laseraltimetrie)*. Ministerie van Verkeer en Waterstaat, Rijkswaterstaat Meetkundige Dienst, Afdeling Geo-Advisering, Delft.

Lemmens, M., Deijkers, H., Looman, P. (1997) Building Detection by Fusing Airborne Laser-altimeter DEMs and 2D Digital Maps. *International Archives of Photogrammetry and Remote Sensing*, Stuttgart, XXXII(Part 3-4W2), pp. 42–49.

Lichti, D., Skaloud, J. (2010) Registration and Calibration. Chapter 3 in Vosselman, G., Maas, H.-G. (Eds.), *Airborne and Terrestrial Laser Scanning*, Whittles Publishing, Dunbeath, UK, pp. 83–133.

Mallet, C., Bretar, F. (2009) Full-waveform Topographic Lidar: State-of-the-Art. *ISPRS Journal of Photogrammetry and Remote Sensing*, 64(1), pp. 1–16.

Meng, X., Currit, N., Zhao, K. (2010) Ground Filtering Algorithms for Airborne Lidar Data: A Review of critical Issues. *Remote Sens*ing, 2(3), pp. 833–860.

Romano, M.E. (2004) Innovation in lidar processing technology. *Photogrammetric Engineering and Remote Sensing*, 70, pp. 1201–1206.

Schnadt, K., Katzenbeißer, R. (2004) Unique Airborne Fiber Scanner Technique for Application-oriented Lidar Products. *International Archives of Photogrammetry, Remote Sensing and Spatial Information Sciences*, XXXVI-8/W2, pp. 19–23.

Shan, J., Toth, C.K. (Eds.) (2018) *Topographic Laser Ranging and Scanning: Principles and Processing.* 2nd Edition, CRC Press, Boca Raton, FL.

Sirota, M., Roth, R. (2017) The Evolution of Lidar: Single Photon Lidar Brings Higher Pulse Rates for Airborne Lidar. *GIM International*, 31(2), pp. 37–39.

Stoker, J.M., Abdullah, Q.A., Nayegandhi, A., Winehouse, J. (2016) Evaluation of Single Photon and Geiger Mode Lidar for the 3D Elevation Program. *Remote Sensing*, 8, p. 767.

Toth, C.K. (2009) Strip Adjustment and Registration. Chapter 8 in Shan, J., Toth, C.K. (Eds.), *Topographic Laser Ranging and Scanning: Principles and Processing.* Taylor & Francis, Boca Raton, pp. 235–268.

Wagner, W., Hollaus, M., Briese, C., Ducic, V. (2008) 3D Vegetation Mapping using Small-footprint Full-waveform Airborne Laser Scanners. *International Journal of Remote Sensing*, 29(5), pp. 1433–1452.

Wästlund, A., Holmgren, J., Lindberg, E., Olsson, H. (2018) Forest Variable Estimation Using a High Altitude Single Photon Lidar System. *Remote Sensing*, 10, p. 1442.

Wehr, A., Lohr, U. (1999) Airborne Laser Scanning: An Introduction and Overview. *ISPRS Journal of Photogrammetry and Remote Sensing*, 54(2–3), pp. 68–82.

Wichmann, V., Bremer, M., Lindenberger, J., Rutzinger, M., Georges, C., Petrini-Monteferri, F. (2015) Evaluating the Potential of Multispectral Airborne Lidar for Topographic Mapping and Land Cover Classification. *ISPRS Annals of the Photogrammetry, Remote Sensing and Spatial Information Sciences*, II-3/W5, pp. 113–119.

Ground-Based Lidar

CHAPTER 6

The chapter on laser light (Chapter 3) focused on the basics of exploiting laser for surveying purposes. Because of the differences in view on the scene and often also the distance between object and sensor, it is convenient to categorise the various Lidar methods into two broad groups: (1) ground-based Light Detection and Ranging (Lidar) and (2) airborne Lidar. This chapter elaborates on ground-based Lidar. Airborne Lidar has been treated in Chapter 5. Ground-based Lidar may be exploited in a static way, which means the instrument remains at a fixed, stationary position during scanning of the scene or parts of it. In static mode, often the laser scanner will be mounted on a tripod. Scanning from a fixed position is called terrestrial laser scanning (TLS). The laser scanner may also be mounted on a car, van, boat, trolley or other ground-based moving platforms, including the human back and a stick held by a human hand. In these dynamic cases, ground-based Lidar is called mobile laser scanning (MLS). Often MLS systems are accompanied by cameras, and if they are, the system is usually called a mobile mapping system (MMS). However, some use the term MLS for a mobile mapping system, especially when the laser is considered the main survey device and cameras as ancillary devices. In this chapter, we adopt the latter view and will term a mobile device equipped with one or more laser scanners MLS, irrespective of whether the system is additionally equipped with cameras or not.

The main laser measurement principles used in ground-based Lidar are (1) time-of-flight, which is pulse-based, and (2) phase shift. Phase-shift systems can capture over one million points per second with precision at the millimetre level. These favourable properties for many applications go at the cost of maximum ranges varying from several tenths to several hundreds of metres depending on the number of wavelengths used. Pulse-based laser scanners may bridge several kilometres – the most advanced systems today are 6 km – and acquire points with precision of (sub-)centimetre level.

This chapter starts with a view on two basic features of ground-based laser scanning, which affect the properties of the point cloud: scan mechanism and scan angle. Next, the basics and characteristics of TLS are elaborated, followed by exposure to MLS systems. As is the case for airborne Lidar, MLS systems capture points while the platform is continuously on the move. As a result, registration and georeferencing of MLS point clouds have much in common with airborne Lidar point clouds, often using – in addition to Global Navigation Satellite System (GNSS) positions and inertial navigation system (INS) measurements – wheel counts. Therefore, we will only briefly touch on the registration and georeferencing of MLS point clouds in favour of providing a brief look-through on their automatic classification, by some called semantic enrichment. With the steady increase of massive point clouds acquired with MLS systems, automatic classification is gaining increasing importance and has become a highly investigated research topic. Finally, an expound is given on indoor MMSs.

6.1 Basic Features

Although TLS operates from a stationary platform and MLS from a moving platform, both have several properties in common, at least partly, including scan mechanism, range measurement principles and the features determining point density. We consider these three aspects in greater detail in this section. Usually, the laser beams of ground-based laser scanners employ the near-infrared (NIR) part of the electromagnetic (EM) spectrum, which means the laser beams are not visible to the human eye.

6.1.1 Scan Mechanism

A laser unit produces individual laser pulses or laser beams at high frequencies. Emitting over one million laser beams per second is not exceptional today. In order to act as a measurement device resulting in 3D coordinates of scene points, the beams have to be distributed over the scene in a regular, controlled fashion. The dispersion of the laser beams can be done by an oscillating or rotating mirror, or by a solid-state apparatus. In contrast to revolving mirror sensors, solid-state sensors have no mechanical parts and their vertical field of view is lower. The angular speed of revolving mirrors, on the other hand, may show anomalies, which reduces the quality of the 3D coordinates of the point clouds. To tackle the occurring disturbances, regular calibration of the laser unit is required, which adds to surveying costs. The most used systems exploit revolving mirror technology. Therefore, we focus on laser scanners based on revolving mirrors in the description of the scanning mechanism.

The dispersion of the beams over the scene is usually done in a regular way. Regular or homogeneous dispersion is achieved through a constant angular velocity of the mirror. In order to capture a scene, the laser beams have to be distributed in two directions, which are more or less perpendicular to each other. For TLS, these will usually be the vertical and horizontal directions. The regular distribution of points in the vertical direction is attained through a rotating or oscillating mirror. The regular distribution in the horizontal direction is either done by a second revolving mirror which is orthogonally placed to the first mirror – usually this will be a rotating mirror to obtain a horizontal field of view (FoV) of 360° – or by incrementally rotating the TLS instrument around its vertical axis. Usually, the scene is scanned in the vertical direction and next in the horizontal direction. Also in most MLS systems, the distribution of the laser beams in one direction is implemented by a revolving mirror. The movement of the platform provides the distribution in the other direction.

In contrast to photographic images, the scene is not captured at an instantaneous moment defined by the exposure time. In principle, the points are emitted and recorded one after the other and the capturing of the scene may take from a few

Figure 6.1 Artefacts may occur when traffic crosses the scene during data capture as shown in this intensity image of a scan of the historical centre of Delft where pedestrians and cyclists cause spikes and occlusion (courtesy: Delft University of Technology, Netherlands)

minutes to up to several hours, depending on laser beam emission rate, the scan angles, and horizontal and vertical FoV. Earlier we noted "in principle" since in today's systems usually several laser beams are emitted simultaneously, for example, 32, 64 or 128 concurrent beams. These sensors are called multi-beam Lidar systems. The number of beams emitted per second, together with the number of concurrent beams, determines the maximum measurement rate, which is usually expressed in kilohertz (kHz). One of the consequences of the "one after the other" emission of beams is that artefacts may occur in dynamic scenes. For example, in surveys where pedestrians or cyclists cross the scene, they may not be recognised as people but as spikes (Figure 6.1). If such artefacts are undesirable, the surveyor has to close off the scene area to all traffic, which might require permission from the local authorities, before pushing the TLS button.

6.1.2 Scan Angle

In addition to the point accuracy, which is at the millimetre level for phase-shift systems and at the (sub-)centimetre level for the pulse-based system as we have discussed earlier, point density is another important quality indicator of a point cloud. Since the laser beams are distributed over 3D space at constant angles and/or increments, the point density decreases with the increasing distance of the object from the laser scanner. Let us call the vertical angular interval α, the distance (range) of the scanner to the object r and the vertical point spacing at distance r PS_r then (Figure 6.2):

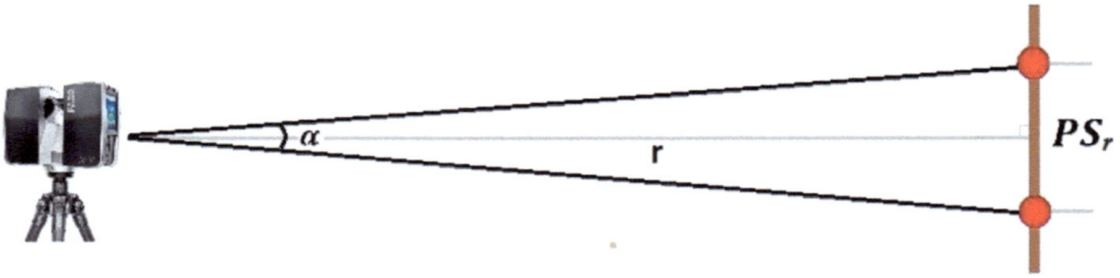

Figure 6.2 The scene is scanned with a constant angular interval; thus, the point density depends on the distance from the laser scanner

$$\tan\frac{1}{2}\alpha = \frac{1}{2}PS_r / r \quad (6.1)$$

Since the angle α is small, we may write: $\tan\frac{1}{2}\alpha = \frac{1}{2}\alpha$ and by rearranging terms, Equation 6.1 reduces to:

$$PS_r = \alpha \cdot r \quad (6.2)$$

Equation 6.2 refers to one-dimensional point spacing. Supposing that the horizontal scan angle is also α, we can compute the point density as $1/PS_r^2$. In order to gain some insight into what this means in reality, let us take ten objects at regular distances from each other: the first at a distance of 10 m from the scanner, the next one at 20 m and the last one at a distance of 100 m. Assume $\alpha = 0.1^o = 0.001745329\,\text{rad}$.

Table 6.1 shows that the point density decreases in a regular fashion with distance from the scanner. When the distance increases from 10 m to 20 m, which means the distance doubles, the point density decreases approximately four times. When the distance increases from 10 m to 100 m, that is, the distance becomes ten times longer, the point density decreases approximately 100 times. So the point density decreases more or less quadratically with distance. A second reason that the point density decreases with distance from the scanner is that the strength of the return decreases with distance. The decrease may be such that the return will be too weak to provide reliable range measurement or no return will be received at all. In the meantime, full-waveform digitisation, initially developed for airborne Lidar (see Section 5.7.1), has found its way to ground-based systems.

Table 6.1 Relation between point density (points/m²) and distance of the object to the laser scanner

Distance (m)	**10**	**20**	**30**	**40**	**50**	**60**	**70**	**80**	**90**	**100**
Spacing (m)	0.017	0.035	0.052	0.070	0.087	0.105	0.122	0.140	0.157	0.175
Points/m²	3,283	821	365	205	131	91	67	51	41	33

6.2 Terrestrial Laser Scanning

TLS has become a standard surveying technology. Many professionals value TLS because of its many favourable properties, including high precision and detailed capturing of the scene. As a result, TLS has become a preferred surveying technique in many earth-related applications including the survey of existing buildings, cultural heritage sites, plants or construction sites (Figure 6.3). The high accuracy paired with measurement speed and high point density has greatly contributed to its profound popularity. Often a TLS system is provided with an onboard calibrated camera, which enables the simultaneous generation of realistic digital 3D models. In this section, we consider the following aspects of TLS:

- Particular features which distinguish the diverse scanners from each other including FoV.
- Aspects involved to obtain 3D coordinates of points from range measurements, co-registration of scans and georeferencing.

Table 6.2 lists some of the main features of eight commercial TLS systems according to the status of September 2018. An impression of the appearance of these systems is given in Figure 6.4.

All systems listed in Table 6.2 do have a complete all-round view, which means a horizontal FoV of 360°. One would expect a larger maximum range than 30 m of the Faro Focus3D X 30 since it is pulse-based. This panoramic scanner is lightweight, easily portable and developed for capturing close-range scenes with high point density.

6.2.1 FoV and Maximum Distance

In Section 6.1.2, we have seen that one of the important features which determine the employability of laser scanners is the measurement principle – time-of-flight or phase shift. Here we consider two other important features: FoV and maximum distance. The part of the scene which can be captured by the TLS instrument during one scan depends on its alignment and FoV (Figure 6.5). In the course of time, the FoV gradually increased. Modern TLS systems have a horizontal FoV of 360°. Since it is not necessary in most outdoor scenes to capture the sky and the ground nearby the scanner, the vertical FoV

Figure 6.3 Leica RTC360 TLS surveying a construction site (left) and detailed view of the instrument, which is lightweight and can capture up to two million points per second

Table 6.2 Main features of eight commercial TLS systems

Scanner	Year of intro	Pulse/phase	Max. measurement rate (Hz)	Max. range (m)	Max. vertical FoV (°)
Faro Focus3D X 30	2015	Pulse	970,000	30	300
Maptek XR3	2017	Pulse	200,000	2,400	100
Riegl VZ-6000	2012	Pulse	300,000	6,000	60
Teledyne Optech Polaris	2016	Pulse	500,000	2,000	120
TI Asahi S-3180V	2013	Phase	>1,000,000	1,800	320
Topcon GLS-2000	2013	Pulse and phase	120,000	350	270
Trimble TX8	2013	Pulse	1,000,000	340	317
Z+F Imager 5016	2016	Phase	1,100,000	365	320

Source: Datasheets of the manufacturers and geo-matching.com

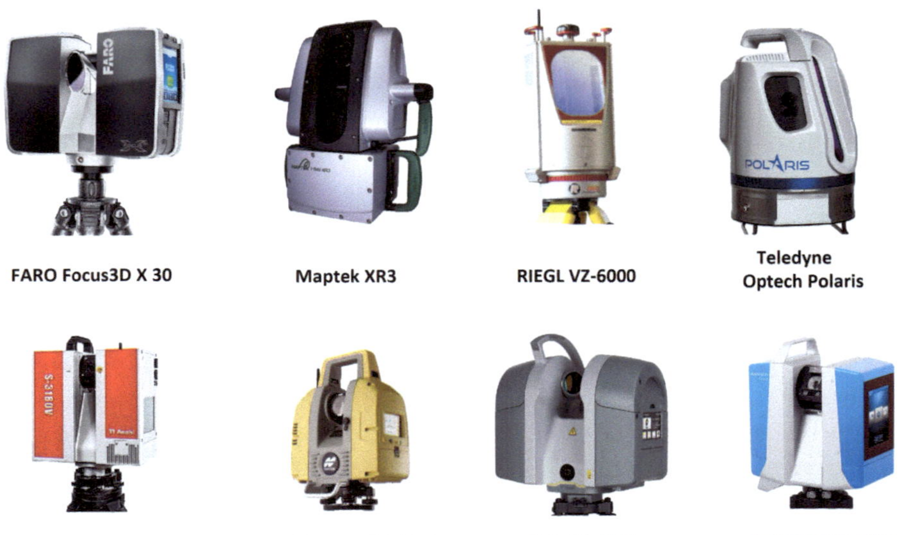

Figure 6.4 View on the outward of the eight commercial TLS systems listed in Table 6.2 (source: manufacturers and geo-matching.com)

usually lies between 60° and 120°. TLS systems developed for indoor and close-range outdoor applications aimed at capturing building, plants and accident scenes, tunnels, and so on may have a considerably larger vertical FoV even up to 320°. Scanners with 360° horizontal FoV and large vertical FoV are called panorama scanners.

The frequency of laser beam emission and the maximum distance are not parameters which can be set independently. The more beams per second are emitted, the less power the pulse will contain and, hence, the smaller the distance the beam can bridge. Of course, the maximum energy of a laser beam is also constrained by eye-safety

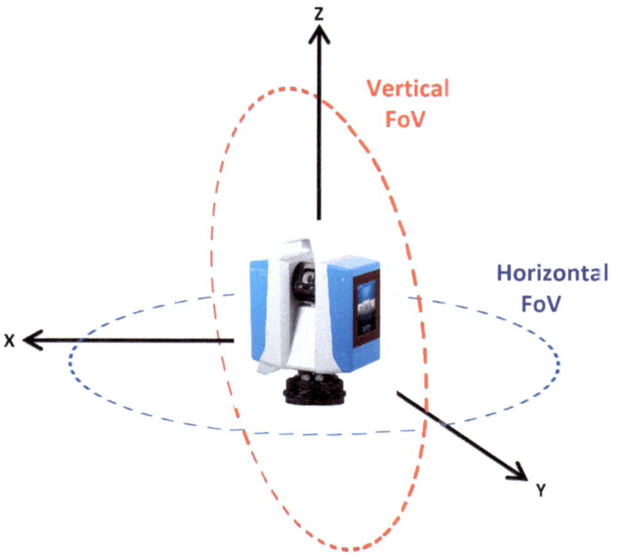

Figure 6.5 Schematic representation of the horizontal and vertical FoV of a ground-based laser scanner

considerations. There is an approximate inverse logarithmic relationship between the laser beam frequency and the maximum range. For example, at a pulse repetition rate of 30 kHz, the RIEGL VZ 6000 has a maximum measurement range for natural targets with a reflectance above 90% of 6,000 m while for a pulse repetition rate of 300 kHz, this maximum range reduces to 3,300 m.

6.2.2 From Ranges to 3D Points

The emitted laser beam measures the distance to an object which the beam meets first. To obtain 3D coordinates, the vertical angle and the horizontal angle have to be measured together with the range. In addition to these geometric parameters, the reflectance value, or intensity, of the return is recorded. The 3D spherical coordinates (r, α, β) are related to the coordinate system of the TLS device. The transformation of the spherical coordinates to 3D Cartesian X, Y, Z coordinates in a more general mapping reference system, such as the geodetic reference system of a country, first requires the transformation of the spherical coordinates to 3D Cartesian X, Y, Z coordinates related to the TLS coordinate system, using the geometric transformation model (GTM) (Figure 6.6):

$$\begin{pmatrix} x \\ y \\ z \end{pmatrix} = r \begin{pmatrix} \cos\alpha\cos\beta \\ \sin\alpha\cos\beta \\ \sin\beta \end{pmatrix} \tag{6.3}$$

When objects are voluminous such as office or university buildings or open-pit mines, several set-up positions are needed to cover the entire object. For example, to capture

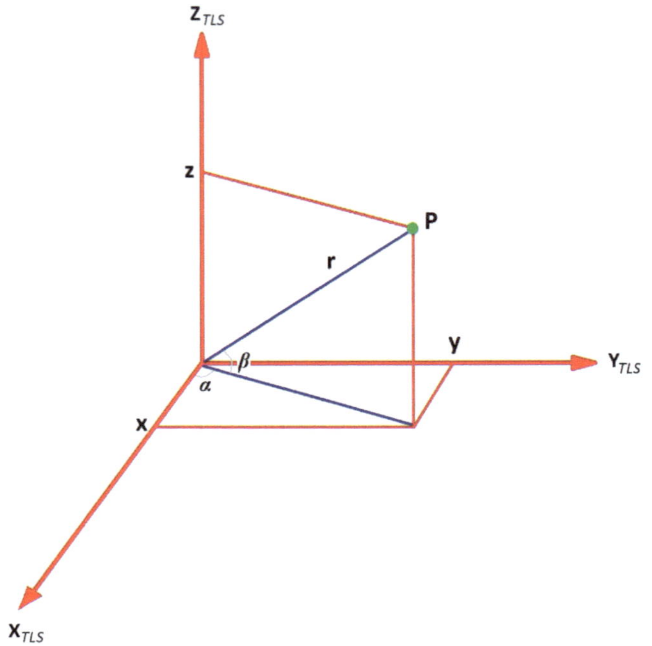

Figure 6.6 TLS centred spherical and Cartesian coordinate systems

the area of 1.5 km by 10 km of an open-pit coal mine in Indonesia with an ILRIS-3D TLS, 40 set-up positions were required (Conforti, 2010) (Figure 6.7).

So surveying a scene usually requires scans from different set-up positions, also called viewpoints. The resulting scans have to be brought into one common coordinate system. This computational process is called registration. Co-registration, usually called registration in ground-based Lidar, is also a necessary step in matching overlapping images. Note that in photogrammetry, registration is indicated as the determination of the exterior orientation parameters of the camera (see Chapter 4). Registration aims at bringing two or more point clouds taken at different times, and/or from different sensors, and/or

Figure 6.7 Optech ILRIS-3D mounted on a tripod with GNSS receiver on top

from different viewpoints in the same common reference system. TLS point clouds of a scene captured from different set-up points are called scans. Registration is an intermediate step to arrive at a point cloud which is georeferenced in a general reference system.

Using tie points in the overlap of adjacent scans, the 3D points, the 3D coordinates, which are related to the individual scans, are brought into one local system that is defined by the project. This part of the process is called registration. Usually, the GTM is based on a rigid body. Since laser scanners measure absolute, metric distances, the scale factor is everywhere equal to one and thus can be omitted in the 3D similarity transformation. As a result, the seven-parameter transformation reduces to six parameters – three shifts and three rotations. Their computation, which is commonly done by least-squares adjustment (LSA), requires the identification of corresponding points in the overlap of the scans.

Usually, a next transformation step is required, in particular when the laser scanning point cloud has to be combined with other data sets covering the same scene or connected to data sets covering neighbouring scenes. In this step, the local coordinate system is transformed into a common national, regional or continental coordinate system. This means that the point cloud has to be georeferenced in a geodetic reference system. This is done by using ground control points (GCP) of which the coordinates refer to the desired reference system using a seven-parameter Helmert transformation also called 3D similarity transformation. As is the case for the registration part also in the georeferencing stage, LSA has to be employed by using more GCPs than strictly necessary to calculate the transformation parameters. Such redundancy adds to precision and the possibility to detect outliers. In summary, georeferencing of a TLS point cloud is carried out in three steps:

1. Transformation of the spherical coordinates of the points in the TLS coordinate system to Cartesian *X*, *Y*, *Z* coordinates
2. Registration: bringing the TLS-centred *X*, *Y*, *Z*, coordinates into a local reference system using tie points in the overlap of scans
3. Georeferencing: transformation of the local reference system to a more general reference system

Step 1 is carried out by employing Equation 6.3. Steps 2 and 3 require the application of a 3D similarity transformation. The general formula reads:

$$\begin{bmatrix} X \\ Y \\ Z \end{bmatrix}^G = \lambda \boldsymbol{R} \begin{bmatrix} X \\ Y \\ Z \end{bmatrix}^L + \begin{bmatrix} X_0 \\ Y_0 \\ Z_0 \end{bmatrix} \tag{6.4}$$

The superscript G refers to the general reference system and L to the local system. λ refers to the scale factor which is one in the registration part; $\boldsymbol{R}$ is a 3×3 matrix containing sine and cosine functions of the three rotation parameters; X_0, Y_0, Z_0 are the three translation (shift) parameters. It is not necessary to perform the three steps in sequence. By combining the formulas, the transformations can be performed simultaneously.

Usually, georeferencing is performed by placing artificial markers, such as retroreflective targets, in the scene. The coordinates of some of these markers are measured with a total station or surveyor-grade GNSS receiver. They act as GCPs for georeferencing. The other markers, which should be placed in the overlap of scans, are used as tie points in the registration step. Sometimes, the surveyor may prefer to measure the coordinates of all artificial markers, which should be regularly placed in the scene to ensure sufficient GCPs are present in each single scan to carry out a 3D similarity transformation. This approach doesn't require overlapping scans, since each scan is transformed individually, and is particularly suited for small projects in which high precision is required and the scene can be captured with a few scans. This is because the number of markers to be placed and measured should be at least three times the number of scans, which would require considerable effort for large scenes and is usually only feasible for smaller scenes.

6.2.3 Tie Points

Registration requires taking provisions before starting the actual scanning. Aligning TLS point clouds – registration – requires sufficient overlap between adjacent scans. The identification of corresponding points in the scan overlaps – tie points – can be done by placing signalised targets, called markers, in the scene at the locations of scan overlaps or by identifying objects or parts of objects, called landmarks, which significantly stand out against the background. The detection of corresponding points can be done manually or automatically.

6.2.3.1 Markers

Signalised targets, called markers here, are used in TLS surveys as tie points or as GCPs for georeferencing. Placing artificial markers in TLS professional surveying projects, in which high requirements are put on accuracy and robustness, is common practice. The large size of a marker and high reflectivity ease the determination of its centre (semi-) automatically or manually from the multiple reflected laser beams. Common markers used in TLS surveys have planar, cylindrical or spherical shapes (Figure 6.8). Software is available which supports the semi-automatic detection of these markers and the computation of the coordinates of their centres. Of course, the size of the markers should be large enough to be hit by many laser beams to enable automatic determination of the centre of the target (Figure 6.9). The advantage of spherical targets is their visibility from any view angle, but the precise measurement of the coordinates of the centroid with a total station is tedious. The manual placing of markers in the scene results in high-precision point clouds but is labour intensive, time consuming and thus expensive while the selection of the marker locations should be carefully done (Theiler, 2015). They should be visible from diverse scan positions and not be placed along a line. Since a TLS survey can last several days, the markers have to be mounted on stable backgrounds and disturbances should be detected by regular inspections. For many applications, points reflected on markers placed in the scene have to be removed from the point cloud during post-processing and they may cause occlusion of relevant scene parts. Occlusion means that a part of an object is not visible from the sensor position due to the presence of another object in the line of sight. Because of these drawbacks, many researchers went in

Figure 6.8 Appearance of targets used as artificial markers

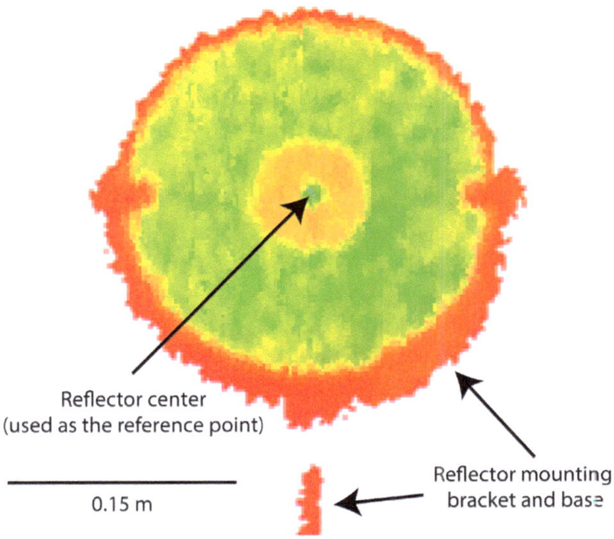

Figure 6.9 Point cloud of a retroreflective target, coloured by intensity, used for registration and georeferencing purposes (source: Wilkinson et al., 2016)

search of registration based on non-artificial markers. These are clearly distinguishable scene points, which are often called landmarks. The choice of the most appropriate landmarks depends on the type of scene. In an oil refinery or other industrial scene, one may (semi-)automatically extract regularly shaped object parts and model them as geometric primitives such as cylinders or spheres. Other methods are based on the detection of interest points and use matching methods similar to the ones used in photogrammetry.

The automated tie point detection methods are still subject to refinement to increase reliability and efficiency (Yan et al., 2017). Finding robust correspondences is challenging due to the following reasons (Rodrigues et al., 2002; Theiler et al., 2015):

- Differences in point density
- Presence of clutter and outliers
- Repetitive adjacent objects
- Insufficient overlaps
- Computational resource requirements

An overview of registration and georeferencing without artificial markers is given by Theiler (2015).

6.2.3.2 Automated Tie Point Detection

Surveyors very well know that careful configuration of a network of markers is tedious and the most time-consuming part of a survey. Triggered by demand of surveyors to relieve their fieldwork efforts, researchers went in search of appropriate automatic registration methods. Many solutions are based on assigning features to points in the overlap of scans and use these features in a subsequent matching step aimed at finding corresponding points (Salvi et al., 2007). Suitable features used in co-registration are treated in Chapter 11. These features should preferably be robust against geometric transformations, particularly they should be rotation and translation invariant. The core of the approach consists of the Iterative Closed Point (ICP) algorithm, which aims at matching points in the one-point cloud with corresponding points in the other point cloud. The search for correspondence is done iteratively using as an optimisation criterion the minimisation of the mean of the squared distances of the corresponding points (Hänsch et al., 2014). When the distance criterion gets smaller than a pre-specified threshold, the iteration stops and the correspondence is settled. To avoid the trap of local minima causing endless iteration, a limit is set on the maximum number of iterations. ICP requires short distances between the corresponding points in the two scans while the differences in rotation should be small. To avoid misalignments, ICP is preceded by a coarse registration step in which rotation and translation invariant features are matched, and approximate transformation parameters are calculated using LSA. The resulting coarse registration is input for the fine registration step based on ICP. Automated registration approaches are not without snags. Finding corresponding points may be hampered when the point densities of the point clouds significantly differ and, in the presence of repetitive structures, which produce similar features at non-corresponding locations, may also cause issues (Theiler, 2015). Differences in point density occur when one scan position is situated close to the object, while the other position is far away.

6.2.4 Measurement of Set-Up Positions

Instead of measuring the coordinates of artificial markers placed in the scene or natural landmarks, also called key or significant points, which act as GCPs, the coordinates of the TLS set-up positions can be determined. One possibility to do so is by signalising the set-up points and then measuring their coordinates using a total station or GNSS receiver. Of course, the latter approach requires line-of-sight to the GNSS satellites. Using permanently signalised set-up positions is feasible when the scene has to be captured regularly, such as open-pit mines for determining the volume of mined gravel, iron ore or lignite (brown coal), or an archaeological site under excavation. Another way to determine the coordinates of the TLS set-up points is by mounting a GNSS receiver on top of the instrument and using differential GNSS to obtain precise measurements (Figure 6.7).

6.3 Mobile Laser Scanner

MLS is increasingly exploited for road inspection, inventory of traffic signs and other road objects, and the creation of maps for road navigation purposes (Guan et al., 2016; Wang et al., 2019). An MLS system is usually mounted on a car, van or other vehicle that can move with traffic speed over roads and highways (Figure 6.10). Around the year 2003, MLS systems, using a car or van as the platform, became operational for surveying and 3D mapping of road scenes and this surveying technology has undergone a steady and rapid development ever since. Today MLS systems are widely used for capturing roads and their vicinity aimed at road inventories on a regular basis.

6.3.1 System Configuration

In contrast to TLS, the platform on which the MMS is mounted is continuously on the move. As is the case for airborne Lidar, registration and georeferencing require continuous measurement of the position and orientation of the Lidar sensors and cameras using ancillary devices. The resulting measurements are linked by accompanying each of them with accurate time stamps. Furthermore, the mutual positions of the diverse sensors have to be determined through a calibration procedure. In order to acquire laser points, images and positioning and orientation data, an MLS integrates one or more laser scanners, one or more digital cameras, a positioning and orientation system (POS) and a control unit. The basic system components of an MLS are indicated in Figure 6.11 on top of a photograph showing the Trimble MX-9. The schematic set-up

Figure 6.10 RIEGL MMS mounted on the roof of car mapping a highway (left) and the historical centre of Krakow (courtesy: RIEGL, Austria)

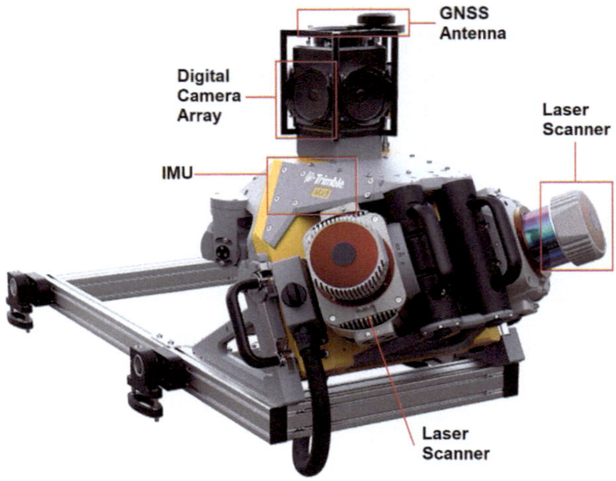

Figure 6.11 Basic system components of an MLS rigidly mounted on a rack; the system shown is the Trimble MX-9

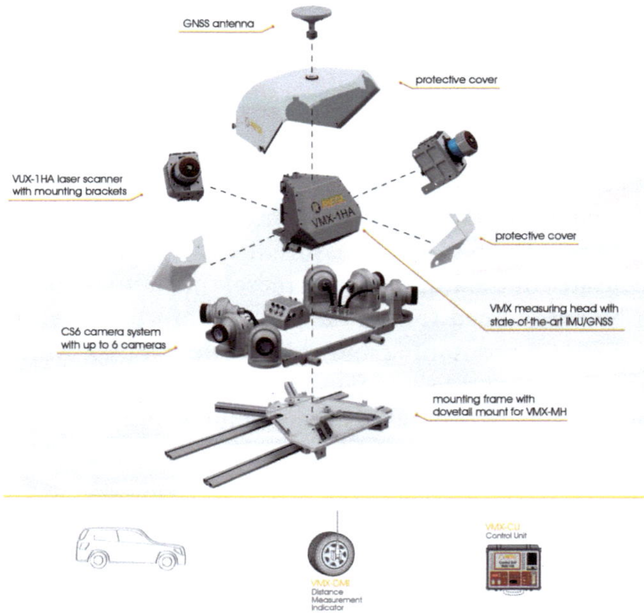

Figure 6.12 Components of an MMS (courtesy: RIEGL, Austria)

of an MLS is shown in Figure 6.12, depicting the components of the Riegl VMX-1HA. Each laser scanner emits laser beams, presently up to over one million laser beams per second, to capture road surfaces and objects above and alongside the road. As discussed in Chapter 3, the distances between sensor and object can be calculated from the pulse travel time, in the case of a time-of-flight system, or from the phase shift. The POS continuously acquires data for calculating the exterior orientation parameters (three coordinates and three orientation angles) using a GNSS receiver and INS, often complemented with a wheel rotation counter. Inertial navigation is used for many different moving platforms and is known under different terms in different disciplines. Within the surveying profession, INS is also called the inertial measurement unit (IMU). Many professional fields in which inertial navigation is exploited have resulted in the development of many types of INS systems. Common types used in MMSs are based on ring laser gyros, fibre-optic gyros and micro-electromechanical (MEM) technology (Puente et al., 2013). The laser scanner primarily measures (1) range – that means the distance from the sensor to the first surface hit by the laser beam, (2) scan angle and (3) intensity of the return. Combining these measurements with those from the GNSS receiver, INS and wheel counter, also called odometer, provide 3D coordinates of millions or even billions of points in a local or national reference system. Usually, the sensors are integrated on one rigid platform. Their mutual offsets have to be quantified through a calibration procedure, which is usually carried out by the manufacturer. Vibrations during the survey, shocks due to holes in the pavement and sudden slowdowns will cause mutual displacements of the sensors and other disturbances. To warrant high-precision surveys all the time, regular recalibration is a necessity. In addition to the 3D coordinates, one or more attributes may be assigned to each point. The attributes may be directly measured by the sensor; this relates particularly to the intensity of the return, computed from a neighbourhood of points or obtained from other sources. Red, green and blue (RGB) values may be assigned to laser points using the simultaneously recorded digital images or image sources having other time stamps (Figure 6.13). From the attributes, features can be derived for registration purposes or for (semi-)automatic classification.

Figure 6.13 MLS point cloud of an urban scene enriched with simultaneously acquired RGB values from images (courtesy: Cyclomedia, the Netherlands)

6.3.2 Positioning and Orientation

The main device for measuring positions is a GNSS receiver. The main device for measuring orientation is an INS. Often an additional measurement device is coupled to this duo: the wheel counter, which provides information on velocity and distance travelled. If the position and orientation are measured sufficiently accurately in a general reference system, registration and georeferencing can be conducted simultaneously. Usually, GNSS and INS are integrated into one system. The ongoing advancements in MEM technology enable the construction of increasingly smaller devices (Figure 6.14).

6.3.2.1 Positioning

The positioning of an MLS is done using GNSS technology. To obtain 3D coordinates with precision at the centimetre level, differential GNSS (DGNSS) should be exploited (see Section 5.5). The majority of MLS surveys concern urban areas and highways. Due to the presence of high-rise buildings (urban canyons), tunnels, high-rise buildings, tree canopy, noise barriers and other obstructing objects, GNSS signals will be frequently blocked during an MLS survey, usually and hopefully for only a short time. Road pavement and building walls will cause multipath, which is also a source of error. GNSS outrage deteriorates positioning precision and thus degrades the quality of the information extracted from the point cloud during further processing. Two remedies can be applied to significantly reduce the degrading effects. A first and obvious solution is by exploring the acceleration measurements along the three orthogonal axes of the INS. Two times integration over time of the accelerations provides positional information. A second solution is using GCPs distributed along the survey trajectory. To achieve the highest accuracy, the two remedies are usually applied in combination.

6.3.2.2 Orientation

An INS is a navigation device that continuously determines velocity, position and orientation of a platform on the move. In contrast to GNSS positioning, which relies on signals emitted by orbiting satellites, INS is an autonomous positioning and orientation solution, which means it does not rely on external sources. Basically, an INS consists of three types of device: three accelerometers, and aligned with these three are gyroscopes

Figure 6.14 Applanix POS LVX, developed for MLS applications, integrates inertial technology with survey-grade GNSS; it weighs 660 g with dimensions: 14.9 cm × 9.3 cm × 4.3 cm

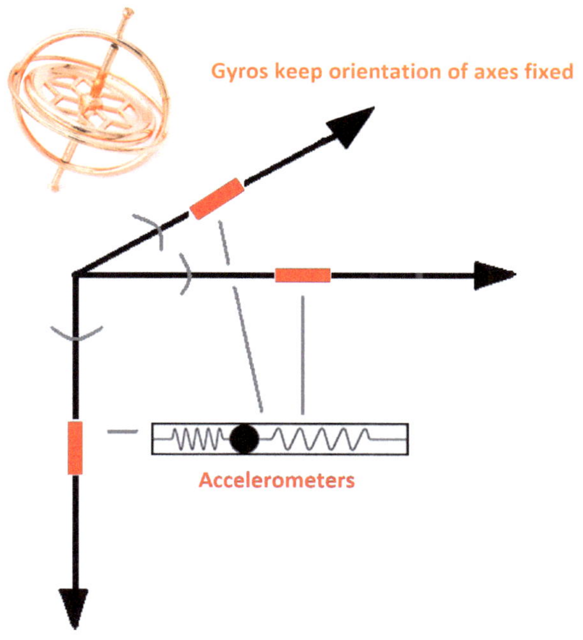

Figure 6.15 Schematic portrayal of the basics of inertial navigation

(Figure 6.15), and to process the measurements, the accelerometers and gyroscopes are connected to a computer equipped with appropriate software (Groves, 2008). The orientation in 3D space of a platform and thus also the orientation of the sensor mounted on that platform can be determined by the gyroscopes, also called gyros. A gyro consists of rotating wheels to maintain a fixed orientation in space. While the platform is moving, the angles between the moving direction and the three perpendicular gyro axes, which maintain their fixed orientation, are measured. These angles indicate the orientation in space of the platform. The three accelerometers mounted on three orthogonal axes indicate the changes in speed. By integration over time, the velocity can be calculated. By integrating over time once more, the position can be calculated from velocity.

The aforementioned description shows that, in principle, an INS system can provide all six parameters of exterior orientation: position and orientation. At first glance, these measurement capabilities seem to make GNSS superfluous. However, the precision of INS positioning rapidly degrades with time due to drift. Not only are the accelerations integrated two times but also the noise, which causes a discrepancy between the actual position and the measured position. These discrepancies rapidly increase over time. On the other hand, INS positions should not be neglected since they are valuable as substitutions during short periods of GNSS outage. Hence, INS positions can fill the gaps due to GNSS signal blocking or multipath. The drift can be compensated for by re-initialisation when GNSS is functioning again. INS sampling frequency is 10–20 times higher than the sampling frequency of GNSS devices. As an autonomous system, INS is immune to any interference from the outside world. How can we integrate the INS, GNSS and possibly wheel counts in the most optimal way?

6.3.2.3 Integration of GNSS and INS

In the aforementioned discussion, we discovered that the characteristics of GNSS and INS devices are complementary and the integration of the measurements of both devices provides a valuable couple for determining the six exterior orientation parameters of sensors mounted in or on a moving platform. In an integrated system, GNSS measurements can compensate for INS drift by re-initialising the INS positions every time a GNSS measurement becomes available. In turn, INS provides intermediate short-term position information and bridges periods of GNSS outage. An optimal combination of the GNSS, INS and wheel counting measurements is usually done through an onboard POS equipped with Kalman filtering software. The basic concept of this filtering procedure, developed by Rudolf E. Kalman in 1960, has been developed further by numerous researchers ever since and its principles and applications have been laid down in numerous textbooks, for example, Brown and Hwang (1997), Grewal and Andrews (2000) and Kim (2011). Kalman filtering provides real-time estimates which are updated while the platform moves and collects new positioning and orientation data. The Kalman filtering procedure takes account of noise and other imperfections.

6.3.3 Ground Control Points

The sensor position is essential for high accuracy, but in urban canyons or under tree canopies, GNSS signals may get blocked or exposed to multipath, which reduces accuracy from a few centimetres to decimetres (Haala et al., 2008). To warrant high-precision end products, GCPs are required distributed along regular distances along the road. Placing and measurement of the GCPs with a total station or a GNSS receiver is a tedious and labour-intensive part of the survey. As a result, researchers are investigating the possibilities of automated GCP determination using ancillary data sources, such as aerial images (Hussnain et al., 2019).

6.4 Automatic Classification

The demand for 3D maps of cities and road networks is steadily increasing, and MLS systems are often the preferred geodata acquisition method for capturing such scenes. Manual processing of massive point clouds is labour intensive and thus time consuming and expensive. This section, which is partly based on Lemmens (2018a, b), briefly touches the state-of-the-art automatic classification and 3D mapping of road objects from point clouds acquired by MMS. Completely automated classification is still in the realm of research laboratory; at best, the proposed methods support human operators to speed up the mapping process (see also Section 1.5).

6.4.1 Need for Automation

Management of roads, their maintenance or reconstruction, requires inventories of pavement conditions, road markings and objects in the vicinity of the road including utility poles, traffic signs, lamp posts and speed guidance boards. The high point density of MLS point clouds allows mapping of traffic guidance arrows and road lines painted on the pavement, vertical road objects, cracks and holes in the pavement, cavity and sagging. Vertical road objects are known as pole-like objects (PLOs), because of their profoundly elongated shape usually extending in the vertical direction. The classification of MLS point clouds is directed towards 3D mapping – that is, the outlining and class assignment of objects. The pipeline of proposed methods usually starts with separating ground points from off-ground points. If the mapping of road surfaces and road markings is not part of the survey, ground points can be removed, which results in a great reduction of points. The next steps depend on the type of scene and the preferences of the researchers. The diverse approaches can be categorised into three groups: point-wise classification, segmentation-based classification and multiscale classification.

Point-wise classification exploits the intensity of the return and/or features derived from a small neighbourhood of points. The geometric features are assigned to each of the individual points, which then are grouped and classified. Segmentation-based methods fit planes, spheres, cylinders or other geometric primitives through neighbourhoods of points. The descriptive parameters of these segments are used as features for further grouping, classification and mapping. The same type of object may appear in road scenes in different lengths, widths and/or heights. The multiscale approach takes account of size variations of objects through combining features computed at various point densities.

Automated classification attempts on MLS point clouds have mainly focused on assigning pre-specified classes to individual points. Point-based approaches are more robust to occlusion and clutter compared to the segmentation-based approaches but the latter are more robust to noise and varying point densities. As a result of their sensitivity to noise and inhomogeneous point densities, point-based approaches may suffer from assigning wrong classes to points. However, segmentation-based methods are sensitive to occlusion and clutter. An example of clutter is the mixing of segments belonging to the one object with segments of another object which is closely attached to the object or overlaps it, which leads to misclassification (Figure 6.16).

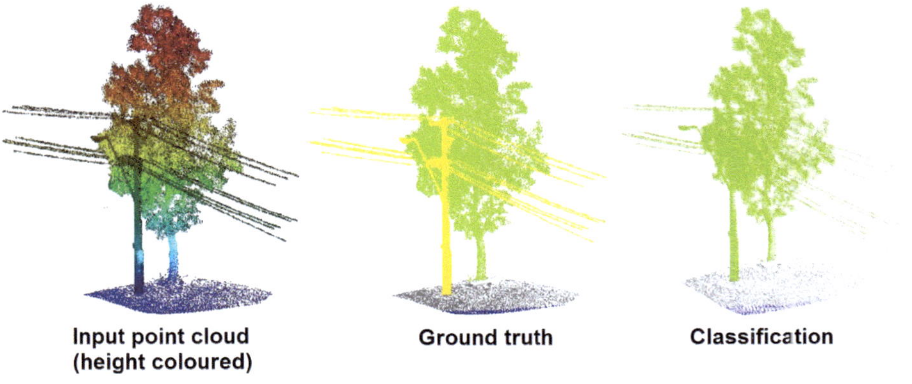

Figure 6.16 The segments of a utility pole partly attached to and overlapping a tree are mixed up with tree segments and misclassified as tree (source: Yang et al. 2017, modified by author)

6.4.2 Local Geometric Structure

MLS point clouds are attribute-poor. In addition to the 3D coordinates usually only the intensity value – often represented as a digital number in the range from 0 to 255 – is available. Many classification methods rely on the use of RGB values from imagery, which may not always be available, and on examining the local geometric structure of neighbouring points. The suitability of the local geometric structure is based on the observation that many objects differ in shape. For example, the majority of buildings can be modelled as an ensemble of planes which – in most cases – intersect perpendicularly at corner lines, while power lines can be modelled as linear elements.

There are two basic descriptors to express the shape of a surface: normal vectors and eigenvalues. Both are assigned to individual points by examining the 3D coordinates of the point under consideration and its adjacent points. If the normal vectors in a neighbourhood point in the same direction, the points most likely form a plane. If they diverge in a systematic manner, they likely form a sphere or a cylinder. When the vectors point in arbitrary directions, the points may be reflected on a fuzzy surface, such as foliage. In addition, eigenvalues of the 3×3 covariance matrix of the 3D coordinates of neighbouring points indicate shape. If one eigenvalue is large and the other two close to zero, the neighbourhood forms a linear element. A plane is indicated by two eigenvalues which have approximately the same value and one eigenvalue close to zero. Spherical and fuzzy surfaces will have three large eigenvalues. It is common practice to derive measures from the eigenvalues which indicate the type of local structure. Examples of such measures are linearity, planarity, sphericity, anisotropy, eigenentropy and local surface variation (see Section 11.6).

6.4.3 Deep Learning

Recent innovations in artificial intelligence include the development of deep learning algorithms based on convolutional neural networks (CNNs). The development of this type of machine learning method has been inspired by the working of the human brain. CNNs have been successfully applied in self-driving cars, robotics and object recognition from images. However, the classification of point clouds appears to be a difficult issue because of the sheer amount of points and the complexity of outdoor scenes. Added to this, the points are not inherently structured as on an image raster while the distribution of points is irregular and non-homogenous. By feeding a CNN with data of an abundance of prototype objects, the algorithm can recognise objects across a broad variety of scenes. However, the training data has to be manually selected, which is more time-costly for 3D models than for 2D models.

6.4.4 Exploiting Scene Knowledge

Scene knowledge can be exploited for checking and improving classification results. The exploitation of scene knowledge can be based on four generic rules:

- *Generic Rule 1*: In 3D mapping, one is not interested in the semantics of individual pixels or points but in what is present on top of the surface of the earth, or bare ground, at specified locations.

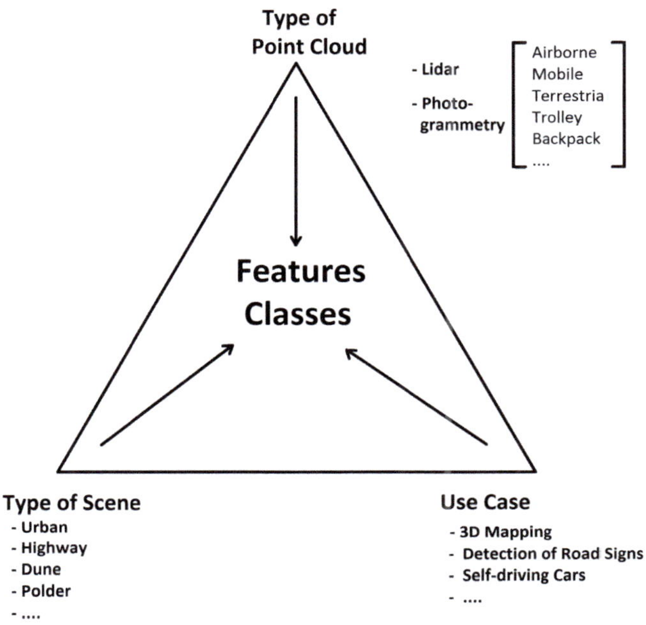

Figure 6.17 Dependence of features and classes on data source, type of scene and use case (courtesy: Lemmens, 2018b)

- *Generic Rule 2*: A point cloud is a blind sampling of the scene, meaning that we have to deal with arbitrary points being part of the surfaces of objects; the determination of object types requires an interpretation stage.
- *Generic Rule 3*: Everything is connected to something else and ultimately to the surface of the earth.
- *Generic Rule 4*: The types of object (classes) which can be recognised in an image or point cloud depend on the (1) type of data source, (2) type of scene and (3) use case.

Figure 6.17 illustrates Generic Rule 4. The data source is key for defining classes. It is obvious that one cannot extract classes from a geodata set which are not implicitly or explicitly present in the data set. For example, a satellite image is not suited for mapping traffic signs. An aerial image provides a different view of a road scene than when the same scene would have been captured by laser scanning from a moving car. Also, the type of scene restricts class definition. It may be a good idea to map banana trees in the city of Abuja, the capital of Nigeria, but not in the capital of Iceland. Also, the task at hand or use case determines the classes to be mapped. A municipality may want to map other object classes compared to a maintenance service of highways.

6.5 Indoor Mobile Laser Scanning

Earlier we considered MLS systems mounted on cars, vans or other vehicles for outdoor surveys. As a result of ongoing miniaturisation of optical and mechanical components

Figure 6.18 MEM technology enables far-reaching miniaturisation of inertial navigation systems

as well as increasing accuracy of micro-electromechanical MEM-based INS, which may fit within the palm of a hand (Figure 6.18), manufacturers increasingly produce small MLS systems, which are mounted on a trolley (Figure 6.19), ensembled in a backpack (Figure 6.20) or fastened on a stick for handheld laser scanning (Figure 6.21). Indoors, GNSS positioning of the MLS is severely hampered or impeded due to the blocking of the GNSS satellite signals.

Other solutions have to be found for registration and georeferencing indoor scans. Appropriate solutions have been developed within the field of robotics. Here, one is faced with the problem of determining the path of a robot through a room, shopping mall or other indoor space and simultaneously acquiring a 3D representation of that space to prevent the robot from colliding with walls or other obstacles. Often no adequate map of the space is available. The solution is based on simultaneously determining the robot's location and creating a map exploiting onboard sensor data. At first glance, this seems an impossible task, a chicken-or-egg dilemma – the robot's location and orientation are essential for generating the map, yet a map is needed to pinpoint the robot's position. The solution lies in "guessing" the position and representing the space based on sensor data and prior knowledge. Next, these guesses are iteratively refined using data collected while the robot is moving using the ICP algorithm, discussed earlier. Simultaneous localisation and mapping (SLAM) solutions are also beneficial for the 3D mapping of indoor and subsurface spaces. The main positioning sensors useful for indoor localisation and orientation are odometers, INS and lasers. Odometers count wheel spins enabling speed and distance travelled to be computed. When they are mounted on two wheels either at the front or at the rear, odometers also provide data about directional changes. INS provides directional and displacement information and lasers detect obstacles in front of the robot. SLAM does not consist of a particular algorithm but is rather an approach in which a diversity of solutions are employed depending on onboard sensors, the nature of the environment and possibilities to connect the trajectories with GCPs. SLAM has

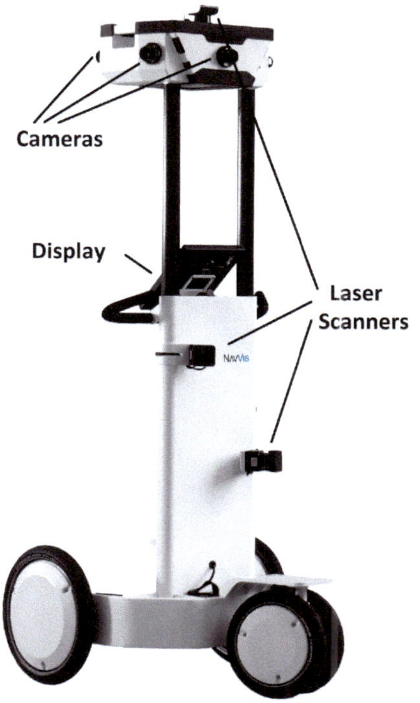

Figure 6.19 NavVis 3D mapping trolley equipped with laser scanners, cameras and a control and data capture unit for indoor 3D mapping (courtesy trolley: NavVis; indication of system components by author)

Figure 6.20 Leica Pegasus:Backpack consisting of five cameras and two laser scanners mounted on a carbon-fibre chassis (courtesy: Leica Geosystems)

Figure 6.21 Handheld GeoSLAM ZEB-REVO mobile 3D scanner

been one of the most prominent successes of robotic research in the past 30 years. It has found its way into many survey applications including MLS. A position paper and tutorial on SLAM is provided by Cadena et al. (2016). Although the problem of SLAM has seen great progress over the last three decades, there remains a need for a lot of fundamental research on methods for creating and maintaining 3D maps from point clouds of indoor scenes.

6.6 Concluding Remarks

Laser scanning has become a daily practice for precise surveying of complex objects and scenes. Looking at the type of platforms, ground-based laser scanning equipment can be categorised into two broad groups:

- Stationary laser scanning, which is best known under the tag terrestrial laser scanning, abbreviated to TLS.
- Laser scanners mounted on ground-based moving platforms, which is best known under the tag mobile laser scanning, abbreviated to MLS.

Ground-based laser scanning has come a long way since the late 1980s. Their point acquisition rate, maximum range and accuracy – that means performance and productivity – have been steadily increased, while the cost of ownership, size and weight decreased. Data is not yet information. The conversion step requires careful processing.

Many researchers devote their scientific work to further automating parts of the processing pipeline including registration and georeferencing up to automated 3D mapping. Before the sheer amount of points can be mapped fully automatically, a long and winding road still lies ahead of us. Nevertheless, the applications of ground-based laser scanning are broadened by the year because continuous advancements release the expertise burden so that a broad pallet of professionals can do the job. Chapter 13 provides a comprehensive exposé of a variety of applications of ground-based Lidar, including inspection of roads, railway tunnels, mapping of technological monuments, 3D mapping of subsurface mines and serious gaming.

References

Brown, R.G., Hwang, P.Y.C. (1997) *Introduction to Random Signals and Applied Kalman Filtering.* 3rd Edition, Wiley, New York.

Cadena, C., Carlone, L., Carrillo, H., Latif, Y., Scaramuzza, D., Neira, J., Reid, I., Leonard, J.L. (2016) Past, Present, and Future of Simultaneous Localization and Mapping: Towards the Robust-perception Age. *IEEE Transactions on Robotics*, 32(6), pp. 1309–1332.

Conforti, D., (2010) Measuring Open Pit Mines: Testing Static and Mobile Laser Scanners. *GIM International*, 24(8), pp. 17–19.

Grewal, M.S., Andrews, A.P. (2000) *Kalman Filtering: Theory and Practice*. 2nd Edition, Wiley, New York.

Groves, P.D. (2008) *Principles of GNSS, Inertial, and Multisensor Integrated Navigation Systems.* Artech House, Boston, London.

Guan, H., Li, J., Cao, S., Yu, Y. (2016) Use of Mobile LiDAR in Road Information Inventory: A Review. *International Journal of Image and Data Fusion*, 7(3), pp. 219–242.

Haala, N., Peter, M., Kremer, J., Hunter, G. (2008) Mobile LiDAR Mapping for 3D Point Cloud Collection in Urban Areas: A Performance Test. *ISPRS Archives of the Photogrammetry, Remote Sensing and Spatial Information Sciences*, 37, pp. 1119–1127.

Hänsch, R., Hellwich, O., Weber, T. (2014) Comparison of 3D Interest Point Detectors and Descriptors for Point Cloud Fusion. *Annals of the Photogrammetry, Remote Sensing and Spatial Information Sciences*, II-3, pp. 57–64.

Hussnain, Z., Oude Elberink, S., Vosselman, G. (2019) Automatic Extraction of Accurate 3D Tiepoints for Trajectory Adjustment of Mobile Laser Scanners Using Aerial Imagery. *ISPRS Journal of Photogrammetry and Remote Sensing*, 154, pp. 41–58.

Kalman, R.E. (1960) A New Approach to Linear Filtering and Prediction Problems. *Transactions of the ASME, Series D, Journal of Basic Engineering*, 82, pp. 35–45.

Kim, Ph. (2011) *Kalman Filter for Beginners: With MATLAB Examples*. Createspace Independent Publishing Platform, North Charleston, US.

Lemmens, M. (2018a) Mobile Laser Scanning Point Clouds: Status and Prospects of Automatic 3D Mapping of Road Objects. *Geomatics World*, 26(2), pp. 16–19.

Lemmens, M. (2018b) On a Knowledge-based Approach to the Classification of Mobile Laser Scanning Point Clouds. *ISPRS: International Archives of the Photogrammetry, Remote Sensing and Spatial Information Sciences*, XLII-4, pp. 343–349.

Puente, I., González-Jorge, H., Martínez-Sánchez, J., Arias, P. (2013) Review of Mobile Mapping and Surveying Technologies. *Measurement*, 46, pp. 2127–2145.

Rodrigues, M., Fisher, R., Liu, Y. (2002) Special Issue on Registration and Fusion of Range Images. *Computer Vision and Image Understanding*, 87(1-3), pp. 1–7.

Salvi, J., Matabosch, C., Fofi, D., Forest, J. (2007) A Review of Recent Range Image Registration Methods with Accuracy Evaluation. *Image and Vision Computing*, 25(5), pp. 578–596.

Theiler, P.W. (2015) Automated Registration of Terrestrial Laser Scanner Point Clouds. PhD Thesis, ETH Zurich.

Theiler, P.W., Wegner, J.D., Schindler, K. (2015) Globally Consistent Registration of Terrestrial Laser Scans via Graph Optimization. *ISPRS Journal of Photogrammetry and Remote Sensing*, 109, pp. 126–138.

Wang, Y., Chen, Q., Zhu, Q., Liu, L., Li, C., Zheng, D. (2019) A Survey of Mobile Laser Scanning Applications and Key Techniques over Urban Areas. *Remote Sensing*, 11, p. 1540.

Wilkinson, M.W., Jones, R.R., Woods, C.E., Gilment, S.R., McCaffrey, K.J.W., Kokkalas, S., Long, J.J. (2016) A comparison of Terrestrial Laser Scanning and Structure-from-Motion Photogrammetry as Methods for Digital Outcrop Acquisition. *Geosphere*, 12(6), pp. 1865–1880.

Yan, l., Tan, J., Liu, H., Xie, H., and Chen, C. (2017) Automatic Registration of TLS-TLS and TLS-MLS Point Clouds Using a Genetic Algorithm. *Sensors*, 17(9), p. 1979.

Yang, B., Dong, Z., Liu, Y., Liang, F., Wang, Y. (2017) Computing Multiple Aggregation Levels and Contextual Features for Road Facilities Recognition Using Mobile Laser Scanning Data. *ISPRS Journal of Photogrammetry and Remote Sensing*, 126, pp. 180–194.

Survey Peculiarities

CHAPTER 7

The creation of point clouds, digital terrain models (DTM), digital landscape models (DLMs) and 3D maps starts with the collection of data on site through a well-designed geodata acquisition process. Data cannot be created out of thin air or as Digital Globe – a commercial provider of high-resolution earth imagery, data and analysis, based in Westminster, Colorado, US – states:

> You cannot create data from nothing and the laws of physics cannot be conquered via software enhancements.

In other words, good information starts with good data and good data starts with a good survey or geodata acquisition process. Indeed, the ultimate aim of collecting point clouds is to extract detailed, reliable and accurate information about a city, building, open-pit mine or other type of object which serves the needs of users who have to solve a particular task at hand. Information is possibly extracted by combining the point cloud with other point clouds, images or maps in a Geographical Information System (GIS). Surveying is not just a mathematical exercise carried out in the real world. The capturing of geodata is a profession in itself and relies for the major part on insight into the characteristics of the devices used to capture the real world: the sensors and measurement equipment and understanding of the task at hand, also called use case. As with any other profession, geodata acquisition also relies much on experience.

Conducting a geodata acquisition exercise in the real world has to cope with physical constraints and relies on knowledge, skills and experience of the survey crew. Any design of a survey is not only constrained by physics but also by legislation and regulations, which may differ from one jurisdiction to another. Providing surveying rules would be unfeasible within the scope of this textbook. The number of sensor types and platforms is extensive and each has its own methods and techniques underpinned by specific approaches to capture a part of the real world. Added to this, a lot of knowledge is concealed within the walls of sensor manufacturers, software developers and surveying firms. Often these vendors provide training courses in order to enable surveyors to get acquainted with their technologies. Nevertheless, it is feasible to consider sampling aspects, accuracy and other aspects which affect the quality of the geodata and derived products in general terms within this book.

We focus our considerations on the generation of digital elevation models (DEMs) and will consider sampling in relation to recognisability and accuracy. A DEM is one of the most important point cloud products of today's automated geodata acquisition techniques (see Chapter 8). Digital photogrammetry (image matching), airborne laser altimetry (Lidar) and SAR interferometry (InSAR) are at present the most important operational automatic data capturing methods for DEM. First, we consider how photogrammetrists adapted the design of a survey to the instrumental constraints, environmental conditions and the human visual system. Next, we have a look at the relationship between the ground sample distance (GSD) and the recognisability of objects. This section is followed by three sections, each discussing a pitfall: point density and recognisability; the space extent of points (points are not points); and finally the effects of terrain characteristics.

7.1 Need for Pilot Studies

In the past, photogrammetrists carried out many tests and pilot studies to adjust and optimally exploit cameras and mapping devices. They did this to design cost-effective workflows aimed at creating or updating topographic maps. Here optimal means the balancing of accuracy, costs of acquisition and processing, costs of collecting missing data in the terrain by topographic surveyors, and recognisability in the images of buildings, roads and all the other objects described in the specifications of the end product: the topographic map. These specifications may differ from jurisdiction to jurisdiction. In aerial photogrammetry, the mapping exercise is carried out by a human operator in the office by using a splendid man–machine interaction approach, which fully exploits the human visual system in a sophisticated way. By offering the left image to the left eye and the right image to the right eye, the human operator observes a 3D image as if he or she were looking through a borehole on the floor of the aircraft (see Section 8.5.1 and Figure 8.12). The tests and requirements resulted in a number of design rules for carrying out the aerial photogrammetric flight. One such rule reads: The scale of the photograph should be no less than two times smaller than the scale of the topographic map, resulting in the rule that if the scale of the topographic map is 1:10,000, the photo scale should be 1:18,000 on average. This average warrants that small fluctuations due to terrain height variations and air turbulence affecting flight attitude will always result in photo scales, which are smaller than 1:20,000. Another rule, which has been induced by the requirement to carry out the measurements in stereo models, states that the along-track overlap should be 60% at least. Although 50% along-track overlap would theoretically be sufficient, a photogrammetric survey is conducted using physical devices operating in a physical environment. As a consequence, air turbulence and misjudgements of time intervals between two consecutive photo exposures may cause gaps in which no stereo vision is possible. Today, the rules for overlaps are 80% or even 90% both along-track and across-track. This rule originates in the need for creating true orthoimages and dense DEMs. Visibility and recognition requirements prescribed that the photos should be taken at high sun angles to minimise the impact of shadows. So summertime would be the best period were it not for deciduous trees and other vegetation, which (partly) occlude many objects when viewing from above. Balancing the needs and limitations, the best period identified was springtime for regions north of the Tropic of Cancer. The best time of the day was identified around noon, provided that there would be no or little haze in the air.

Today, many pilot studies are conducted to find out the optimal geodata acquisition strategy to capture cities, forests and other landscapes by laser scanning. Usually, the use of laser scanning in a professional field is induced by specialists who become enthusiastic about the possibilities of the novel technology, and want to find out whether this advanced technology, which is brand new for them, can support their tasks. For example, foresters are highly interested in airborne Lidar because the resulting point clouds can be studied in an office environment and the laser pulses do not only reflect on the tree canopy but are also able to reach the ground. This provides extremely valuable information to foresters against costs which are just a fraction of applying conventional, field-based surveys. The results of the tests find their way into scientific journals, which focus on forestry, and may inspire colleagues to also exploit the possibilities of airborne Lidar. Usually, this is the way how a geomatics technology increases its popularity throughout the years.

7.2 GSD and Recognisability

A photogrammetric or Lidar point cloud is the result of a sampling process of object space, for example, consisting of ground, buildings, roads and trees. When conducting a laser scanning or a photogrammetric survey, the question pops up: "What should be the point density, in case of laser scanning, or what should be the ground sample distance (GSD), in case of photogrammetry?"

Formally GSD is defined as the distance between the centres of two adjacent pixels measured on the ground. In an image with a GSD of 0.15 m, the centres of adjacent pixels projected on the ground are 15 cm apart. GSD may also be understood as the side length of the square which one pixel covers in object space, which means a GSD of 0.15 m corresponds to a ground area of 15 × 15 cm. This is often also called spatial resolution. The larger the GSD, the lower the spatial resolution, and the fewer details can be distinguished.

Some may be inclined to think that the higher the point density or the smaller the GSD, the more useful information may be extracted from the geodata set. Although oversampling seems to be a good idea, it requires extra surveying time, thus contributing to costs of geodata acquisition and necessary computer resources. Furthermore, a higher than required number of points for the task at hand slows down data processing and when the local point density reaches its application-specific optimum, redundant points which occupy (exhaustive) storage space without adding relevant information (Puttonen et al., 2013). Oversampling may also result in erroneous analysis results, leading to wrong conclusions. First, we have to recognise that features at the surface of the earth can be described at a multitude of scales. In general, for a certain application domain, only a limited range of scales will be adequate. Too few details may give results which might be useless for the task at hand. Too many details may conceal the bigger picture so that one no longer sees the forest through the trees.

An obvious example is a valley. On top of the slope of the valley, many details may be present that are not of interest when one wants to compute, for example, the erosion in terms of soil loss per square metre per year. However, when gradients are computed that do not represent the trend of the slope but the small-scale fluctuations on top of the valley slope, erroneous results will be derived. So, at too fine a scale, one is swamped with extraneous detail. In contrast at too coarse a scale, important features may be missed. When a slope is represented by some sample points badly distributed over the valley, water run-off may be computed which has nothing to do with any reality. Other scenes where point densities resulting from laser scanning may be too high are urban areas for which locations of objects and their rough outlines have to determined without aiming at accurate object modelling, and forests where reflections on irregularly shaped vegetation causes too many reflections of laser beams (Puttonen et al., 2013).

Stating that a higher point density accomplishes a better terrain description is based on a faulty modelling of the part of the real world within the task at hand. A complicating factor is that for many applications, the optimal GSD is unknown. This is because the representation of the earth's surface in digital format, the mathematisation of physical real-world phenomena and the use of computational models to describe earth-related processes are relatively recent advancements. To arrive at deep insight from which practical rules of thumb can be derived a lot of tests have to be conducted, as has been the case for applying photogrammetry for topographic mapping demonstrated in the above discussion.

Within photogrammetry, a lot of research has resulted in a few rules of thumb on the relationship between GSD and the recognisability of objects. These rules have been developed with topographic mapping as an application domain in mind. Assuming that the area covered by one pixel is similar to the spatial resolution, one may conclude from the Nyquist–Shannon sampling theorem that the size of an object should be around one-tenth of the pixel to ensure that its mapping will not be affected by the position and orientation of the object with respect to the sampling scheme (Blaschke, 2010). This means that a proper outlining of an object from an image requires that it should be covered by at least ten pixels along the longest side. This is illustrated in Figure 7.1 showing three situations. In the first situation, the object is completely encapsulated by the boundaries of the pixel. At best, an indication of what is present in object space can be

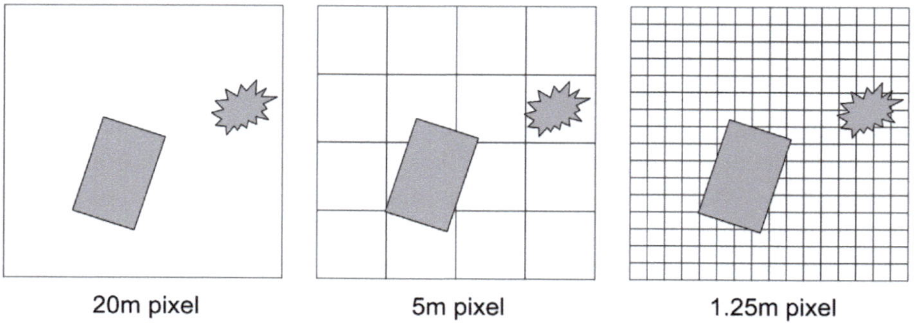

Figure 7.1 Dependence of identification and outline of an object on ground sample distance (GSD) (source: Blaschke, 2010)

obtained, although the resulting mixed pixel may likely impede proper classification. In the second situation, the object and GSD are similar in size, which may result in a more reliable classification of the pixel, but will not contribute to a proper outlining of the object. In the third situation, the GSD is significantly smaller than the size of the object, allowing a reliable outlining.

The Nyquist–Shannon sampling theorem, mentioned earlier and also called Nyquist theorem, aims to provide a rule for digitising analogue time signals so that the original signal can be reconstructed from the sampled version without loss of information. The principle is based on the fact that any analogue signal can be decomposed into many waves with subsequently smaller wavelengths or larger frequencies. Most signals will have a minimum frequency and a maximum frequency. The difference between the maximum and the minimum is called bandwidth. The Nyquist theorem states that the analogue signal can be reconstructed if the sampling rate is two times the bandwidth. Suppose that the bandwidth of an analogue signal is 10,000 Hz, then the signal can be perfectly reconstructed when sampled with a time interval of 1/20,000 s. Often the minimum frequency will be close to zero. In these cases, the bandwidth equals the maximum frequency in the signal. That is why the Nyquist theorem is usually reformulated as follows: If the highest frequency in a signal is F_{max}, the signal can be reconstructed when sampled with a frequency of $2F_{max}$, which is called the Nyquist frequency, or a time interval of $0.5/F_{max}$. This is the formulation in which the Nyquist theorem is known best.

When collecting a laser point cloud, it might seem a good idea to apply the Nyquist theorem to determine the point sampling interval or point density. The reasoning may go as follows. Suppose one wants to map fences, which might be a necessary activity for a particular task at hand. The specifications state that the height of fences, which have to be mapped, should be at least 1.5 m, the length at least 5 m and the width at least 0.75 m. To apply the Nyquist–Shannon sampling theorem, one further recognises that the width of the object has the smallest extension, resulting in the conclusion that the scene has to be sampled with a point interval of 0.75/2 = 0.375 m. Since there is a straightforward relationship between point sampling interval, also called point spacing, and point density, the resulting point density becomes 7 points/m^2. Till so far the way of reasoning. But makes it sense?

Applying the Nyquist theorem for determining the point cloud density for mapping topographic objects is associated with three main pitfalls when applied without further considerations. These pitfalls concern the pattern of the points, the dimension of the points and the terrain characteristics. In the sequel, we will consider these three aspects in greater detail.

7.3 Point Density/Point Pattern

How many points are necessary to reconstruct, for example, a roof from an airborne laser scanning point cloud? From a mathematical point of view, it suffices to use three points to fit a plane through. However, a reliable reconstruction requires a good distribution of the points over the plane. The points should not clutter or clump together in the centre or in one of the corners. The points should also not lie on the same line. Furthermore, outliers may be present, caused by the presence of small structures – small relative to the size of the roof – such as a chimney or a small dormer, which may affect the orientation of the roof in space. This discussion shows the dilemmas in the design of a Lidar or photogrammetric survey.

The point density pitfall originates in using the notion of point density as a substitute for sampling interval. Laser scanners scan line by line. This means the sampling interval in the two directions may differ significantly. The sampling interval for the one direction may be sufficient to capture the object but may be insufficient for capturing objects of the same size but orientated in the perpendicular direction (Figure 7.2). Furthermore, oscillating scanners generate irregular samples as a result of their design as shown in Figure 5.4 in Chapter 5. A point cloud generated by dense image matching may theoretically provide one height value per pixel. However, due to object characteristics in particular the photometric properties, also photogrammetric point clouds, will show an irregular, inhomogeneous distribution of points. For laser point clouds, the properties of the surfaces will also affect the distribution of points. A mobile laser scanner will have an inhomogeneous distribution of points due to the fixed scan angle: The larger the distance between sensor and object, the large the point spacing will be and thus the coarser the point cloud (see Section 6.1.2). The point density is not only a crucial scanning parameter, which determines how well information can be extracted from a laser point cloud, but also the pattern of the footprints (Ulrich, 2013). Specifying point density in points per square metre assumes that the points are regularly distributed over object space. This is because point density is an overall average measure which provides no information about the actual distribution of points over space. As a result, the information content of data sets with similar point densities differs depending on the point pattern.

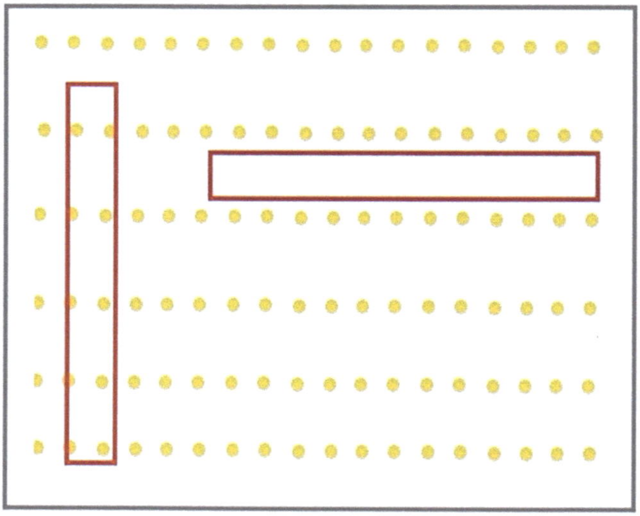

Figure 7.2 The sampling interval of laser scanners may differ in scan line direction and the moving direction of the platform; the same object lying in the one direction may be sufficiently covered, while this may not be true for the other direction

7.4 Points Are Not Points

A second misconception seems to be induced by the term point. The points in a laser point cloud are acquired by physical devices which operate in a physical environment. As a result, a point in a laser point cloud is not a point in the mathematical sense, having no dimension. Due to beam divergence combined with flight height, the pulse will have a certain circular extension, called footprint, when hitting the object surface (see Section 3.6 on Beam Divergence). When the beam divergence is 0.3 mrad, the diameter of the footprint will be 30 cm at a flight height of 1 km. When the beam divergence is 0.8 mrad, this value will be 80 cm. Beam divergence may introduce errors in the laser point. However, this property can also be beneficially exploited for improving the capturing of specific types of object spaces. For capturing the built environment, it is desirable that the edges of buildings, roads and so on are captured as sharp as possible (Xu et al., 2008). This can be achieved by using small footprints. In modern laser scanners, the user can often adapt the footprint within certain ranges – either step-wise or continuously – to meet the needs best. Furthermore, the footprint is also determined by flight height: The higher the aircraft flies, the larger the footprint will be. A higher flight height introduces also more loss of pulse energy on the way from sensor to object and back. In forest areas with tall canopy foliage, an increase of the footprint results in better penetration of the foliage so that the pulse can reach the ground (Hopkinson, 2007). This property is beneficial for creating DTMs of the forest area and computing the height of the forest stand if the sensor is able to capture the first and last return. However, a large beam divergence means that the pulse contains less energy per unit of area. So, the chance that energy is received back from the part of the pulse that reached the ground surface, diminishes with increasing beam divergence. This is why foresters prefer small footprint airborne Lidar point clouds. The reduction of pulse energy with increasing flight height affects also the point pattern, because objects with low reflection properties may be captured at low flight height, but not at high flight height. The relationship between pulse strength and footprint size has to be considered when adjusting the footprint as a survey parameter in flight planning.

Adopting the Nyquist theorem as the leading guideline for determining the point density of an airborne Lidar survey aimed at capturing a powerline would result in the following reasoning. Since the diameter of the individual powerlines is 10 cm, the sampling interval should be 5 cm, resulting in a point density of 400 points/m^2. However, even when the point density is just a few points per square metre, the powerline will be hit by laser pulses although the diameter of the individual powerlines is much smaller than the sampling interval. This is not a flaw in the Nyquist theorem, as some are inclined to think, but is because of the phenomenon, described earlier and called beam divergence. For laser point clouds, the use of the Nyquist theorem is misleading since laser points do not return as points but exhibit a footprint of which the diameter depends on the beam divergence and the distance from the sensor to the object. When the fence in the above-mentioned example would be a hedge, it may even happen that no signal may be received from the top of the hedge because the pulse will penetrate the coarse distribution of leaves and branches.

7.5 Terrain Characteristics

Usually, the measuring process is the best understood part of automatic DEM generation. Prior to operation, a quantification of the measurement accuracy can be gained by calibration. With error propagation theory, quantifications of the accuracy of the elevation values can be obtained a priori to carrying out actual measurements. Once acquired, the DEM can be compared with sample values obtained with a measurement technique, which is an order of magnitude better than the survey technique. For validation purposes, this approach is well accepted (see Section 7.6). The nature of measuring processes causes that all observations are inevitably contaminated by error; errors are an empirical reality. Besides inaccuracies and errors introduced by the measurement devices operating in a physical environment, uncertainty in DEMs is also introduced by the characteristics of the terrain surface.

It is a general experience that when sampling a terrain for the second time with the same measuring devices and the same point density, the difference between the two resulting DEMs is much larger than one would expect on the basis of the measurement accuracy of the equipment. Basically, this effect is due to ground cover characteristics. This means that the quality of a DEM does only depend partly on the characteristics of the measurement technique or the setting of the sensor and survey parameters. For a larger part, the quality depends on the characteristics of the ground surface and the objects on top of it. The uncertainties introduced by terrain characteristics concern primarily microrelief, coverage, slope and reflectivity (Lemmens, 1999). These four aspects are considered in the sequel in greater detail.

7.5.1 Microrelief

Within the present discussion, we assume that the DEMs are generated from airborne Lidar point clouds. When capturing parts of the earth's surface, microrelief refers to the fluctuations in height values within the region for which the height value of the sample point is assumed to be representative. When firmly standing with both booths on the ground, the land surveyor carries out an interpretation process during the point sampling of the terrain while operating a Global Navigation Satellite System (GNSS) receiver, a total station or a levelling instrument. The same is true for the photogrammetric operator who extracts 3D points from a stereo image pair using a digital photogrammetric workstation (DPW) or, in former days, an analytical plotter. The surveyors and operators choose locations in which the height is close to the average terrain height value within the vicinity of the point (Figure 7.3). In other words, representative points are gathered. When using automated DEM capturing techniques, the interpretation part is lost. One may say that the terrain is captured blindly. Consequently, sample points will in general not be representative of their vicinity. They may even be returned from mobile objects such as cars, dogs and cattle. The effect is that repetitive automatic sampling will show larger differences in height values than one would expect on the basis of the accuracy of the measurement technique.

7.5.2 Coverage

Automated DEM capturing techniques sample the terrain blindly; the sampling is non-discriminative. As a result, the sensors may capture a surface, which may not correspond to what the user wants. The unwanted objects of which the surfaces may be included in

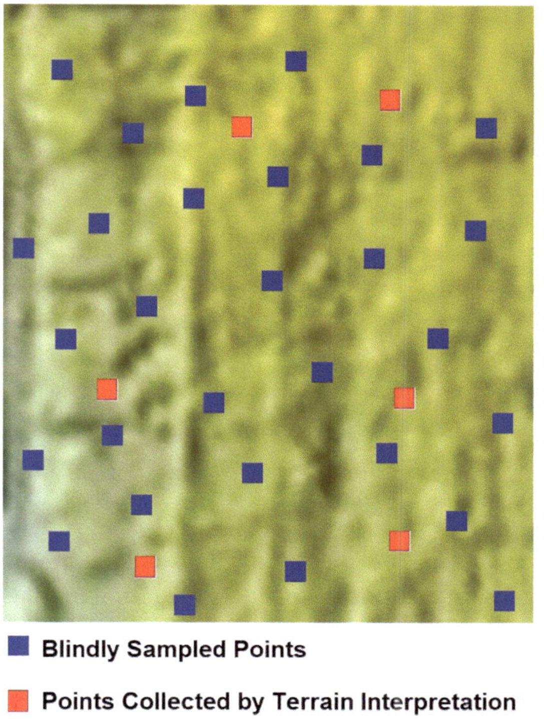

Figure 7.3 Microrelief affects the accuracy of point collection by blind sampling methods compared to knowledgeable interpretation by surveyors

the DEM are houses, trees, hedges and cars. The actual result is a digital surface model (DSM). In other words, automated techniques are not able to distinguish among the many types of objects present on the surface of the earth. This induces the need for the removal of these unwanted objects, which requires an interpretation process. Since automated data capture is usually associated with huge amounts of data, attempts have been undertaken to remove unwanted objects by automatic filtering operations. Since the mid-1990s, a tremendous amount of effort has been spent on developing ground-filtering techniques, which can automatically remove unwanted objects in order to obtain a bare-ground DEM (see Chapter 10). However, none of the plethora of ground-filtering algorithms is able to generate a DTM, which fulfils the needs of the task at hand. This observation includes that an airborne Lidar point cloud or a point cloud generated from photogrammetric image matching technique is not fit for use in many GIS applications. At best, it is a quasi-DTM which still needs a lot of manual editing before the information that the GIS analyst wants has been revealed.

7.5.3 Slope

Slope causes two main types of error. The first type refers to errors in the horizontal position of the sample points. Errors in the planar coordinates cause height errors of which the value depends on the slope angle. For airborne Lidar DEMs, such types of error can be introduced by pointing jitter, that is, the pointing accuracy of the scanning mirror. Another type of error depends on the slope within the region of measurement. When using airborne laser altimetry, the terrain slope causes a part of the footprint to hit the ground earlier compared to other parts. Consequently, one part of the pulse is already scattered back while the other part still has to hit the ground. This induces the problem of which part of the recorded back-scattered pulse refers to the ground. When applying photogrammetric image matching techniques, errors may occur when the slope within the region of measurement is not adequately incorporated into the matching model. For example, the slope is modelled as a ramp edge, while the actual slope contains second- and higher-order terms.

7.5.4 Reflectivity

DEM generation methods in which the points are automatically sampled rely on the reflectivity properties of the terrain surface. Water bodies cause problems for both Lidar and image matching. Water bodies absorb the major part of the energy of laser pulses in the near-infrared part of the electromagnetic (EM) spectrum while the reflected part may be subject to specular reflection yielding reflection of the signal in the direction away from the recording platform. Specular reflection may also cause corner reflection by multipath effects when buildings and other discontinuous objects are in the neighbourhood of the reflected signal, resulting in elevation dips. Using image matching, water bodies, as well as sand bodies like beaches and dunes, show little or no texture impeding reliable and accurate image matching. The above-mentioned reflectivity characteristics of terrain surfaces result in systematic errors (SEs) and data gaps in the DEM. In addition, reflectivity characteristics of the terrain may cause that the signal-to-noise ratio of the recorded data is low, producing higher than normal random height errors in the data.

7.5.5 Concluding Remarks

The above-mentioned considerations reveal that differences in terrain characteristics cause inhomogeneous quality of the DEM. The height of sampled points on, for example, hilly terrain covered with forest will show a lower accuracy than those on paved parking lots. To arrive at a proper quality description, points should be differentiated according to the type of terrain they are captured from. Consequently, terrain classes should be assigned as additional attributes to the data points.

7.6 Survey Design Issues

To obtain a sense of the issues a surveyor may get confronted with when conducting a laser scanning survey, we present here a short discussion on preparing and conducting a terrestrial laser scanner (TLS) survey. We present this discussion also because one may get inspired by the product brochures of TLS manufacturers in which the operation of their instruments in the field is presented as a plug-and-play game. Of course, TLS is able to capture quickly all types of complex scenes rapidly and relatively easily with high details. This can be done by just placing the instrument in front of the object, levelling it roughly, entering the boundaries of the part of the scene to be captured and pushing the button. It is as easy as that, at least as far as the instrument side of the survey is concerned. The snag lies not in operating the instrument but in the prelude to this, in scene monitoring during capture and in integrating the resulting point cloud in other geodata sets. We focus here on issues occurring while conducting the on-site survey.

Before placing the instrument, the operator has to ensure that no objects intervene between the instrument and the scene that might occlude essential parts of it. This means no cars, lampposts, traffic signs, vegetation, people or cows. Potential sources of failure may also be hidden in the scene; lapses may result especially from the reflectance characteristics of the surfaces of objects. Hitting a surface, a laser beam may interact with it in three ways: It may be reflected, absorbed or transmitted. Only reflected beams will reach the instrument and thus be of value for the task at hand, but one reflection is better than another. Ideally, a surface behaves as a diffuse reflector, so that the resulting reflections have similar strength in all directions. In this case, most of the signal returned from the surface reaches the instrument. But when parts of an object have a specular surface, the reflection is deflected and little or no signal is returned to the instrument. Therefore, mirrors, shiny metal and brackets of neon lights present in the scene have to be removed or covered. The strength of reflection, recorded in addition to the time of flight of the laser beam, provides the operator with a helpful means of detecting lapses in reflectance during the processing of point clouds. The intensity of the return is an important factor in determining the accuracy of TLS point measurements. Furthermore, laser beams may be transmitted through windows; this can occur with buildings, and results in the recording of objects on the other side of the glass or no returns at all. To avoid problems in reconstructing the final scene, windows have to be covered, or special caution must be exercised in processing the point cloud. Highway police, when using scanners to record the site of an accident, spray powder on the windows of the demolished cars, which are involved in the accident, to prevent the scanner "looking" through the glass. Pedestrians tend to cross a scene whilst it is being captured, and the same is true of dogs, cars and cyclists. These moving objects occlude the objects of interest. Crossing of pedestrians and bikers, while the scene is being captured, causes the presence of spikes as a result of the scanning characteristics (see Figure 6.1, Chapter 6). The surveyor must thus close off the scene area to all traffic before pushing the button or accept that artefacts are present in the point cloud. In practice, closing the scene to moving objects often proves to be cumbersome.

7.7 Validation and Accuracy

Validation aims at determining whether the acquired DEM, which is delivered by the survey contractor, obeys the product specifications. Major specifications include systematic error (SE), random error, average point density and maximum size of no-data areas. The maximum size of no-data areas is a tricky quality criterion since water bodies will be the main source of no-data areas. To validate this criterion, the DEM should be combined with a map or aerial images outlining water bodies.

A measure of elevation accuracy of DEMs is usually provided by the root mean square error (RMSE) and SE statistics. Both are usually computed by comparing a small subsample of the elevations of DEM, which are well distributed over the area, with elevation reference values. The reference elevations have to be measured with precision, which is about an order of magnitude better (say ten times) than the geodata acquisition technology with which the DEM has been obtained. Often reference values are measured in the field using GNSS devices. The RMSE is expressed as:

$$\mathrm{RMSE} = \sqrt{\frac{\sum_{i=1}^{n} \left(Z_i - z_i\right)^2}{n}} \tag{7.1}$$

Z_i refers to the reference values, z_i to the corresponding DEM elevations and n is the number of points involved in the validation. The SE, also called mean error or bias, is the difference between the mean of the point cloud values and the mean of the reference elevation values:

$$\mathrm{SE} = \frac{\sum_{i=1}^{n} \left(Z_i - z_i\right)}{n} \tag{7.2}$$

If the SE is zero, the RMSE statistic is essentially a standard deviation. This is based on the assumption that errors in the DEM are randomly and normally distributed (Fischer and Tate, 2006). However, if the SE is not equal to zero, the RMSE will be biased, which violates the assumption that the RMSE may be used as a substitute for the standard deviation. A measure which resembles better the standard deviation (σ) is to correct the DEM height for the systematic error by subtracting the SE from the DEM height and dividing by $n - 1$ as is the case for the standard deviation (Fischer and Tate, 2006):

$$\sigma = \sqrt{\frac{\sum_{i=1}^{n} \left(Z_i - \mathrm{SE} - z_i\right)^2}{n-1}} \tag{7.3}$$

Since it will only occur occasionally that both the DEM point and the reference point will be located at the same point in the horizontal plane, one has to carry out an interpolation to guarantee that the DEM point and the reference point do have the same planar coordinates. Chapter 9 elaborates on various methods which are suited for the interpolation of DEMs.

In practice, the collection of reference values is often done by measuring regularly distributed points on validation fields, such as parking lots and football grounds, well-distributed over the survey area. The measurements are usually conducted with high-performance equipment based on differential GNSS. When the differences between DEM elevations and reference elevations are within predefined limits, according to the used error criterion, the DEM is approved and accepted. Nevertheless, caution is required, since one is often inclined to consider the validation criteria as an average description of the quality of the entire data set. Since the validation fields are usually the most and best measurable parts of the area, the validation error statistics will provide a

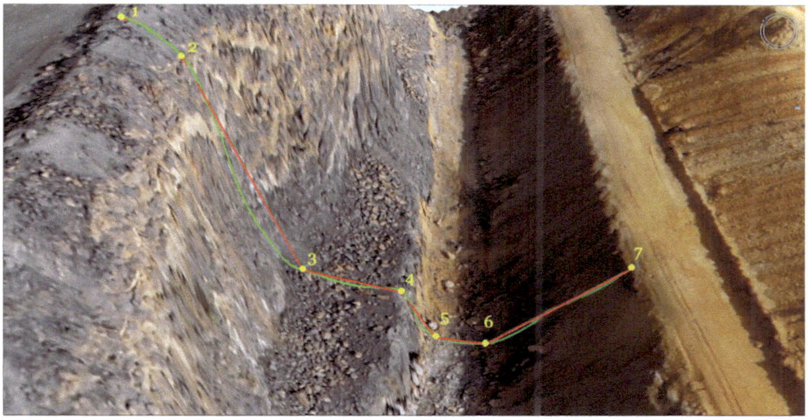

Figure 7.4 Location precision versus point density; the red profile is measured in the field, the green one by virtual surveying (source: Op 't Eyndt and Volkmann, 2013)

too optimistic view of the quality of the entire point cloud. The quality of the point cloud is inherently inhomogeneous because of the large influence of the terrain characteristics, which may vary locally, as discussed earlier.

To trace outliers, visual inspection of the end product is often carried out. When the DEM is photogrammetrically created using image matching techniques, the height points can be draped over the stereo image to inspect the presence of mismatches in a stereoscopic view. Advanced graphical techniques for inspection of DEMs when no reference is available have been developed too. Such a validation procedure is helpful to remove erroneous results. Visual quality control is insufficient when the data are wrong or are corrupted by error (Burrough and McDonnell, 1998).

It has been recognised early on that single point accuracy, expressed in error types like SEs, random errors and percentage of outliers, alone is not sufficient to describe the quality of a DEM. Also the point density in relation to the terrain fluctuations plays an important role. When the terrain surface in an area is perfectly horizontal, one height point will suffice to describe the elevation of the entire area. However, when relief fluctuations are present, more height points are necessary to obtain a representation, which is sufficiently accurate. The more rapid the fluctuations occur in the terrain, the higher the point density has to be. If the point accuracy is high but the point density is low while the terrain fluctuations are high, interpolated points will show low accuracy; the terrain is under-sampled (Figure 7.4). The choice of the point density depends not only on the terrain fluctuations but also on the required level of detail necessary within the application domain.

7.8 Final Considerations

This section presents some basic considerations on flight planning software tools, the feasibility of simultaneously mounting different sensors on an aeroplane and the appearance of objects in different point cloud types. This section discusses also surveying in harsh environments and ignorant laymen who may expect too much from point cloud acquisition and processing technologies.

7.8.1 Software Tools

Manufacturers of Lidar scanners and photogrammetric cameras deliver software tools to design the survey. When aircraft are the carrying platform, the software is often called flight planning software.

With today's positioning and orientation systems (POS) onboard, based on GNSS and inertial navigation systems, the flight can be steered by these systems allowing autonomous capturing of the scene. To ease the life of the survey operator, some manufacturers have developed specifications for different survey types. For example, Pix4D flight planning software supports the highly automated capturing of images from unmanned airborne system (UAS) platforms (Figure 7.5). The UAS operator can select a flight configuration, define the shape of the area, the size of the area to be captured and the flight parameters such as the image overlap, camera angle or flight height. The flight can be monitored from a computer device showing what the camera captures, the flight altitude and speed (Figure 7.6). The software may also enable checking the completeness of the survey on site so that omissions can be corrected while still being present on site. The creation of an accurate DEM requires a significant amount of manual editing in the post-processing stage to minimise the number of errors. For this purpose, editing tools are indispensable.

Figure 7.5 User interface for planning a UAS photogrammetric survey (courtesy: Pix4D, Switzerland)

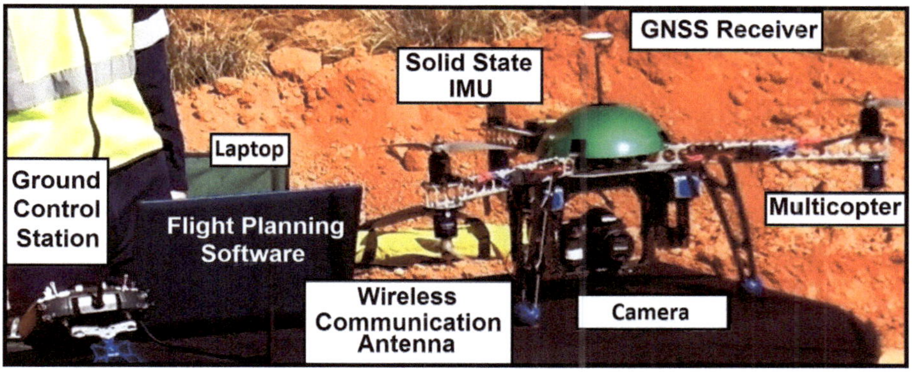

Figure 7.6 Components of a UAS equipped with a camera device

7.8.2 Simultaneous Acquisition

When designing an airborne survey, it may seem a good idea to capture both imagery and Lidar point clouds during the same flight by mounting the two sensors side by side in the same aircraft. However, the optimal flight parameters may differ for the two sensor systems. For example, the optimal flight height for the laser scanner may result in unfavourable extents of occlusion and lean-over in the images. In addition, the requirements for the distances between subsequent flight lines may differ. A distance, which may be optimal for one sensor, may be too small for the other sensor. Therefore, the distance between flight lines is determined by the weakest link. As a result, the simultaneous collection of imagery and Lidar data during the same flight may become more expensive than afterwards combining airborne Lidar point clouds and imagery captured during separate flights. Compared to Lidar, images have another weak point: The quality of images depends on weather conditions. Particularly haze and clouds should be absent as much as possible while the sun should illuminate the landscape from high angles to minimise the length of shadows. Lidar, as an active system, is largely independent of weather conditions except for clouds, and can even be used during nighttime. The time span available for Lidar flights is thus much bigger than for capturing images and waiting for good weather may be more expensive than acquiring Lidar point clouds and images during separate surveys. However, it should be recommended to add a consumer-grade camera onboard the aircraft to support the processing of airborne Lidar point clouds. The aim of the camera is not to be used as a measuring instrument but as an additional tool to identify which objects might have caused deviant behaviour of points. For example, the origin of data gaps may be identified in the images.

7.8.3 Harsh Environments

In many countries, in sub-Saharan Africa, peasants individually cultivate land parcels based on use rights. The land is owned by the community the peasants live in. The use rights are not formalised in official registers but based on customary law, memorised by the elders and chiefs, and orally transferred from one generation to the other. The value of agricultural land located close to cities is steadily increasing as a result of advancing urbanisation, and strangers settle in the midst of former close-knit communities. To secure tenure, the rights have to be formalised and registered in land administration systems and on cadastral plans. The creation of cadastral plans is highly supported by UAS photogrammetry; property boundaries can be easily delineated on orthoimages acquired by UAS. By combining orthoimages with the DSM, a digital landscape model can be created. Conducting surveys in countries near the Equator is often accompanied by harsh weather conditions. Heat and humidity may make the sensitive devices inoperative and are severe threats. Between the end of November and the middle of March, the Harmattan blows dust from the Sahara desert to West Africa, which may disrupt the UAS's mechanical parts, while distrustful natives may obstruct surveying (Schröder, 2015).

7.8.4 Laymen's Ignorance

The human visual is so well developed that laymen hardly grasp that the projection of a point cloud on a computer screen is just a visualisation. The point cloud itself does not contain information about the mutual spatial distribution of the points. The structure, which the human brain observes, is something that is a projection of the human brain on what the eyes see. It is an interpretation, which is unconsciously performed and thus concealed. Indeed, the human brain, averse to chaos, immediately starts to determine structures in whatever signals the eyes offer to the brain. And the brain reacts by imposing structure on groups of points. This is very hard to explain to people who observe not the individual points on the screen, but fuse points together to objects of which they know how they appear in reality. However, the clusters of points, which seem to form a group that belongs to the same object for the human brain, are stored in an unstructured way on a computer-accessible medium. The points don't know that they belong together nor do they know that they are reflections of the same objects. This phenomenon is extremely difficult to apprehend by laymen. "Why should something a human being can do so easily not be done by an advanced computer?" The layman may even state: "I clearly perceive a car and I don't see any reason why we would not be able to programme a computer to do the same. I think it should be rather easy to do because computers are so powerful today." In this context, I would like to quote the words of David Marr. He was a British neuroscientist and physiologist and an early and most creative computer vision researcher. His seminal book – *Vision* – focused on the automatic extraction of visual information from images. He sighs in this book (Marr 1982, p. 16):

> the idea that extracting edges and lines from images might be at all difficult simply did not occur to those who had not tried to do it. It turned out to be an elusive problem.

7.9 Concluding Remarks

The above-mentioned discussions demonstrate that the information present in point clouds can't be properly exploited without a basic understanding of the characteristics of the sensors which acquired the point cloud. Each type of survey requires a thorough preparation and planning stage in which the survey parameters are defined depending on the needs of the task at hand. There are no rules of thumb which can be blindly applied to any type of object space, any type of geodata acquisition technology and any type of use case, be it topographic mapping, 3D city mapping or outlining of general boundaries for cadastral purposes.

The distribution of objects and natural phenomena over the surface of the earth is inhomogeneous. When one walks through a scene, one will meet many different objects of which the spatial arrangement varies all along the route. Objects vary in size/volume, shape and orientation. This variation in spatial structure demands thought about which objects one wants to capture and provides selection criteria on minimum size/volume and so on. The notion of geographical variation has to be recognised during the design of a survey. The design of a sampling strategy will far and foremost depend on the application domain, which means the task at hand.

This chapter has demonstrated that geodata acquisition is accompanied by many pitfalls and that each step within the workflow, from design to data processing, requires a high degree of caution, insight, skills and thorough craftsmanship in the first place. Indeed, geodata acquisition is a profession in itself. To quote myself (Lemmens, 2016):

> One thing is for sure: how advanced a device may be, if not in the hands of a skilled person the output will be trash.

References

Blaschke, Th. (2010) Object Based Image Analysis for Remote Sensing. *ISPRS Journal of Photogrammetry and Remote Sensing*, 65(1), pp. 2–16.

Burrough, P.A., McDonnell, R.A. (1998) *Principles of Geographical Information Systems*. 2nd Edition, Oxford University Press, New York.

Fisher, P.F., Tate, N.J. (2006) Causes and Consequences of Error in Digital Elevation Models. *Progress in Physical Geography*, 30, pp. 467–489.

Hopkinson, C. (2007) The Influence of Flying Altitude and Beam Divergence on Canopy Penetration and Laser Pulse Return Distribution Characteristics. *Canadian Journal of Remote Sensing*, 33, pp. 312–324.

Lemmens, M.J.P.M. (1999) Quality Description Problems of Blindly Sampled DEMs. In Shi, W., Goodchild, M.F., Fisher, P.F. (Eds.), *Proceedings of the International Symposium on Spatial Data Quality '99*. Hong Kong Polytechnic University, Hong Kong, pp. 210–218.

Lemmens, M. (2016) EDM without Reflectors. *The American Surveyor*, September 2016, pp. 1–3.

Marr, D. (1982) *Vision: A Computational Investigation into the Human Representation and Processing of Visual Information*. W. H. Freeman and Company, San Francisco.

Op 't Eyndt, T., Volkmann, W., (2013) UAS as a Tool for Surveyors: From Tripods and Trucks to Virtual Surveying. *GIM International*, 27, pp. 20–25.

Puttonen, E., Lehtomäki, M., Kaartinen, H., Zhu, L., Kukko, A., Jaakkola, A. (2013) Improved Sampling for Terrestrial and Mobile Laser Scanner Point Cloud Data. *Remote Sensing*, 5(4), pp. 1754–1773.

Schröder, M. (2015) UAS Experiences in Hot and Dusty Africa: Heat, Dust and Distrustful Indigenous Peasants. *GIM International*, 29(12), pp. 21–23.

Ullrich, A. (2013) Sampling the World in 3D by Airborne LIDAR: Assessing the Information Content of LIDAR Point Clouds. *Photogrammetric Week '13*, pp. 247–259.

Xu, B., Li, F., Keshu, Z., Lin, Z. (2008) Laser Footprint Size and Pointing Precision Analysis for Lidar Systems. *The International Archives of the Photogrammetry, Remote Sensing and Spatial Information Sciences*, XXXVII(Part B1), pp. 331–336.

Part III
2.5D Mapping of Point Clouds

Digital Elevation Models

CHAPTER

8

If you have ever cycled through hilly or mountainous landscapes, your body, heartbeat and breathing felt the impact of rapid changes in terrain elevations. Indeed, while cycling, you have to cope with the terrain. The shape of the earth's surface is a key factor for determining many physical processes, not only the steaming of your own body and the splashing of sweat drops while you swing the pedals around, but also the flow of rainfall water, the steady downward movement of fertile soil called erosion, the melting of glaciers and the beech of seawater on dunes. To obtain a deeper understanding of these phenomena, geoscientists rely on accurate and detailed terrain descriptions in the form of samples of elevation represented numerically so that they can be processed by a computer using proper algorithms embedded in software. These numerical representations are usually called by the overarching term: digital elevation model, abbreviated to DEM. Also, engineers rely on DEMs for carrying out their design, construction and maintenance work. In road construction, for example, volumes and masses can be calculated from an accurate, reliable and detailed DEM to support earthworks. By comparing time series of DEMs, captured weekly, monthly or yearly, mining engineers are able to calculate ore or gravel extraction of an open-pit mine. Time series of DEMs are also a valuable data source for monitoring dune erosion. Combining DEMs of the bare ground surface with DEMs of a canopy, forest engineers can calculate the height of individual trees, the density of the canopy, biomass and the stand volume amongst others. This is vitally important for monitoring the production potential of extensive cultivated forest areas in countries such as Russia, Estonia and Canada. Digital representations of elevation also form the basis for geomatics products such as orthoimages, which are often used as backdrops or reference maps in GIS analyses.

The importance of DEMs in the various geosciences and practice is reflected by the many textbooks written on digital terrain representations, including El-Sheimy et al. (2005), Li et al. (2005), Peckham and Jordan (Eds.) (2007) and Florinsky (2016).

This is the first chapter of Part III of this book. Part III contains three chapters and entirely focuses on digital terrain representations. The emphasis on this topic is justified because of its huge importance for environmental analyses, governmental planning and monitoring, construction engineering and supporting the quest of gaining insight into the physical processes that the earth endures. This chapter starts with a characterisation of DEM point clouds – the section heavily relies on Lemmens (2014). The second chapter of Part III addresses the topic of computing height values of unmeasured points on the terrain surface from measured points, a process called interpolation. The final chapter considers the binary classification problem of separating bare ground points from building, vegetation and other above-ground points. This process is usually called ground filtering.

The present chapter aims at providing basic insight into how terrain surfaces are represented in numbers so that calculations can be carried out on computers using GIS software or programming languages such as Matlab or Python, which do have extensive libraries for processing point clouds. As discussed earlier, DEMs are a basic data source for many applications in geosciences and engineering, including agriculture, forestry, the environmental sciences, and mining and civil engineering. As a result, no uniform definitions of the diverse types of terrain surface representation exist. Section 8.2 briefly considers the various definitions and presents the DEM taxonomy adopted in this book. Next, this chapter has a look at DEM representations in the form of triangular irregular networks (TINs), grids and quadtrees. The latter are discussed in the framework of progressive sampling. This chapter ends with a short description of a data format specifically developed for the storage of point clouds: the LAS data format.

8.1 DEM Point Clouds

A point cloud is a set of data points represented in a preferred coordinate system. The data set consists of measurements taken at discrete points of a curved 2D surface in 3D space. This 2D surface may be smooth or may contain discontinuities such as cliffs and building facades. The number and density of the points should be such that the 2D surface can be reconstructed – that is, at every point on the surface, a value can be calculated from the measured values in the vicinity using one of the many interpolation methods developed in course of time. In principle, the curved 2D surface can represent any instance such as soil pollution, forest biomass, rainfall, terrain elevation or the seabed. In the field of geomatics, the phenomenon will usually be the terrain surface or the seabed in the form of a DEM. The elevations can be acquired by a land surveyor standing with both boots firmly on the ground while operating a Global Navigation Satellite System (GNSS) receiver, total station, levelling instrument or terrestrial laser scanner. Compared to these techniques, DEM acquisition using airborne Light Detection and Ranging (Lidar) is able to acquire detailed and high-accuracy DEMs highly automatically.

Today, large areas are captured by airborne Lidar and airborne photogrammetry through highly automated sampling. In addition to these geodata acquisition techniques which conduct measurements while the surveyor or operator is present in or above the terrain, another point cloud source consists of maps on which elevation contours are drawn. Often these sources will be topographic maps. Using a digitising tablet, the point cloud can be created in the office. The mapping of seabeds or the bottoms of other water bodies is usually done with multibeam sonar. In the pre-computer era, terrain surfaces were mainly represented by contour lines or by relief shading. Especially, the Swiss topographic maps are famous for the artistic quality of the relief shading. This doesn't mean that contours are not used anymore. On the contrary, today many contour maps are automatically generated from DEMs.

8.1.1 DEM Nucleus

A DEM point cloud consists of a set of planar locations indicated by their (*X*, *Y*) coordinates to which elevation values have been assigned: one value per (*X*, *Y*) location. The coordinate system may be a national or continental geodetic reference system or a local system specifically developed for the task at hand. In a fully 3D map of each planar location, multiple *Z*-values may be assigned. Because only one *Z*-value is assigned per *X*, *Y* position, a DEM is often called a 2.5D representation or a 2.5D map (see Section 1.2). Added to the elevation values, other attributes may be attached to the *X*, *Y* component, such as reflection intensities of the laser return or the values of the red, green and blue (RGB) bands of a colour image (Figure 8.1). The RGB values may be recorded by a camera, which accompanied the laser scanner during the survey. In a photogrammetric DEM, generated using image-matching techniques, the RGB values stem directly from the images. Airborne Lidar is able to record more than one return per outgoing pulse, resulting in multiple *Z*-values per *X*, *Y* component. If the laser scanner is equipped with the facility of full-wavelength digitisation, hundreds of returns can be recorded per outgoing pulse. However, when constructing a DEM from these Lidar point clouds, only one of these values is preserved. In a forest area, for example, this may be the returns stemming from reflections on the bare ground or from reflections on the canopy. When preserving the returns which reflected on the canopy, the resulting surface is often called digital canopy model (DCM). The *X*, *Y* component and its attributes form the nucleus of the point cloud, and the number of nuclei of a DEM point cloud may run into billions.

8.1.2 Labels

The attributes may not only concern measurements but may also be labels representing classes, which are useful during analysis and mapping in the later processing stage. For

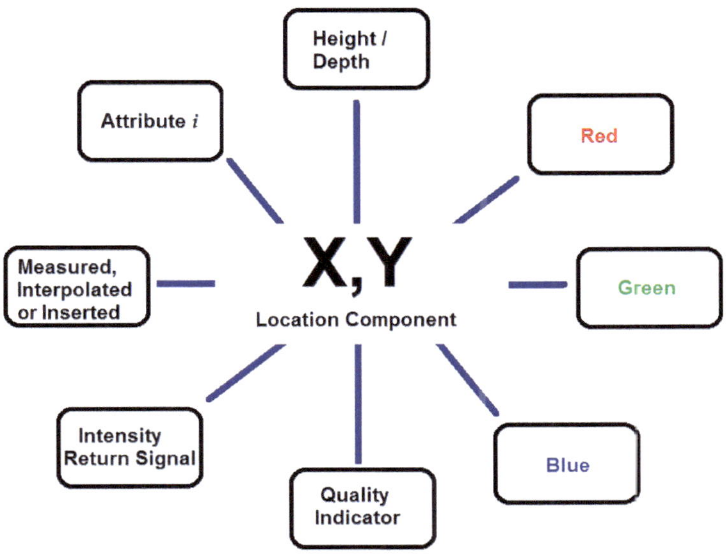

Figure 8.1 Diagram representing the nucleus of a DEM point cloud consisting of a location component (*X*, *Y*) and a variety of attributes (source: Lemmens, 2014)

example, the nationwide DEM point cloud of the Netherlands, called AHN3, distinguishes, in addition to ground-level points, four other point classes such as buildings; bridges, locks and other constructions; and others. One often forgets that information without quality description is actually poor information, when it may be called information at all. Therefore, quality indicators should also be assigned either to the individual points or to the point cloud as a whole. One of the quality indicators should indicate whether the points are measured, interpolated or inserted by a human operator based on generic or detailed knowledge of the terrain. The collection of DEM point clouds is not an aim in itself, but the starting point for creating a broad spectrum of products, which are valuable information sources for multiple application areas as discussed in Section 1.1.

8.2 Terminology

Digital terrain surface representations are exploited by professionals in virtually all geo-disciplines which describe and analyse physical phenomena taking place on, below or above the surface of the earth. They are also used by civil engineers and architects to design, construct and maintain buildings, bridges and other constructions. The military appreciates DEMs because they enable the automatic guidance of missiles at low flight altitudes over long distances. Furthermore, they are important for volume calculations in the mining industry. Since they are used in so many disciplines, the terminology has been developed over time and is far from unified. In the early days of computers, they became important tools in cartography and geomatics amongst others to obtain a numerical representation of terrain elevations at planar locations distributed over the area. These representations were called digital terrain models (DTMs) (Miller and Laflamme, 1958) and used for creating elevation contours. In other disciplines, what Miller and Laflamme have named DTM is called digital elevation model. Indeed, the terms DEM and DTM are often considered to be synonyms and used interchangeably.

8.2.1 Various Definitions

The US Geological Survey (USGS) defines a DEM as a "digital file consisting of terrain elevations for ground positions at regularly-spaced horizontal intervals." Maybe this definition is influenced by Frederick Doyle (1920–2013), who was a senior advisor for cartography at the USGS and 1980–1984 ISPRS President. One of his papers starts with the words (Doyle, 1978): "A Digital terrain model is an ordered array of numbers that represents the spatial distribution of terrain characteristics." On their GIS Dictionary website, Esri defines DEM as "the representation of continuous elevation values over a topographic surface by a regular array of Z-values, referenced to a common vertical datum," and digital surface model (DSM) as a "representation of continuous elevation values over a topographic surface, including vegetation and man-made features, by a regular array of Z-values, referenced to a common datum." Note that the term *datum* is equivalent to what we indicate in this book as *coordinate system* or *reference system*.

The definitions mentioned earlier are aimed at professionals working in the United States and do not only include a description of the type of terrain surface involved but also indicate how the Z-values should be arranged in the computer, namely in a regular array. In Europe, INSPIRE (2009) defines a DEM as a generic term for DTM and DSM. In this context, a DTM represents the 3D shape of the earth's bare surface, excluding as possible any other features placed on it. A DSM describes the 3D shape of the earth's surface, including all static features placed on it. In addition, INSPIRE (2009) notes: "in a DSM temporary phenomena should not form part of the surface, but due to the technical difficulties in removing them some of these features may also be present in the surface." The latter refers to dense point clouds produced by airborne Lidar and photogrammetry. The automated ground-filtering methods used to separate the bare ground from everything above the bare ground are not perfect. Recurrently, points on cars, trains, cows and other mobile objects will remain undetected during automated ground filtering or are overlooked by human operators during manual editing.

8.2.2 DEM Taxonomy

The non-universal use of the terms DEM, DTM and DSM induces the need to uniquely define the terms within the context of this book to ease communication. Basically, three types of terrain surfaces can be distinguished. The first one is the umbrella surface which includes road surfaces, roofs, foliage and other non-moving objects (Figure 8.2). Throughout the world and in various disciplines, there is a major consensus that this surface should be called a digital surface model (DSM). Next, there is the representation of the bare ground. In the past, some people, especially those with a background in civil engineering, have suggested using the term digital ground surface (DGS) to indicate this type of elevation surface (Grist, 1972). The term DEM just includes the word elevation but doesn't specify to which surface the elevations refer – it can refer to the bare ground, the umbrella surface or bare ground including terrain features. Therefore, the term DEM is well suited as an overarching idiom for indicating all types of terrain surface representations, regardless of whether the elevations are stored in a grid or the form of a point cloud. This is the definition of DEM we adopt in this book. Figure 8.3 schematically indicates the relationships among the diverse terrain surface representation.

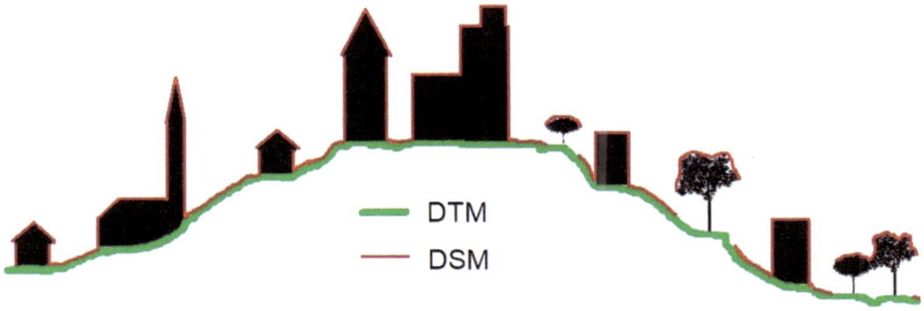

Figure 8.2 Schematic indications of the difference between DSM and digital elevation model (DTM)

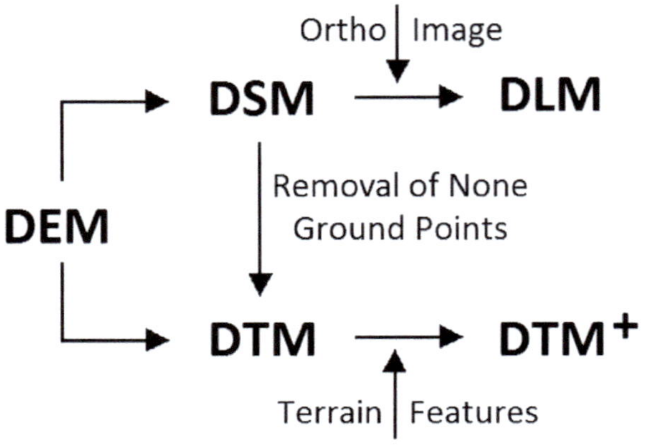

Figure 8.3 Scheme showing the relationship among the diverse terrain surface representations

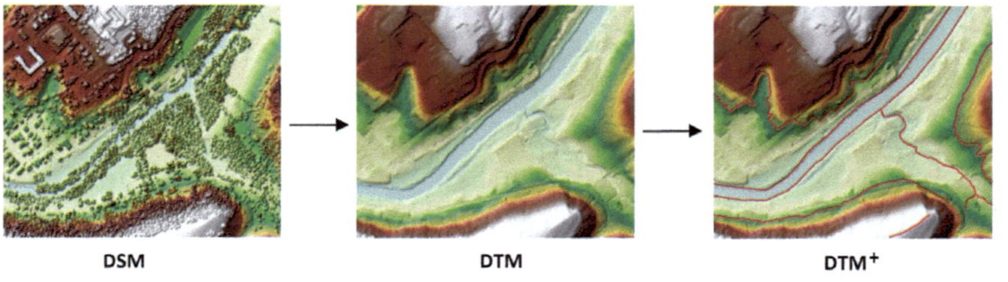

Figure 8.4 Visualisation of the differences between terms DSM, DTM and DTM$^+$ used in this book (source individual pictures: Esri website)

The introduction of a new taxonomy doesn't make much sense and neither does the changing of the definition of existing terms, which are widely used and strongly embedded within the core of a discipline. Therefore, the definition of the umbrella surface, which we use in the framework of this book, is the terrain surface including all static objects placed on the bare ground. We reserve the term DTM for representation of the bare ground surface. Often one wants to augment DTMs with terrain features such as ridges and streams. We will call this type of surface: DTM$^+$. Figure 8.4 visualises the difference between DSM, DTM and DTM$^+$. Others want to visually augment the representation by draping orthoimages over the DEM. Some call this representation a virtual landscape model. We prefer to name such representation a digital landscape model (DLM). Figure 8.5 shows an example of a DLM of a quarry located near Kassel, Germany. The quarry, covering an area of 16.5 ha, was generated from 330 images with ground sample distances (GSDs) as small as a few centimetres. A DLM provides not only a nice visualisation of the landscape but is also used by surveyors to carry out terrain measurements in the office, called virtual surveying (Op 't Eyndt and Volkmann, 2013). The text box provides a further account of virtual surveying.

Figure 8.5 DLM representation of a quarry generated by draping an orthoimage over a DSM, both created by UAS photogrammetry (source: Op 't Eyndt and Volkmann, 2013)

VIRTUAL SURVEYING

Unmanned airborne systems (UASs) equipped with a camera are able to create highly automatically DLMs, which is a computer representation of the actual terrain. Usually, the area covered during a UAS survey is counted in dozens of hectares, which means that the size of the site is relatively small. The photogrammetric UAS survey may take one day or less. After its creation, the land surveyor can display the DLM on a computer screen in order to measure points of interest by mouse-clicking as if he were placing a survey rod over a terrain point. In contrast to photogrammetric stereo models, DLMs allow the landscape to be viewed from many viewpoints. The operator can move through the terrain quickly and easily and even leap over hedges, houses and scrubs. Zooming and roaming enable the reliable interpretation of terrain details. As a result, productivity is improved. In addition, virtual surveying improves safety when open-pit mines, quarries, busy roads with heavy traffic and other hazardous sites have to be surveyed.

8.3 Triangular Irregular Network

The basic idea underlying the generation of TIN is to approximately reconstruct the continuous surface by connecting the individual DEM sample points. As a result, a TIN represents the continuous surface covered by a DEM point cloud as a set of connected triangles; there are no overlaps and no gaps. The vertices of each triangle are spanned up by three DEM points. All points are part of two or more triangles.

8.3.1 Two Basic Constraints

DEM points which have not been detected as vertices – thus are not part of a triangle – do not contribute to the representation of the terrain. This would be a waste of valuable and costly information. Indeed, the acquisition of elevation values is often an expensive exercise, although modern Lidar and photogrammetry are able to collect billions of points highly automatically within a time span of a few hours. To avoid waste of information, an important constraint put on the generation of the triangles is that within each triangle no elevation point is allowed to be present (Figure 8.6). When developing an algorithm based on this constraint, the result may be that no points are located within the interior of any of the triangles indeed, but that many triangles are elongated (Figure 8.7, left). In elongated triangles, at least two of the three vertices are far apart while the third vertex is close to the edge connecting the two distant vertices. This is not a favourable property when reconstructing a terrain surface – it results in an unrealistic surface which may have little resemblance with the actual terrain. In addition, elongated triangles violate the fact that points which are in the vicinity of each other will likely show more resemblance than points, which are far away from each other. To avoid elongated triangles, a second constraint is introduced, reading: The vertices of a triangle should be as close as possible to each other. This constraint may also be formulated as follows: Triangles should have a compact shape. From a geometric point of view, the compactness constraint encompasses maximisation of the minimum angle of each triangle. This constraint embraces that the edges should be equilateral as much as possible. In an equilateral triangle, the three edges have the same length.

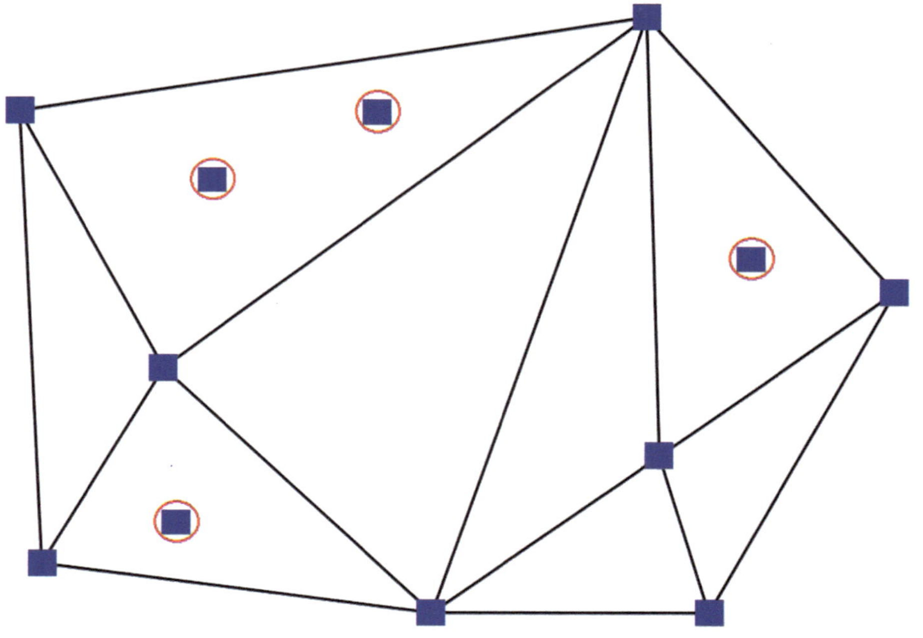

Figure 8.6 Wrong construction of triangles; a major constraint in the construction of a TIN is that no points are allowed to lie within the interior of the triangles

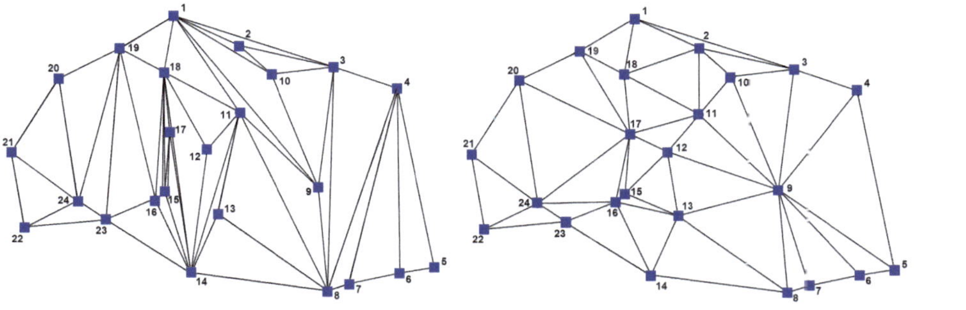

Figure 8.7 Two types of triangulation applied to the same point cloud; the TIN on the right side is based on the Delaunay triangulation

8.3.2 Delaunay Triangulation

An approach which has such a favourable property is by conducting a Delaunay triangulation, named after Boris Delaunay (1890–1980), a Russian mathematician and student of George Voronoi. Through the three vertices of a triangle, just one and only circle can be drawn. This unique circle is called the circumcircle (Figure 8.8). The Delaunay triangulation is defined by the constraint that the circumcircle of any triangle in the point cloud should be empty; that is, none of the circles may contain one or more points. The centres of the circumcircles are the vertices of the Voronoi tessellation (Figure 8.9). It is often said that the Voronoi diagram is the dual graph of the Delaunay triangulation. Algorithms to create a Delaunay triangulation and a Voronoi tessellation are implemented in GIS software packages and are also available in programming languages such as Matlab. Some interpolation methods rely on first creating a TIN from the DEM (see Section 9.5). TINs provide a realistic surface representation when the DEM point cloud is acquired by digitising contour lines from, for example, topographic maps. The distribution of such points is usually irregular because the digitising interval along the contours will usually be (much) smaller than the contour interval on the map. More

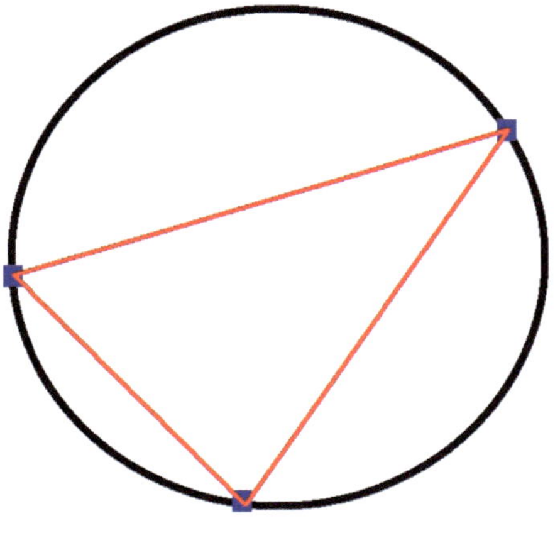

Figure 8.8 Circumcircle through three points

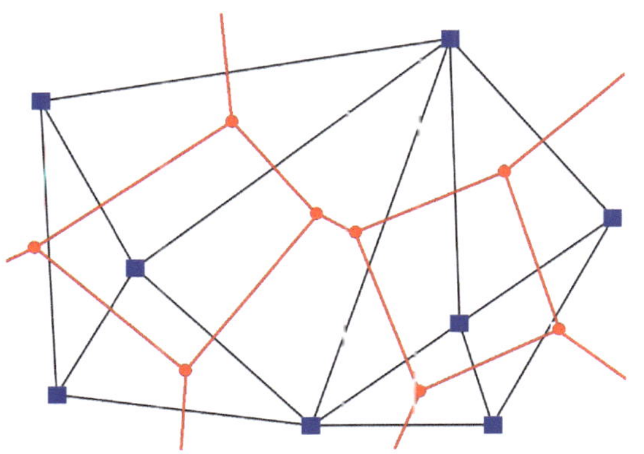

Figure 8.9 Dual graph: the centres of the circumcircles shown in red are the vertices of the Voronoi tessellation

specifically, when the points are irregularly distributed, a TIN is able to reconstruct surfaces reasonably well.

8.4 Gridding

A point cloud representation of measured terrain elevations contains the original information on the terrain shape. Conducting computer operators on the original measurements will only improve the quality of the information if knowledge about the terrain is implemented in these operators. This is for example the case when applying ground filtering – that is, the removal of outliers, or unwanted points, particularly non-ground points (see Chapter 10). Nevertheless, sometimes operations are carried out to obtain another representation without improving the information content. Point clouds stored in the LAS format (Section 8.6) are troublesome to work with in a GIS environment, especially if the point clouds are bulky. The massive amount of points may slow down processing during analysis. Added to this, many analysts, who explore geodata, are often used to working with raster data such as images. Processing of raster data is well supported by GIS software. Therefore, many analysts prefer the use of raster representations. The transformation from point cloud to raster – gridding or rasterisation – results in a data format which is similar to an image. As a result, the most common data structure to represent a DEM is the square grid, consisting of regularly distributed grid points to which elevations are assigned. An image raster cell is called pixel. The storage of the elevations of the DEM and the Lidar reflectance values results in two data layers.

The creation of a grid representation from DEM point clouds can be done in several ways. One of the most basic approaches is to create a square grid by dividing the reference surface into non-overlapping, fully connected tiles. In other words, a raster is superimposed over the reference surface which will usually be a plane. Only DEMs of large extent, for example covering an area of 100 km or more, require an ellipsoid as reference. The size of the raster cells has to be chosen such that it corresponds to the density of the DEM point cloud, if the point cloud consists of quasi regularly distributed points. For example, when the point density is 1 point/m^2, then a proper raster interval seems to be 1 m. Some raster cells will contain more than one point. To calculate the elevation of this raster cell, one can average the elevations or take their median value, the minimum, the maximum or another statistical value.

8.4.1 Basics of Gridding

Gridding requires the setting of the size of the grid cells, which is usually uniform over the entire DEM area. The choice of the size of the grid cell requires attention. A naïve approach would be to compute the average point density by dividing the total number of points present in the data set by the area covered by the point cloud, without checking the homogeneity of the point distribution. The outcome may read as n points/m^2. Next, the size of the square grid cells can be computed as $n^{-1/2}$. However, the points in raw, unprocessed point clouds are usually inhomogeneous, which means they are irregularly distributed over space. This is due to the characteristics of the sensors and the properties of the objects present in the scene (Figure 8.10). For example, a ditch, water reservoir or

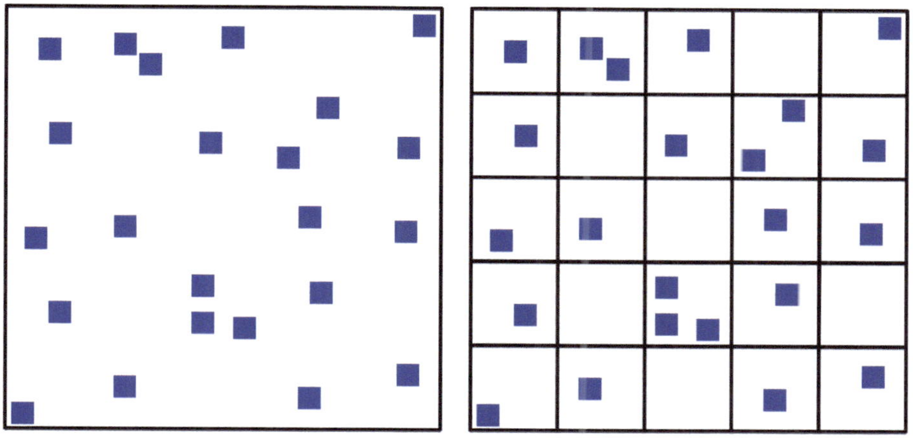

Figure 8.10 Superimposing a raster over a DEM point cloud of which the points are irregularly distributed results in empty raster cells and cells with two or more elevation values

other water body will give no or only a few returns when capturing the scene with airborne Lidar. As a consequence, many raster cells of areas covered by water bodies will have no value, and they will be empty. These cells can be assigned a no-data flag, or an elevation value can be computed from the elevations of the surrounding cells. On the other hand, some raster cells will have two or more elevations within their borders. The averaging of two or more elevations within one raster cell, as well as the computation of an elevation for empty cells, can be conducted by a moving 3 × 3 kernel, also called a window. This kernel checks whether two or more values are present in the centre cell of the kernel and if so computes the average or median of these values. If the centre of the kernel is empty, the average or the median of the surrounding elevations are computed and assigned to the centre of kernel.

If the raster cell covers a zone between two landscape types, averaging may result in an elevation which may be too high or too low. Figure 8.11 shows a 5 × 5 patch taken from a DEM raster. The raster cell with coordinates (2, 2) contains no elevation value but is surrounded by cells with elevation values. Computing the average results in an elevation value of 158 m. Applying the median results in (144 + 146)/2 = 145 m. The median forces the cell to belong to either one or the other landscape type, while averaging warrants a gradual transition.

8.4.2 Gridding by Interpolation

The superimposition of a raster over a DEM point cloud results in empty cells and cells with more than one point. The above discussion has shown that averaging or taking the median may bring resolution. However, it may frequently occur that also all or nearly all adjacent cells are empty. In Figure 8.11, the (2, 2) raster cell is surrounded by raster cells for which elevations are available. In contrast, the empty (4, 4) raster cell is surrounded by just two raster cells with elevations. Estimating its elevation from just a few surrounding values would generate unreliable results.

	1	2	3	4	5
1	142	144	183		
2	131		207		180
3	135	146	176		
4		141			
5				174	

Figure 8.11 Gridding of an irregularly distributed point cloud may result in many adjacent empty raster cells making the use of a moving averaging kernel unreliable

To resolve this problem, which will frequently occur in irregularly distributed point clouds, the computation of the elevation at a grid point is usually done by exploiting one of the interpolation techniques considered in Chapter 9. A popular interpolation scheme for this purpose is inverse distance weighting (IDW). This is because the concept can be easily understood and the method warrants that measured points which are close to the grid point have a bigger influence on the computation of the elevation of that grid point than on points farther away.

8.4.3 From Grid Coordinates to 3D Coordinates

The transformation from the grid coordinate system of each cell $(i, j), i = 1,\ldots,N; j = 1,\ldots,M$ to 3D coordinates is straightforward when the coordinates of the centre of the top-left grid cell (X_0, Y_0) are known in a ground-based reference system and the ground spacing of the grid cells is known: (ΔX, ΔY):

$$\begin{aligned} X &= X_0 + \Delta X \cdot i \\ Y &= Y_0 + \Delta Y \cdot j \\ Z &= Z(i, j) \end{aligned} \tag{8.1}$$

Compared to the storage of a DEM in the form of a point cloud, a grid representation requires a modest amount of computer memory. A further saving in storage requirement can be obtained by considering that the elevations within the raster will vary within a certain range: $\Delta Z = Z_{max} - Z_{min}$, where $Z_{max} = \max(Z(i,j))$ and $Z_{min} = \min(Z(i,j))$ with the domain of (i,j) as defined above. Then Z_{min} can be subtracted from all the height values:

$$Z'(i,j) = Z(i,j) - Z_{min} \tag{8.2}$$

and stored as a single number valid for the entire raster. The difference between the actual elevation and minimum elevation ($Z'(i,j)$) can be stored in a representative number of bits as if the attribute value was a grey value in a digital image. Suppose that the elevation values are stored at centimetre level and the elevation range covered within the entire raster, with the raster coordinates running from $i = 1, \ldots, N$ and $j = 1, \ldots, M$, is 4,030 cm, then a 12-bit quantisation suffices since 12 bits can store $2^{12} = 4{,}096$ different values. Accordingly, the computation of Z in Equation 8.1 changes to:

$$Z = Z_{min} + Z'(i,j) \tag{8.3}$$

8.5 Progressive Sampling and Quadtrees

Up to the year 2000, photogrammetry was the most important geomatics measurement technology for creating DEMs. At that time, the analytical plotter was in use as the major measurement device (Figure 8.12). This precision instrument basically consists of two digitising tablets connected by a computer. The movement of the two tablets in the two orthogonal x and y directions is steered by the computer. In the two tablets,

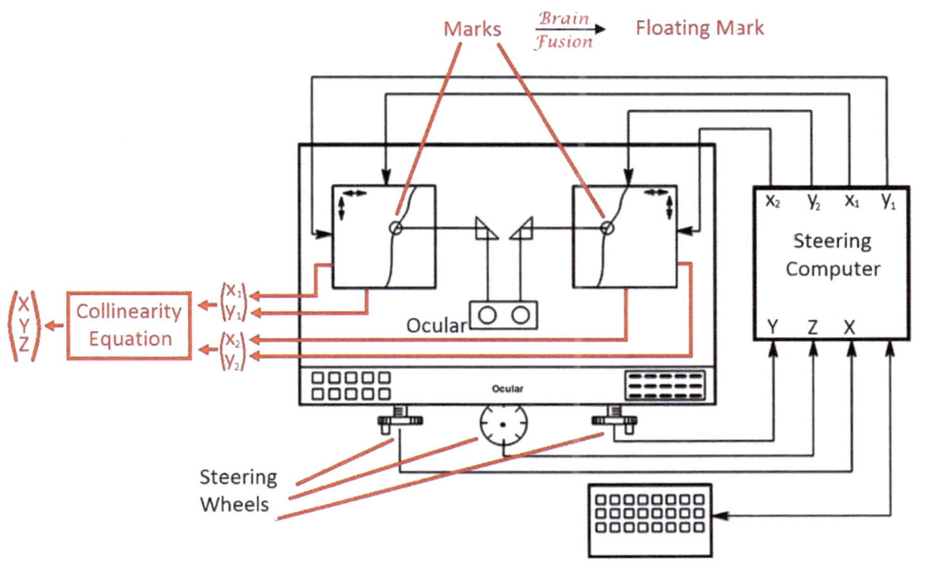

Figure 8.12 Schematic overview of the components of an analytical plotter used for DEM generation

overlapping aerial photographs are placed. Next, the two orthogonal image coordinates – indicated in Figure 8.12 as (x_1, y_1) and (x_2, y_2) – of corresponding terrain points are digitised. The corresponding points are points in the overlaps which are projections in the two images of the same terrain point. After the collection of the image coordinates, the computer transfers the (x_1, y_1) and (x_2, y_2) image coordinates to the (X, Y, Z) coordinates of the terrain point in a preferred geodetic coordinate system. The computation is conducted by applying the collinearity equation, which is implemented in software. The collinearity equation is the central equation of photogrammetry which relates planar image coordinates with the 3D coordinates of points in object space (see Section 4.3.1). After computation, the 3D coordinates are stored on computer memory.

8.5.1 Progressive Sampling on an Analytical Plotter

To ease and speed up the measuring process, the image coordinates of corresponding points are not measured separately and consecutively but simultaneously. The analytical plotter is constructed in such a way that a sophisticated use is made of the human visual system which is able to create a stereo model (essentially a 3D view). This means that the brain fuses the two images, which cover the same part of the terrain taken from different viewpoints, into one, single 3D image. The two overlapping images are observed by the human operator through the viewing system – that is, ocular – of the analytical plotter: One image is presented to the left eye and the other image to the right eye. As a result, the human operator sees a virtual 3D representation of the terrain. In the lens system of the ocular, measuring marks are attached. The function of these marks is similar to the cursor of a digitising tablet. By brain fusion, the human operator perceives the two marks as one mark that floats above the terrain. By turning the Z-wheel of the analytical plotter, the operator moves the floating mark in a vertical direction until it lands on the terrain surface.

The acquisition of a DEM using an analytical plotter was done by defining a fixed grid interval. After input of the interval, the computer steered the digitising tables to the corresponding positions. Next, the human operator let the floating mark land on the terrain surface and pushed a button to collect the image coordinates from which the 3D coordinates of the terrain point were computed. This approach considerably improved productivity compared to methods which were conventional at that time. A major disadvantage was oversampling in many parts of the area. This is because the distribution of the sampling points on a regular grid is not related to local terrain fluctuations. The use of a regular grid requires that the grid interval is adjusted to the smallest terrain fluctuations one wants to capture over the entire survey area. The sampling interval for flat terrain is the same as for terrain with a lot of relief. As a result, in flat regions, the number of elevations will be unnecessarily high which contributes to computer storage requirements, computational burden and waste of expensive labour time. Why store a bunch of elevations over an area in which the elevations are not significantly changing? At an early stage of creating DEMs by photogrammetric means, this disadvantage has been recognised by photogrammetric researchers.

A solution to this dilemma was developed by Makarovic in the early 1970s (Makarovic, 1973). He introduced a method, called progressive sampling, to adjust the sampling interval to the local terrain variations. A prediction of terrain variations was computed from previous measurements. The larger the variations, the smaller the interval was set. In other words, progressive sampling aims at reducing the number of elevation sampling points without compromising accuracy by only conducting measurements where the terrain surface is significantly changing. Significantly, this aspect is related to the task at hand. Since the sampling interval is adjusted while the data collection process is progressing, progressive sampling is also called adaptive sampling. Progressive sampling can also be conducted afterwards, when the data is already collected. This idea can be implemented by using a quadtree structure.

8.5.2 Quadtree

A quadtree is a way of organising data in a database such that the data can be quickly accessed for retrieving the data (Burrough et al., 2015). Organising the data in a quadtree prevent the entire database from having to be searched for a prespecified type of data, such as all data with a certain label located in an area represented by a polygon (Longley et al., 2015). The basic idea for using a quadtree as a data structure is to split up the bounding box of the area covered by the geodata into smaller quadrants until each quadrant is homogeneous according to a predefined criterion (Samet, 1989, 1990).

To demonstrate the creation of a quadtree, let's consider the example of a raster in which the cells have either the value 0 or 1. The bounding box of the area is subdivided into four quadrants of equal size. Each quadrant is being examined in the presence of zeros and ones. If both values are present, the quadrant is further split up into four subquadrants. If only zeros or ones are present, the quadrant is filled with uniform values and no further splitting is required. This process is recursively repeated until only quadrants remain which contain uniform values. The smallest possible quadrant is the single raster cell. The data structure is hierarchical – at a certain hierarchical level, all quadrants have the same size. In heterogeneous areas, in which the values fluctuate rapidly, the recursive subdivision will end up in small quadrants, which lie at deeper hierarchical levels while in homogeneous areas the quadrants will be large and the hierarchical level shallow. The quadrants at the higher hierarchical level are sometimes called parents, and at the lower hierarchical level, children. In this example, the criterion of the division was homogeneity in binary values. Another criterion may be the presence of data. In a DEM point cloud, elevations may be sparse in some regions, while in other regions plenty of elevations have been measured. In this case, each quadrant is being examined based on the presence of elevations. If no elevations are present in a quadrant, the subdivision stops. The quadtree is also a useful data representation for DEM point clouds especially when the elevations are irregularly distributed over the area.

8.5.3 Progressive Sampling on a Quadtree

To implement the idea of progressive sampling in a quadtree data structure, not the presence of elevations within the quadrant, but their degree of similarity, is examined. A suitable similarity criterion is the variance, or its square root, called standard deviation, of the elevations of the points present in the quadrant. If the standard deviation is below a certain threshold, the quadrant is considered to be homogeneous and no further

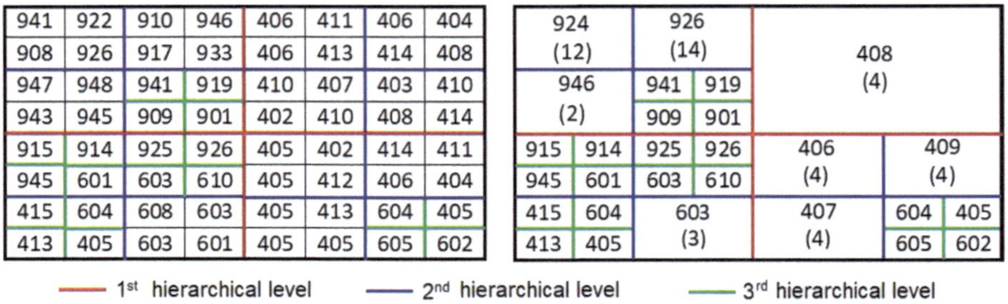

Figure 8.13 Building a quadtree from a raster representation of a DEM

splitting is needed. Figure 8.13 shows an example of a quadtree created from a raster DEM; the similarity criterion used is the standard deviation. The three-digit numbers are the elevations computed for the quadrants by averaging all the elevations covered by the quadrant. The numbers are presented dimensionless, the reader can think about centimetres, inches, feet or metres. The numbers between brackets indicate the standard deviation. The threshold on the standard deviation was set to 15, which means when the standard deviation is bigger or equal to 15, the quadrant has to be further subdivided. Another useful decision criterion could be the difference between the maximum and minimum elevation of points present in the quadrant. It is not necessary to transfer the point cloud to a raster representation, before building the quadtree. The quadtree can also be built up by examining the elevations of the points present within each quadrant.

8.6 LAS

The previous sections considered several data structures for storing point clouds, including TINs, grids and quadtrees. Usually, the points are transferred to these data structures in a post-processing stage. Immediately after the acquisition, the majority of point clouds are stored in a LAS format. This is also the format in which many laser scanning and photogrammetric surveying companies deliver point clouds to their clients. This data storage format, which is primarily meant as an exchange format, is the topic of this section.

8.6.1 Need for a Standard

The development of the LAS format has a practical and pragmatic background. The need for standardisation of exchange formats of massive data sets has been already experienced since the early 1970s. Indeed, for several decades in a row, users of remotely sensed data sets, in particular satellite images, got annoyed by the many proprietary data formats that various space agencies and earth-observation companies created. Each vendor developed its own file format(s) to best support the characteristics of the multispectral images produced by their sensors. Also, many manufacturers of laser scanners developed their own data format, unconnectable with systems developed by others, which hampered the transfer of point clouds from one processing system to the other. With the rise of laser scanning sensors, history got in danger of repeating itself.

To prevent a proliferation of data formats for airborne Lidar point clouds, the American Society of Photogrammetry and Remote Sensing (ASPRS) developed a standard exchange format in the early stages of the advancement of laser scanning technology. The rationale for developing this format was not only to support users' independence of proprietary formats, which prevents data being easily taken from one system to another, but also to provide an alternative to the ASCII file format used by many companies. The storage in ASCII format gives slow access to the data, results in excessive large file sizes even for modest amounts of data and requires exorbitant processing time as the numbers stored in the form of ASCII characters have to be first converted into numerical values before calculations can be conducted. It is as if the data is stored in Roman numbers on which first a translation to Arabic numbers has to be carried out before calculations can commence. Numbers in LAS format are stored in binary form to save space and to avoid the processing time needed to convert numbers written in ASCII code into a form on which computers can carry out calculations. Initially, the LAS format has been developed for exchanging laser scanning point clouds. However, the LAS format has become a generic format for the storage of point clouds captured by a variety of geodata acquisition technologies. The LAS format is a publicly available standard developed and maintained by ASPRS intended for use without license.

8.6.2 Tools Exploiting LAS

ASPRS continuously adjusts the LAS standard to reflect new advancements in sensor technologies. For example, the introduction and success of full-waveform digitisation resulted in a modification of the LAS format so that at the time of writing 15 returns per outgoing pulse can be stored. In the meantime, companies have developed and are continuing to develop processing and compression tools to speed up access time, processing and analysis of point clouds and to reduce storage space needed. Rapidlaso, based in Germany and headed by founder and owner Dr Martin Isenburg, for example, has created LAStools and LASzip. LAStools consists of modules for classifying, tiling, converting, filtering, TIN creation, contouring, rasterisation and many other indispensable tools while the open-source LASzip compressor has become the de facto industry standard for compressed Lidar data (rapidlasso.com).

8.7 Concluding Remarks

In many countries, the acquisition of point clouds for creating DEMs is often initiated by governmental agencies. For example, many countries in Northwest Europe, including Austria, Finland, Germany, Sweden, the Netherlands, Sweden and Switzerland, do have a nationwide DEM available acquired by airborne Lidar. The Netherlands started already in 1996 with the collection of dense DEMs using airborne Lidar. Meanwhile, the third edition is available as open data. In many countries, nationwide laser DEMs are available to everybody in the open data framework. The benefits of airborne Lidar for creating high-resolution, nationwide DEMs are well recognised and include high point accuracy coupled with high point densities up to dozens of points per square metre. In addition, Lidar allows highly automated processing, which brings costs substantially down.

References

Burrough, P.A., McDonnell, R.A., Lloyd, C.D. (2015) *Principles of Geographical Information Systems*. 3rd Edition, Oxford University Press, New York.

Doyle, F.J., (1978) Digital Terrain Models: An Overview. *Photogrammetric Engineering and Remote Sensing*, 44(12), pp. 1481–1485.

El-Sheimy, N., Valeo, C., Habib, A. (2005) *Digital Terrain Modeling: Acquisition, Manipulation and Applications*. Artech House, Norwood, MA.

Florinsky, I. (2016) *Digital Terrain Analysis in Soil Science and Geology*. 2nd Edition, Academic Press, Oxford.

Grist, M.W. (1972) Digital Ground Models: An Account of Recent Research. *The Photogrammetric Record*, 7(40), pp. 424–441.

INSPIRE (2009) Infrastructure for Spatial Information in Europe D2.8.II.1 Data Specification on Elevation – Draft Guidelines, http://inspire.ec.europa.eu/documents/Data_Specifications/INSPIRE_DataSpecification_EL_v3.0.2.pdf

Lemmens, M. (2014) Point Clouds (1): The Functionalities of Processing Software. *GIM International*, 28(6), pp. 16–19.

Li, Z., Zu, Q., Gold, C. (2005) *Digital Terrain Modeling – Principles and Methodology*. CRC Press, Boca Raton, FL.

Longley, P.A., Goodchild, M.F., Maguire, D.J., Rhind, D.W. (2015) *Geographic Information Science and Systems*. 4th Edition, Wiley.

Makarovic, B. (1973) Progressive Sampling for Digital Terrain Models. *ITC Journal*, 28, pp. 111–130.

Miller, C., Laflamme, R.A. (1958) The Digital Terrain Model: Theory and Applications. *Photogrammetric Engineering*, 24(3), pp. 433–442.

Op 't Eyndt, T., Volkmann, W. (2013) UAS as a Tool for Surveyors: Virtual Surveying Instead of Tripods and Trucks. *GIM International, UASSpecial*, pp. 26–29, https://www.gim-international.com/content/article/uas-as-a-tool-for-surveyors.

Peckham, R.J., Jordan, G. (Eds.) (2007) *Digital Terrain Modelling: Development and Applications in a Policy Support Environment. Lecture Notes in Geoinformation and Cartography*. Springer, Berlin, Heidelberg.

Samet, H. (1989) *Spatial Data Structures: Quadtrees, Octrees, and Other Hierarchical Methods*. Addison-Wesley, Reading, MA.

Samet, H. (1990) *Applications of Spatial Data Structures: Computer Graphics, Image Processing, and GIS*. Addison-Wesley, Reading, MA.

Interpolation

CHAPTER

9

Interpolation is widely applied to point clouds to reconstruct terrain surfaces or other continuous earth-related surfaces. This chapter provides the basics of interpolation aimed at reconstructing the continuous terrain surface from elevations, which have been automatically sampled using advanced geomatics data acquisition technologies. The resulting point clouds are dense and should be approached as big data. The literature on interpolation presents and examines many methods, usually developed with a certain application or group of applications in mind. The interpolation methods implemented in geographical information system (GIS) packages aim to serve a broad spectrum of geoscience professionals.

Indeed, the various geosciences have developed an abundant amount of interpolation methods to address the problem of reconstructing continuous surfaces from a sampled collection of points. As a result, a lot of publications have emerged with mining, soil pollution, watershed analysis and other specialised applications in mind. Li and Heap (2008) investigated the performance of over 40 interpolation methods and noted that their list is not exhaustive. The sampling may be sparse or dense, the measurements may be accurate or uncertain, the surface may be smooth or ragged with abrupt changes. All these factors influence the choice of the interpolation method and have contributed to the plethora of methods. Some may perform better for a certain application than others at the cost of abundant computation time which is unpractical for large point cloud data sets. Whether the performance is good or bad may be affected by the setting of tuning parameters, which have to be predefined by the user, and may require time-consuming tests and a lot of experience and craftmanship from the human operator.

Each interpolation is based on underlying assumptions about the density of points, the distribution of the points over the surface, which means the sampling pattern, the type of phenomenon, computational limitations and other factors. It is important that users keep these factors in mind when choosing an interpolation method from the broad pallet available within their GIS software. A bad evaluation criterion is whether the result is visually appealing. In general, physical phenomena do not behave nicely. So, a nice, appealing output should make the truthful analyst, who is proud of his craftmanship, suspicious. Appealing visualisations may make governmental authorities and the general public enthusiastic but they do not contribute to uncovering the truth and nothing but the truth. Words can conceal hidden (political) agendas, as can do statistics and as can do maps. See, for example, Darrell Huff's insightful book *How to Lie with*

Statistics, first published in 1954. Covering topics such as "correlation does not imply cause," it has long been a widely used standard book. Also using maps it is easy to lie, as Mark Monmonier convincingly demonstrated in his book *How to Lie with Maps*, first published in 1991. Ludwig Wittgenstein (1889–1951) stated in his Tractatus Logico-Philosophicus: "Everything that can be thought at all can be thought clearly. Everything that can be said can be said clearly." And "Whereof one cannot speak, thereof one must be silent." Paraphrasing the language philosopher brings us to the following statement:

> Everything which can be expressed by visual means should be expressed clearly.
> Something that cannot be visualised, that must be left invisible.

This chapter focuses on interpolation techniques applied to digital elevation models (DEMs). A DEM is one of the most important point cloud products of today's geo-data acquisition techniques which enable the acquisition of point clouds by highly automated means. DEMs, of which the diverse definitions have been discussed in Chapter 8, have become a fundamental source for many mapping products, including orthoimages, digital landscape models and 3D city models. DEMs are indispensable for earth-related research and applications, including deformation and flood-risk monitoring, volume estimation in the open-pit mining industry and forest management. Digital photogrammetry (image matching) (see Chapter 4), airborne Light Detection and Ranging (Lidar [see Chapter 5]) and SAR interferometry (InSAR) are the most important operational automatic data-capturing methods for topographic DEM creation at present.

The underlying principles of interpolation have been treated in a number of textbooks, including Journel and Huijbregts (1978), Burrough (1986), Isaaks and Srivastava (1989), Cressie (1991), Walden and Guttorp (1992), Wackernagel (1995), Goovaerts (1997), Lloyd (2010) and Chilès and Delfiner (2012). Understanding these principles often requires an in-depth insight into mathematics and statistics from the reader. It is our aim to put emphasis on the concepts of the diverse interpolation methods rather than scrutinising the mathematics. Hence, the amount of mathematics distributed over the pages of this chapter will be limited, while the mathematical language used will be as simple as possible but not simpler than that. Many publications are available on spatial interpolation, and to cover the broad subject in detail would require a book in itself. This is all the more true because plenty of interpolation methods often appear in several variations. Therefore, we limit ourselves to a panoptic number of local interpolation methods, which are commonly used, readable and available in GIS software packages. Particularly, the focus will be on inverse distance weighting (IDW), splines, Kriging, triangulated irregular network (TIN), natural neighbour (NN), the DEM interpolation method developed at Australian National University (ANU) known as ANUDEM and bilinear interpolation. These are the interpolation methods most often encountered in GIS packages and used for DEM interpolation. In addition, we will briefly consider two non-conventional interpolation methods: (1) one based on fitting a polynomial through a set of points based on least-squares adjustment and (2) the other one based on deriving distance weights from the 2D Gaussian bell-shape function, which is well known in statistics.

9.1 Basics of Interpolation

Interpolation aims at predicting the values of a certain variable of interest at unsampled locations based on measured values at points within the area of interest (Burrough and McDonnell, 1998).

On the one hand, there are the measured points, which do have known values, on the other hand, there are an infinite number of points with unknown values. For some of these points, one wants to know the value to proceed with the task at hand. In this chapter, we will refer to a location at which the elevation has to be computed as an unknown point and locations at which the values are known through measurements as measured points.

Interpolation methods may be grouped into two broad classes: local and global approaches. Local methods compute the value of an unknown point from the measured or known values of neighbouring points. Global methods use all measured points in the data set. This is impractical for large point clouds containing millions or billions of points. Therefore, we limit our considerations to local interpolation methods. Local interpolation methods are also more in accordance with Tobler's law.

9.1.1 Tobler's Law

Many phenomena in life which are located close together will show similar characteristics. This is true for the income level of people living in a block of dwellings, for terrain elevations and for other geographical phenomena. In other words:

> Everything is related to everything else, but near things are more related than distant things.

This is the so-called first law of geography, also known as Tobler's law, named after the geographer and cartographer Waldo Tobler (1930–2018), who formulated the law for the first time. This first law of geography is the foundation of local approaches to spatial interpolation.

To obtain a value of an unknown point, the measured points in the vicinity of the point are exploited. The closer the unknown point lies to a measured point, the more similar its behaviour will be. Hence, it is feasible to assign to nearby points a higher importance, expressed as a weight, in the computation of the unknown value, than points which are farther away.

The plethora of interpolation techniques available in GIS software has been initially developed for certain application domains in which reality is sampled in a particular way. For example, the South African mining engineer D.G. Krige was interested in predicting the presence of gold in explored areas based on known deposals of the metal. The method he developed and published in a South African journal (Krige, 1951) has been picked up by the French mathematician Georges Matheron (Cressie, 1991). Matheron (1963) was one of the first books on geostatistics in which also the approach of Krige was considered. He named the method of making reasonable predictions of the most likely distribution of minerals based on samples from a few boreholes after his developer *Kriging*. Particularly, the method has been developed for the coarsely sampled data set.

Today, Lidar and photogrammetric point densities are becoming denser by the year. Local interpolation requires the specification of a neighbourhood around the unknown point. The next section elaborates on neighbourhood selection.

9.1.2 Definition of Neighbourhood

The neighbourhood around an unknown point may be defined according to a distance metric or a numeric metric (Figure 9.1). In the numeric metric, the number of measured points is predefined: n measured points which are closest to the unknown point are selected. In a distance metric, the neighbourhood is selected by defining a range: All known points lying within the defined range are incorporated in the interpolation. In two dimensions, the definition of a range may be done by setting the radius of a circle or the side length of a square as shown in Figure 9.1. In three dimensions, the range may be set by the radius of a sphere or the side length of a cube. Other more complex shapes are possible although they are seldom used. The distance metric approach introduces the selection of the range, which means how large the neighbourhood of influence should be. The numeric metric introduces the issue that the neighbourhood of influence varies with local point density while the number of measured points may have nothing to do with the terrain characteristics and only depend on the sampling strategy or characteristics of the sensor.

The numeric and distance metric can be combined. Within the neighbourhood a prespecified number of points, for example, 12, which are closest to the unknown point, may be selected (Figure 9.1 D). This introduces a mechanism of point selection which depends on the local point density, which may be caused by the properties of the object but also by the sampling method. In an airborne topographic Lidar point cloud, water bodies may significantly reduce the point density at these locations due to the fact that near-infrared is absorbed by water. An airborne Lidar, which scans the terrain based on an oscillating mirror, causes a sawtooth sampling pattern. This may cause irregular distribution of points over the neighbourhood while the points are mainly aligned with the scan direction. The computation of the elevation of an unknown point close to the scanline may be mainly computed from measured points within the same scan line. This effect is demonstrated in Figure 9.2. The limitation to a fixed number of points within a neighbourhood is primarily induced by computational effort considerations and particularly makes sense when the point densities considerably vary over space.

The selection of the neighbourhood should be carefully considered. Using a distance metric, the selection of the range may be defined as a function of the interpolator used.

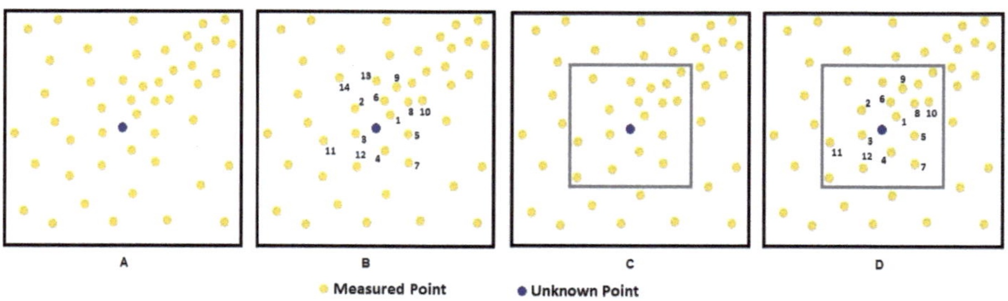

Figure 9.1 Neighbourhood selection: (A) distribution of measured points and inserted unknown point; (B) numeric metric; (C) distance metric; (D) hybrid metric

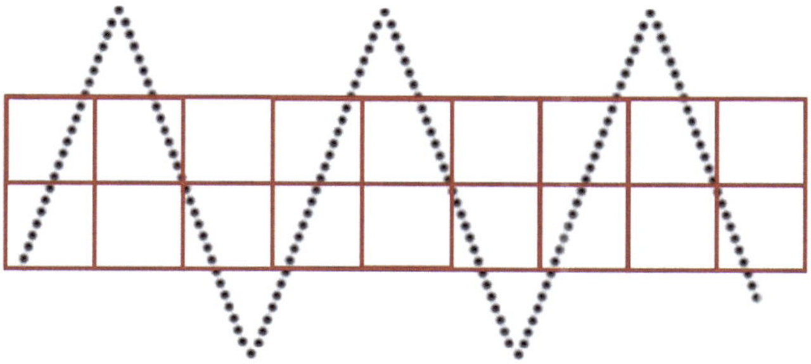

Figure 9.2 Due to the sawtooth scan pattern of laser scanners using an oscillating mirror, an interpolation neighbourhood may mainly contain points along the scan line

In the case of a numeric metric, n will be a fixed number. Several methods have been developed to determine the weights. The most simple approach is by introducing a numeric metric with $n = 1$ and a weight of 1. This approach results in the nearest-neighbour interpolation. Just the value of the nearest known point is selected and assigned to the unknown point. The algorithm is very simple to implement and may be feasible for dense and quasi-homogeneous point clouds.

9.1.3 Why Interpolation?

Point clouds in the context of the present book consist of blindly sampled points of continuous surfaces of objects. Reconstruction of the surfaces from the measured points is done by computing the value of an unknown point from one or more measured points surrounding the unknown point.

Spatial interpolation has to be conducted for a variety of GIS analysis tasks. These tasks may be divided into four main groups:

- For a project, one needs bare ground elevations at certain locations, which are not present in the data sets. The purchase of a digital terrain model (DTM), in the form of a point cloud, brings relief. The GIS software allows computation of elevations at unknown points using measured points in the neighbourhood.
- The analyst finds out that combining the DTM with multispectral aerial images adds to the reliability of the result. This requires rasterisation or better gridding of the DEM: For each raster cell, elevations have to be computed from the measured points.
- Next, the GIS analysts find that the DTM is referenced in the world geodetic reference system 1984 (WGS84) and the aerial image in a national geodetic reference system. To transfer the gridded DTM to the national frame, a geometric transformation has to be conducted. To compute the elevations in the new raster resulting from the transformation, an interpolation has to be carried out, for example a bilinear interpolation.
- The accuracy of an airborne Lidar DEM point cloud produced by a contractor commissioned by a governmental agency needs validation – that is, a check on the extent to which the DEM meets the specifications. Two specifications concern

maximum values set on the root mean square error (RMSE) and the systematic error (see Section 7.6). A second contractor gets tasked to conduct Global Navigation Satellite System (GNSS) surveys to collect reference elevations on parking lots and football fields well distributed over the survey area. The planar locations of the GNSS points do not coincide with the DEM points. To bring them to the same location, interpolation is necessary.

9.1.4 Weights

Tobler's law, discussed earlier, may be mathematically implemented by introducing distance-related weights:

$$Z(x,y)=\sum_{i=1}^{n} w_i Z(x_i,y_i) \tag{9.1}$$

where $\sum_{i=1}^{n} w_i = 1$ and $Z(x,y)$ refers to the elevation value which has to be computed at position (x,y) and $Z(x_i,y_i); i=1,\ldots,n$ refer to the n measured points in the neighbourhood. Writing the above-mentioned equation in vector notation results in the following expression:

$$Z(x,y)=\boldsymbol{w}^T \boldsymbol{z} \tag{9.2}$$

where **z** is the vector with the elevation values at the measured points and **w** is the corresponding vector with weights. If the uncertainties in the elevation values of the measured points are known and expressed in the form of a covariance matrix, **C**, the variance of the elevation value of the known point can be computed by error propagation:

$$\mathrm{Var}\big(Z(x,y)\big)=\boldsymbol{w}^T \boldsymbol{C}\, \boldsymbol{w} \tag{9.3}$$

Often the uncertainties introduced by the measurement equipment are known or can be computed from validation by confronting the measurements with reference points. For example, the random noise of a DTM acquired by airborne Lidar may be quantified as 15 cm in terms of RMSE. Supposing that the measurements are independent, which means there is no correlation so that the off-diagonal elements of the covariance matrix will be zero, and that all measured points have the same standard deviation (usually set equal to the RMSE), the standard deviation of the interpolated point can be computed. Let us consider that, as an example, 12 measured points – standard deviation σ – are involved in the interpolation and that every point has the same weight, which means de facto an ordinary averaging is carried out. The weights will be 1/12 for each known point resulting in $\mathrm{Var}(Z(x, y)) = 1/12\sigma^2$. As a result, the standard deviation of the interpolated point will be 0.29σ. This would imply that the quality of an interpolated point is better than the quality of the measured points. Indeed, as well known, averaging reduces the impact of random noise and averaging – that is what we did.

However, the reasoning underpinning the above discussion only includes the measurement part as a source of uncertainty. This is just a small part of the sources that affect the quality of interpolation. There are many more causes of uncertainty (Fisher and Tate, 2006). The single point accuracy of an interpolated elevation will only depend for a minor part on the measurement accuracy and depends for the major part on disturbing factors caused by the terrain slope, the microrelief, terrain cover, reflection characteristics of the objects and other effects (Lemmens, 1999). So, it doesn't make much sense to claim that the accuracy of an interpolated point would be better than the accuracy of the measured points. The only interpolation method which takes the uncertainties introduced by terrain characteristics into account is Kriging. As a result, Kriging is the only method which allows computation of a quality statistics, which makes sense, for the interpolated points.

The weights may be derived from the distance, r_i, from the unknown point to the measured point as follows:

$$r_i = \sqrt{(x_i - x)^2 + (y_i - y)^2} \tag{9.4}$$

Use of only the distance in the computation of the weight results in IDW interpolation. A basic assumption underlying IDW is that the correlation between height diminishes with distance. In addition to the distance, a smoothing and or tension parameter may be introduced which leads to spline interpolation.

9.2 Inverse Distance-Weighted Interpolation

IDW is one of the oldest interpolation methods. Proposed and developed in the mid-1960s by Shepard (1968), the method is commonly used for a variety of sampled data sets, including DTMs. Today, IDW has proven to be useful for reconstructing surfaces from DTMs and this is underpinned by the assumption that measured points close to the unknown point should have a larger influence than points farther away – the closer an unknown point is located to a measured point, the more similar it will be. Many GIS users apply this method because the basic concepts of IDW are easy to understand and its simplicity and availability in GIS software contribute to its popularity.

The unknown attribute value $Z(x, y)$ is computed as a distance-weighted average of values at measured points lying within the neighbourhood surrounding the unknown point. The weights given to the measured points in the vicinity of the unknown point are determined as an exponential function of the reciprocal of the distance to the unknown point:

$$w_i = 1/r_i^q / \sum 1/r_i^q \tag{9.5}$$

The power exponent q, which has to be tuned by the user, mainly determines the performance because it balances the influence of nearby points compared to points farther away. With the increase of the power exponent, measured points farther away from the unknown point get less influence. The most popular choice for the power exponent is 2 and the resulting method is often called inverse square distance weighting. When $q = 0$, IDW reduces to a moving-average interpolation in which all measured points have equal weight.

Whether the relationship drops slowly or rapidly with distance is somewhat concealed in the power exponent. Setting the power to 1 corresponds to a slow drop of relationship, a high power to a rapid drop. To get an impression of the drop-off rate, Figure 9.3 depicts the decline of influence with increasing distance between an unknown point and the measured point for power exponents ranging from 1 to 4.

IDW requires the definition of the neighbourhood of the unknown point. When introducing a distance metric, the size of the influence sphere should be in balance with the power exponent. A low power means that the neighbourhood should be larger than when setting a high power. When a point lies five times farther away from the unknown point than the closest point, its fractional influence is just 4% when the power is 2. The power exponent can thus be used for determining the cut-off range: Points lying outside this range have so less influence on the elevation of the unknown point that it would be a waste of computer resources to include them in the computation. In turn, when the size of the cut-off range has been defined by the user as well as the fractional influence, the relevant power exponent can be computed. The relationship between the cut-off range (R_0), fractional influence (p) and the power exponent (q) is:

$$R_0^{-q} = p \tag{9.6}$$

For example, if the user wants to set $q=2$ and $p=10\%$, $R_0=3.16$. If the user wants to define the influence sphere as well as the fraction, q becomes:

$$q = -({}^{R_0}\log p) \tag{9.7}$$

For example when $R_0=5$ and $p=4\%$, q becomes 2.

The computed values are always within the range of the measured values, indicating that the elevation computed for the unknown point will be never larger or smaller than

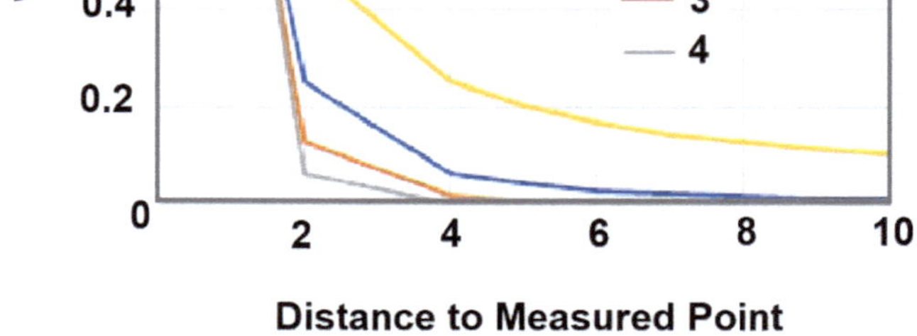

Figure 9.3 Inverse distance-weighted interpolation: drop-off of the influence with distance from the unknown point as a function of the power exponent (q)

the elevation of the measured points in the neighbourhood. In a sparse DTM peaks, riffs, chines, razor-edges and dips may not be represented and IDW will not be able to reconstruct them in contrast to spline interpolation. IDW does not look at the point pattern – that is, the distribution of the measured points over space. In some cases, this means that a directional dependency or bias will be introduced (Robeson, 1997). If the distribution of measured points is also sparse, the terrain surface may be badly reconstructed. However, IDW works well for dense DTMs and if the terrain is smooth.

9.3 Spline Interpolation

Spline interpolation is also called radial basis functions' interpolation because the method exploits a linear combination of several independent basis functions, one for each measured point. Radial basis functions were initially developed for curve and surface fitting.

A spline is a polynomial type of function which is not forced to exactly fit through the measured points. From a statistical point of view, this means that the measured elevations are uncertain, partly because they are contaminated by random noise. So the measured points are allowed to deviate to some level from the reconstructed surface. The size of the deviation may be set by the user by introducing a smoothing parameter. The setting of this smoothing parameter may be defined based on the accuracy of the measurements. The smoothing parameter can be defined not only on the basis of knowledge about the accuracy of the data, but also on scene knowledge combined with point density.

Spline interpolation exploits a linear combination of N basis functions (one for each measured point in the neighbourhood). The selection of basis functions determines the properties of the spline interpolation. For two-dimensional applications, basis functions are functions of planar coordinates plus one or more additional parameters which determine how well the functions fit the measured points. The additional parameters release the basis functions from the constraint to exactly go through the measured points.

The major basis functions are spline with tension, completely regularised spline, multiquadric spline, inverse multiquadric spline and thin-plate spline (Johnston et al., 2001). As examples, we provide the basis functions for the multiquadric and the thin-plate spline. The multiquadric basis functions proposed by Hardy (1990) and widely used in surface-fitting can be written in a generalised form (Wang and Liu, 2002) as:

$$\varphi(r_i) = (r_i + \sigma)^q \tag{9.8}$$

If $q = 0.5$, Equation 9.8 becomes Hardy's original multiquadric function and if $q = -0.5$, Equation 9.8 becomes the reciprocal multiquadric spline. The parameter r defines the distance from the unknown point to the measured point and is computed from the planar coordinates of both according to Equation 9.4. The additional parameter σ in Equation 9.8, also called tension, smoothing, shape or regularising parameter, is provided by the user and is thus a tuning parameter. Note that if the smoothing parameter σ is neglected – that is, set to zero – and negative values are assigned to the power exponent q, the spline approach based on multiquadric function reduces to IDW interpolation.

The parameters r and σ appear also in the basis functions of the thin-plate spline:

$$\varphi(r_i) = (r_i.\sigma)^2 \ln(r_i.\sigma) \tag{9.9}$$

For each measured point, there is one basis function and the functions are combined by computing a weighted average and adding a bias parameter β:

$$Z(x,y) = \sum_{i=1}^{n} w_i \varphi(r_i) + \beta \tag{9.10}$$

The weights and the bias parameter are determined from the data and the basis functions. A small value of the smoothing parameter results in a spline surface for which the measured points do not deviate much from the reconstructed surface while a large value results in a spline which lies farther away from the measured points and thus provides a smoother solution.

To get a more tangible impression of splines, one may have a look at where they come from. Initially, splines were physical devices made of rubber which could be flexibly curled through duck points so that the total curvature of the surface was minimised. The minimisation process depended on the tension a human hand put on the sheet.

In addition to a smoothing parameter, also a tension function can be exploited in spline interpolation. The tension function controls the distance over which measured points have an influence on the reconstructed surface.

The elevation computed for the unknown point does not necessarily fall within the elevation range of measured points. This means that the elevation may be larger than the maximum elevation of the measured or smaller than the minimum elevation. As a result peaks, riffs, chines, razor-edges and dips may be reconstructed even for sparse DTMs. This is in contrast with IDW, which is not able to reconstruct these landscape components.

9.4 Kriging

Kriging is widely used for DEM interpolation. It has a sound underpinning in statistical theory and the implementation is convenient (Meyer, 2004). Kriging provides the best linear unbiased estimators of elevations at unknown points. The solution is optimal in a least-squares sense and the weights are based on the inverse of the covariance matrix.

The principles underpinning interpolation by Kriging are described in many textbooks. Basically, Kriging exploits Equation 9.1. In contrast to IDW, the weights are not determined by "guessing" the relationship between the unknown point and the measured points by introducing a power exponent on the inverse of the distance. Kriging aims at quantifying this relationship based on sound statistical considerations. Kriging is a generic name for a family of generalised least-squares regression algorithms (Daya, 2015). This family is called Kriging in recognition of the work of Daniel Krige (1919–2013), a South African mining engineer and pioneer in geostatistics. We will only briefly consider the concepts of this interpolation method and do not go into the mathematics and statistics underlying the approach. A detailed description can be found in Burrough et al. (2015). This book, which is widely available and should have a prominent place in the bookcase of every geoscientist, dedicates an entire chapter to Kriging interpolation.

In their search for mineral deposits, such as copper, zinc, palladium, cobalt and coltan, but also gold, mining engineers often have to rely on samples from just a few boreholes. The estimation of the presence of minerals for the mining industry from sparse samples requires not only thorough geological knowledge but also sophisticated prediction techniques. This is the origin of Kriging interpolation and now is widely used in geography and spatial data analysis.

The weights are based not only on distance as is the case for IDW, but also on the quantification of the spatial relationships among the measured points. These relationships can be numerically expressed in a covariance matrix or a variogram. This means that the spatial pattern or arrangement of the measured points is also taken into account (Johnston et al., 2001). So the weights depend on a fitting model for the measured points, the distances of the measured points to the unknown points and the spatial pattern of the measured points.

In contrast to other interpolation methods, Kriging is able to calculate the precision of the computed value of the unknown point using error propagation methods. The method requires a lot of computer resources and computation time depending on the number of measured points involved and computer speed. Nevertheless, Kriging has become an important part of spatial interpolation methods within GIS and is frequently used to compute terrain surfaces represented in a grid format. The application of this interpolation method requires a thorough scrutinising of the spatial behaviour of the phenomenon and Kriging requires a skilled analyst with geostatistical knowledge. If the neighbourhoods of interpolation do not sufficiently overlap, for example when they have just one point in common, a terrain surface reconstructed by ordinary Kriging (OK) can be patch-wise discontinuous (Meyer, 2004). This is especially true when the points are located at the boundary zone of two landscape types; in particular, flat terrain adjacent to terrain will rapidly change elevations. This effect is demonstrated in Figure 9.4.

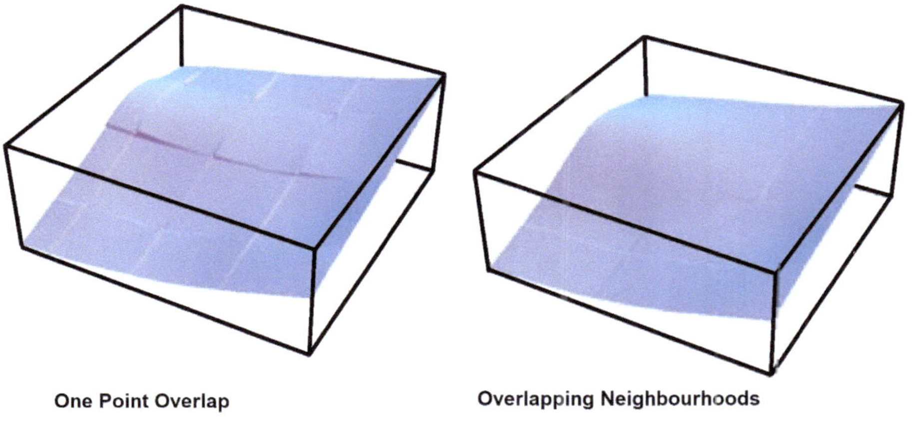

Figure 9.4 The same DEM patch interpolated with neighbourhoods sharing one point (left) and sharing sufficient points; the discontinuity, running from left to right in the middle, is clearly visible (source: Meyer, 2004)

9.5 TIN Interpolation

TIN modelling is a simple way to construct a surface from measured points. After conducting a TIN operation, the terrain surface is represented by a network of connected, non-overlapping triangles much in the same way as land surveyors did in the past to create accurate nationwide geodetic coordinate systems. Usually, the term network is replaced by mesh in the context of TIN modelling. It is a particularly useful technique for irregularly spaced points, for example when the elevation values are acquired through the digitisation of contour lines. An introduction to the creation of TINs is provided in Section 8.3.

The vertices of the triangles, which are created using Delaunay triangulation, consist of measured points. The edges of each triangle connect nearby points. Three points are selected such that the triangles do not overlap and the network is as compact as possible. This is achieved by introducing the constraint that the circumscribing circle of any triangle contains no measured points except the three vertices defining the triangle. The elevation value of the unknown point is found by intersecting the vertical line through (x, y) with the plane spanning up the TIN triangle (Figure 9.5). Because the values at the three vertices of the triangle are known and the distances between vertices can be calculated, a simple linear equation can be used to calculate an interpolated value for any position within the boundary of the TIN. The mathematical relationship between the Z value and the planar coordinates (x, y) is given by:

$$Z(x,y) = ax + by + c \tag{9.11}$$

The three parameters (a, b, c) can be exactly solved, no least-squares adjustment needed, by the three points spanning up a triangle in a TIN. Next, the Z values at an unknown point at location (x, y) can be calculated from Equation 9.11. TIN Interpolation is based on a numerical metric consisting of three points for each interpolation. Using only three

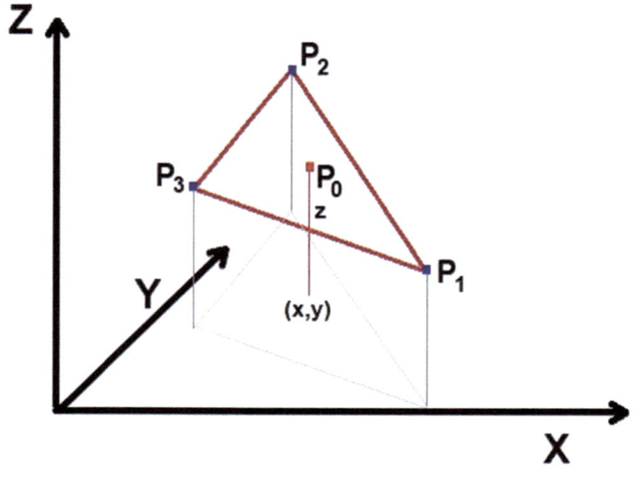

Figure 9.5 Sketch of the principle of TIN Interpolation

points per unknown point makes TIN interpolation fast. On the other hand, the three points spanning the triangle are not necessarily closest to the unknown point (Figure 9.6). TIN guarantees that the unknown point is encapsulated by the three known points. The price to be paid for this beneficial property is that the final result may not be as accurate as other methods (Mitas and Mitasova, 1999). Since the reconstructed surface consists of a patchwork of triangles, the reconstructed surface may look jagged (Figure 9.7). The

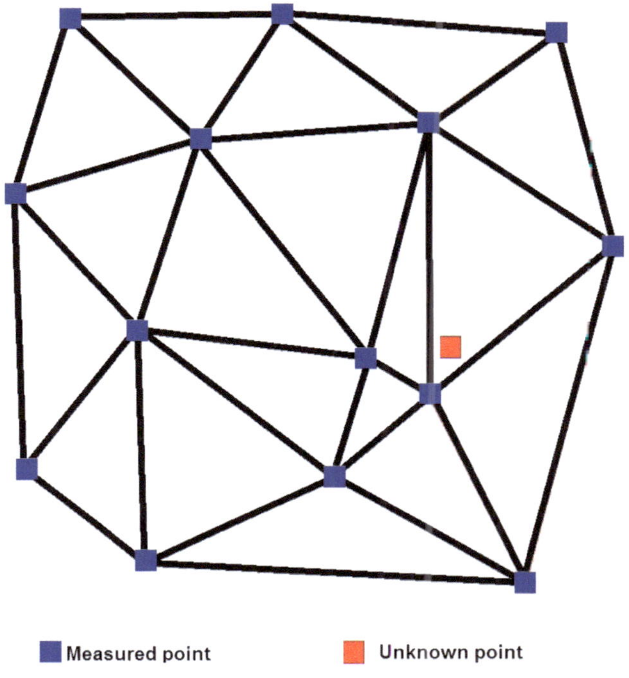

Figure 9.6 TIN interpolation does not necessarily take the closest point, but warrants that the unknown point is surrounded by three known points

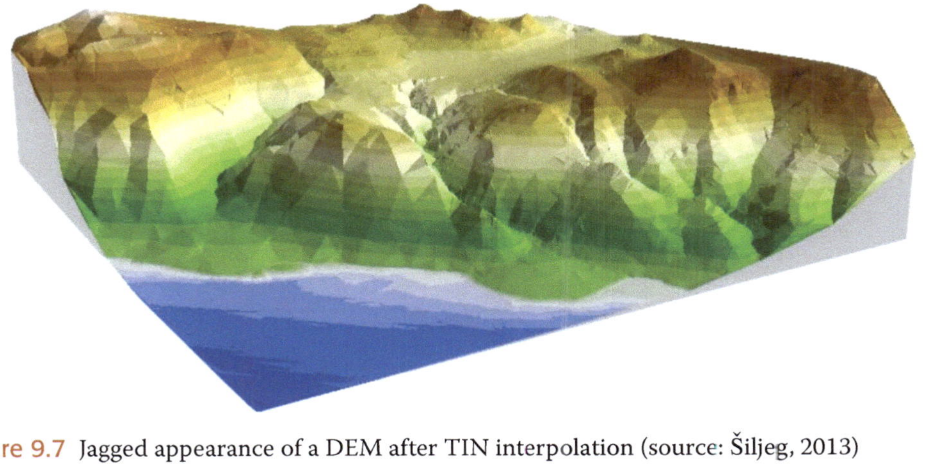

Figure 9.7 Jagged appearance of a DEM after TIN interpolation (source: Šiljeg, 2013)

transition from one triangle to the other is usually not smooth but may look toothed. This is especially true when the slopes at the edge of two adjacent triangles are not aligned but point in different directions. Discontinuities and other landscape structures can be easily incorporated (Mitas and Mitasova, 1999). The method is well suited for reconstructing surfaces from irregularly spaced measured points, like points extracted from contour lines.

9.6 Natural Nearest Interpolation

Natural nearest (NaN) interpolation is based on weighting the elevation values of measured points in the neighbourhood of the unknown point according to Equation 9.1. The weights of the NaN interpolation, developed by Sibson (1981), are not based on distance as in the case of IDW but on area. The question becomes now: how to compute the size of the area and its fractional contribution? The computation of the area is based on Voronoi tessellation of the measured points. Since Voronoi polygons, also known as Thiessen polygons, cover the entire surface without gaps or overlaps, an unknown point will always be located within one and only one Voronoi polygon. The insertion of an unknown point will result in a new Voronoi tessellation. By draping the new Voronoi polygon, created by the insertion of the unknown point, over the original polygons, the overlaps of the new Voronoi polygon with the surrounding original polygons can be computed. Let's call this area B_i. The larger the proportion of this area in relation to the size of the original polygon A_i, the larger the influence of the elevation of the measured point spanning up this polygon should be. In other words, the proportion of the area that is cut off by the new Voronoi polygon determines the weight (Figure 9.8). By substituting the distances in Equation 9.1 by the proportion of the areas, the NaN interpolation weights become:

$$w_i = B_i / A_i \tag{9.12}$$

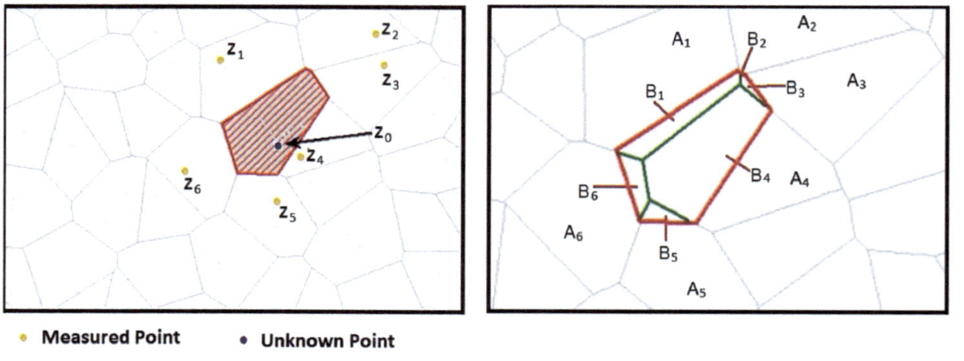

Figure 9.8 Sketch demonstrating the principle of NaN interpolation. The red polygon is newly created after insertion of the unknown point (Z_0) (left); $B_i; i = 1,\ldots,6$ are the areas of the overlaps of the new polygon with the surrounding original polygons

The area of a polygon can be computed from the well-known formula:

$$A = \frac{1}{2}\sum_{i=1}^{n} x_i (y_{i-1} - y_{i+1}) \tag{9.13}$$

The definition of the points of the polygons is circular, meaning that $(x_{n+1}, y_{n+1}) = (x_1, y_1)$. Here, the coordinates $(x_i, y_i); i = 1,\ldots,n$ refer to vertices of the Voronoi polygon.

NaN interpolation doesn't require any user input on the size and shape of the neighbourhood, the number of known points to be included in the interpolation or any other tuning parameters. NaN performs equally well for regularly as well as irregularly distributed known points (Watson, 1992). As is the case for IDW, NaN does not create peaks, dips or other artefacts; interpolated elevations are never smaller or larger than the measured points used in the interpolation. The proportional use of the overlap areas warrants that nearby measured points have a larger influence than those farther away. The method adopts the local distribution of known points, since all Voronoi polygons surrounding the unknown point are exploited provided that they are directly adjacent to the inserted Voronoi polygon.

9.7 Bilinear Interpolation

Grid generation using local interpolation, of whatever type, is based on weighted averages of values at nearby points. The assumption is that the elevation at each grid point is likely to be similar to other values in the vicinity. Gridding may have been carried out by the producer of the DEM and the user may have no access to the original point cloud. To combine the raster DEM with other geodata sets, which are georeferenced with respect to another coordinate system, a geometric transformation has to be carried out. Every geometric transformation of raster data or gridded data induces the need for starting a resampling procedure. This need results from the fact that the transformation of integer input coordinates of grid points will not necessarily lead to integer output coordinates, meaning that the position in the output image will not coincide with the grid point. To let the results of the transformation coincide with the grid point, an interpolation is necessary.

The procedure starts with defining a rectified output grid that is initially empty. For each output grid point, the position in the input grid is computed by performing the inverse geometric transformation. The two calculated coordinates of the position in the input will usually not exactly coincide with a grid point in the output grid. The unknown value at this position in the input grid is computed by interpolation. Next, this value is assigned to the grid point in the output raster. The simplest interpolation method is nearest-neighbour interpolation, which we treated earlier: The value of the point projected in the input grid is taken of which the x, y coordinates are closest to the computed coordinates. Another often used interpolation scheme is based on bilinear interpolation. A third interpolation scheme is bicubic spline interpolation, which is based on fitting a two-dimensional third-order polynomial through the values of the neighbourhood. Here, we will look at the bilinear interpolation scheme in greater detail. Consider the

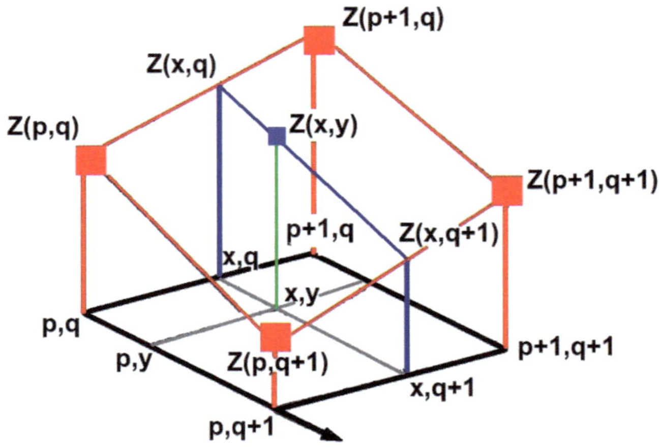

Figure 9.9 Diagram demonstrating bilinear interpolation

values of four grid points arranged in a 2 × 2 neighbourhood. Call these values: $z(p, q)$, $z(p + 1, q)$, $z(p, q + 1)$ and $z(p + 1, q + 1)$ and call $dx = x - p$ and $dy = y - q$, then (Figure 9.9):

$$z(x,y) = z(x,q)(1-dy) + z(x,q+1)dy$$

$z(x, q)$ and $z(x, q + 1)$ can be derived as:

$$z(x,q) = z(p,q)(1-dx) + z(p+1,q)dx$$

$$z(x,q+1) = z(p,q+1)(1-dx) + z(p+1,q+1)dx$$

Leading to:

$$z(x,y) = z(p,q)(1-dx)(1-dy) + z(p+1,q)dx(1-dy) + z(p,q+1)(1-dx)dy + z(p+1,q+1)dxdy \quad (9.14)$$

9.8 ANUDEM

The DEM interpolation method developed by the ANU – known as ANUDEM – has been designed for generating hydrologically correct DTMs. In addition to interpolation the software, which is widely used, incorporates constraints on landscape features. ANUDEM resulted from the research effort of Hutchinson (1989). Further developments of ANUDEM are discussed in Hutchinson et al. (2011). At the core, a thin-plate spline interpolation is used. The smoothing parameter, within the method called roughness penalty, is modified to allow the terrain to follow ridges, streams, cliffs and other abrupt changes in terrain elevation. The method employs geomorphological knowledge to avoid elevations which are unlikely, given the process which created the landscape.

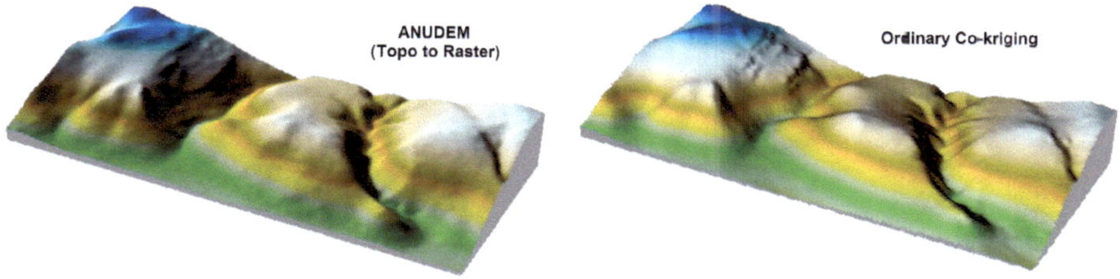

Figure 9.10 Comparison of the result of ANUDEM interpolation (left) with ordinary co-Kriging applied on the ravines of Mernjača and Mednjača, covering an area of 230 ha, in the Lake Vrana national park; the grid distance is 5 m for both representations (source: Šiljeg, 2013)

An example of such knowledge is that running water, which is a major cause of erosion, creates landscapes which may have many hilltops but few sinks. The geomorphological knowledge is implemented as constraints which ensure the generation of a connected drainage pattern. In addition, the intelligent exploration of a priori knowledge during interpolation also brings about the fact that less measured points are needed which reduces the costs of acquiring reliable input data. The input data may include elevation contours – in the past, the most common way of representing elevation – DTM point clouds, streamlines, sinks, cliff lines and polygons describing (lake) boundaries. The output is a gridded DTM presented in raster format. The ANUDEM interpolation method has been implemented in the ArcGIS toolset under the name Topo to Raster. Figure 9.10 shows the outputs of ANUDEM interpolation and ordinary co-Kriging.

9.9 Non-Conventional Interpolation

Most of the interpolation methods, including their variants, described in the previous sections are implemented in GIS packages. Therefore, they are often used for a variety of applications including DEM interpolation. Because they are readable and available, they are also often used in pilot studies aimed at finding out the best interpolation methods for a certain DEM acquisition technique, point density, terrain shapes and land cover. This section threats two more methods which may be beneficially used for the interpolation of DEMs. The first method is based on approximating the local terrain surface by a mathematical surface defined by higher-order polynomials. The second approach is based on a weight decay with distance based on the well-known Gaussian bell-shaped exponential function.

9.9.1 Local Polynomial Interpolation

One of the most intuitive approaches to interpolation is by fitting a curved surface through a neighbourhood of points. Mathematically, such a surface can be defined by a two-dimensional polynomial of order n. A two-dimensional polynomial of order n can be written as:

$$h(x,y)=\sum_{i=0}^{n}\sum_{j=0}^{n}a_{ij}x^{i}y^{j};i+j\leq n \tag{9.15}$$

If the surface may be assumed to be piecewise sufficiently smooth, a second-order polynomial will often suffice. A second-order two-dimensional polynomial may be mathematically formulated as:

$$Z(x,y) = a_0 + a_1x + a_2y + a_3xy + a_4x^2 + a_5y^2 \tag{9.16}$$

The six coefficients a_i, $i = 0, \ldots, 5$ can be obtained by defining a linear system of equations, using well-distributed measured points in the vicinity of the point with an unknown value. To obtain an exact solution at least six measured points are required. Usually, one will take more points and compute the six coefficients by a least-squares adjustment procedure. Once the coefficients have been calculated, the value Z at location (x, y) can be computed. This interpolation has several disadvantages. Firstly, least-squares adjustment consumes a lot of computational power, which makes the method unattractive when many unknown points have to be calculated. Furthermore, the method violates the assumption that the values of nearby points show more resemblance to the value of the unknown point than points far away. This disadvantage could be forestalled by assigning weights to the observation depending on the distance to the unknown point. However, this approach complicates computation further. Nevertheless, this approach can be applied for conducting ground filtering immediately followed by the generation of a grid DTM (see Chapter 10).

When assuming that the mathematical surface represents the local terrain surface sufficiently well, the approach can also be used for removing outliers. This polynomial approach allows to carry out a simultaneous outlier detection and interpolation (SODI) procedure. Outliers are points which deviate too much from the reconstructed surface according to a prespecified criterion. They may be caused by the reflection of laser pulses on dogs or bicycles or by multipath effects. Outlier detection requires that the elevation difference between each measured point used in the calculation of the surface and the surface is computed.

A feasible approach for removing outliers from a data set is to conduct this process by removing outliers one after the other, starting with the largest outlier. This process developed by Baarda (1968) is called data snooping and goes as follows. Compute the distances (Δh_i) between the n measured points and the computed surface, select the largest distance (Δh_{max}) and check whether this distance is larger than a threshold (τ) based on a prespecified decision criterion, which may be founded on sound statistical reasoning:

$$\text{if } \Delta h_{max} = \max\left|\Delta h_i\right|(i = 1, \ldots, n) > \tau$$

If the largest distance is larger than the threshold, the measured point is considered to be an outlier and removed from the data set. Since the outlier has impacted the computation of the distances of all measured points to the surface, the coefficients of the polynomial have to be computed again, but now without the measurement identified as an outlier. After readjustment, the distances are computed again and checked against the threshold. The process of removing outliers continues until no outliers are found anymore. The setting of the threshold (τ) may be derived from the known standard deviation (σ) of the elevation measurements. When one has an idea about the severity or impact of non-detection of outliers or removing points which are no outliers, a confidence level may be introduced based on the impact considerations. There is always a balance between the impact of removing measurements which are no outliers and non-detection of outliers. Feasible settings of the threshold lie between 2σ and 3σ. The approach assumes that the amount of outliers is modest, in the order of a few percentages. When the number of outliers is larger, computation time would increase beyond practical application especially when the DTM contains millions or even billions of points.

9.9.2 Gaussian Interpolation

The two-dimensional Gaussian bell-shaped exponential function is used in image processing to reduce the impact of noise and also to create image pyramids. Image pyramids are a sequence of images derived from the original image. Each image at a higher level in the pyramid has a coarser resolution – and thus a larger GSD – than the image at the lower level. Image pyramids are amongst others used in the image-matching process for creating point clouds from overlapping images. The same type of function can also be used for computing unknown points from measured points.

Tobler's law can be mathematised by recognising that many natural phenomena show an exponential decrease of the influence. One of the most popular functions to describe exponential decay in space or time is the Gaussian bell-shaped function which is well known in statistics (Figure 9.11). In digital image processing, the Gaussian function is used as a smoothing filter to reduce noise and to create the scale space which is among others used in image matching. From the Gaussian function, the weights of neighbouring known values can be computed which are based on distances between the planar position of the unknown points and the known neighbouring points. Defining the coordinates of the unknown point as (x, y) and those of the known points as $(x_i, y_i), i = 1, \ldots, n$, then the differences can be computed $\Delta x_i = x_i - x; \Delta y_i = y_i - y$

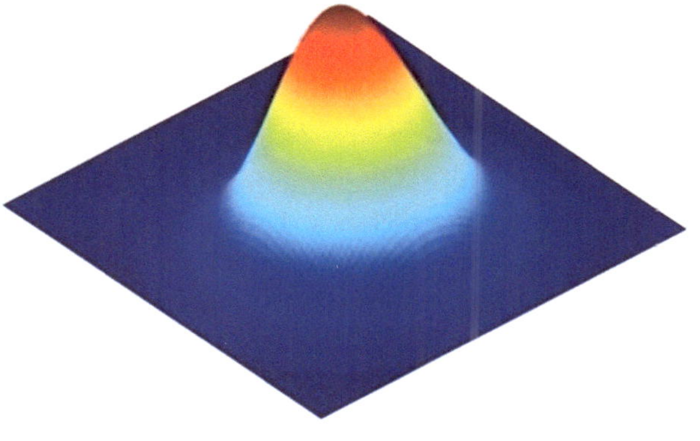

Figure 9.11 Two-dimensional Gaussian function

1.0	1.6	1.9	1.6	1.0
1.6	2.6	3.1	2.6	1.6
1.9	3.1	(·)	3.1	1.9
1.6	2.6	3.1	2.6	1.6
1.0	1.6	1.9	1.6	1.0

1.0	2.7	3.7	2.7	1.0
2.7	7.1	9.8	7.1	2.7
3.7	9.8	(·)	9.8	3.7
2.7	7.1	9.8	7.1	2.7
1.0	2.7	3.7	2.7	1.0

Figure 9.12 Gaussian interpolation kernels for computing the elevation of empty raster cells from elevations in surrounding cells

Sigma can be considered as a parameter that defines the area of influence, the larger sigma is, the larger the influence area. The weights in Equation 9.1 can now be determined from the circular Gaussian function:

$$w_i = \frac{1}{2\pi\sigma^2} e^{-\left[\Delta x_i^2 + \Delta y_i^2\right]/2\pi\sigma^2} \tag{9.17}$$

σ can be considered to be a factor, which defines the sphere of influence of the Gaussian function and thus introduces a distance metric. This approach leaves the user with the tuning of the σ-parameter, which is an important tuning parameter in addition to the setting of the kernel size. Both parameters are dependent and their settings should be balanced. To ease computation, especially for elevation arrays in which the heights are arranged in a grid structure, the Gaussian function can be transferred to a Gaussian kernel. As stated, the tuning of the σ-parameter is important. For example, setting σ to 1 will result in the 5 × 5 kernel as shown in Figure 9.12, left. Note that the centre of the kernel is empty. For this raster cell, the elevation value has to be computed. When setting σ to 0.7, the 5 × 5 kernel will look as shown in Figure 9.12, right. The values within the kernel define weights and these should sum up to 1, as indicated in Equation 9.1. This means that the outcome of applying the above kernel must be divided by the sum of weights. Since not all grid points will contain a height value within the area covered by the kernel, the sum of weights will not be a fixed value but has to be computed for each unknown point separately.

9.10 Impact of Terrain Shapes

In the treatment of the diverse interpolation methods in the previous sections, we have also considered performance characteristics. Different interpolation techniques may result in different surface representations as demonstrated in Figure 9.13. This section focuses on the influence of terrain structure on the interpolation result. In the previous sections, it has also become clear that the many investigations attempting to quantify the accuracy of DEM generation are inspired by the fact that there are so many sources of DEM point clouds, so many application areas and so many different terrain structures. The terrain shape of a dune area differs from a river polder area or hilly terrain.

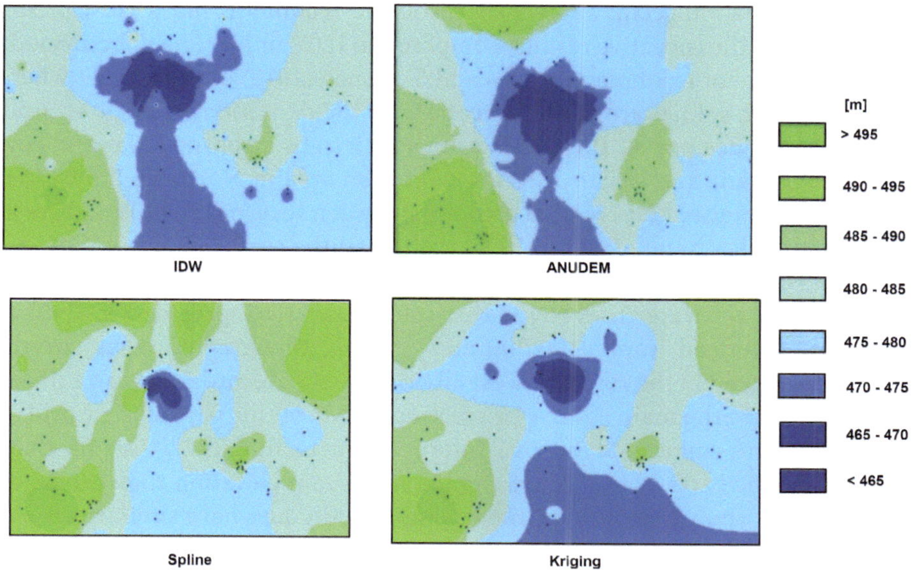

Figure 9.13 Different interpolation methods applied to the same point cloud may generate different surfaces

An area covered by production conifers in Estonia differs a lot from an area covered by tea plantations in Kerala, India. How does terrain shape impact the result of an interpolation exercise? First, we consider that all interpolation methods are affected by abrupt changes in terrain elevations, which means the presence of discontinuities in the terrain surface. Next, we will summarise several studies on how terrain shape and terrain coverage affect the accuracy of some of the interpolation methods treated in this chapter. Finally, remedies will be discussed to limit the impact of sudden changes in terrain shape.

9.10.1 Performance

Guo et al. (2010) investigated the accuracy of NN, IDW, TIN, splines, OK and universal Kriging (UK) on Lidar DEMs in relation to the terrain variability and point density. They found that IDW, NN and TIN are more efficient for generating DEMs from Lidar point clouds. Kriging methods are more reliable if accuracy is the most important performance criterion.

Bater and Coops (2009) examined amongst others IDW, natural neighbourhood and two spline interpolation methods on a small footprint airborne Lidar DEM, captured over areas with diverse types of vegetation on Vancouver Island, British Columbia, Canada. To examine the effects of terrain fluctuations and ground cover on interpolation accuracies, the area was partitioned based on point density, terrain slope, type of vegetation and normalised difference vegetation indices (NDVI) derived from satellite imagery (see Section 11.7.1). The researchers found that the regularised spline and IDW interpolation methods generated the most extreme outliers of up to ±6 m on slopes. Overall, NN interpolation provided the best results with a minimum of effort.

Spline interpolation seems to be a sophisticated technique and it is surprising that research found rather negative results, compared to IDW or NaN neighbourhood, when applied to DEMs. For example, Garnero and Godoneb (2013) observed that where IDW and NaN succeed, splines are unable to cope with the denser data sets up to 1 point/10 m^2 and even produce peaks. The DEMs were acquired by airborne Lidar and covered a mountainous area in the Italian Alps and an urban area in Northern Italy. Originally, the DEMs were acquired for creating orthoimagery at a scale of 1:5,000 and bare ground DEMs. On the other hand, Pavlova (2017) compared the performance of IDW, Kriging, ANUDEM, splines, NN and TIN, which are all interpolation methods implemented using ArcGIS 10 software. They found that splines and IDW provided the best results. The DEM was derived from 1:100,000 and 1:50,000 topographic maps covering the Omsk region located in the south of West Siberia, Russia. According to the author, topographical maps are the main source for creating DEMs of this part of the world.

Given the performance of various interpolation techniques, it seems that some methods can cope better with abrupt changes in terrain elevation than others. On the other hand, tests show that various interpolation techniques have similar performance, provided that the terrain fluctuations within the neighbourhood of interpolation are sufficiently smooth. Hence, it is important that the unknown point and the measured points used in the interpolation belong to the same landscape type (Lemmens, 1999). For example, when the elevation of an unknown point located in a valley is computed from four valley points and three slope points, the interpolated elevation may be too large. In the next section, this topic will be scrutinised in greater detail.

9.10.2 Discontinuities

All interpolation methods assume, implicitly or explicitly, that the terrain surface varies smoothly within the neighbourhood of interpolation. This means that interpolation methods do usually not cope with abrupt elevation changes within the vicinity of the unknown point. However, DEMs may contain abrupt terrain changes introduced by buildings, bridges and other man-made constructions in built-up environments. Also in natural, mining and agricultural areas, sudden jumps in elevation values may occur, introduced by cliffs, quarries, open-pit mines, river dikes and so on. The effect on the interpolated value of the unknown point is demonstrated in Figure 9.14.

When the geomorphology of the terrain varies, this means that there is a lot of relief present, one may get an overestimation or underestimation of the interpolated height values. This is demonstrated in Figure 9.15, where the red point (number 0) is the unknown point. The elevation has to be computed from seven surrounding measured points (1, …, 7). The unknown point lies in the valley, and its actual height is 143.17 m.

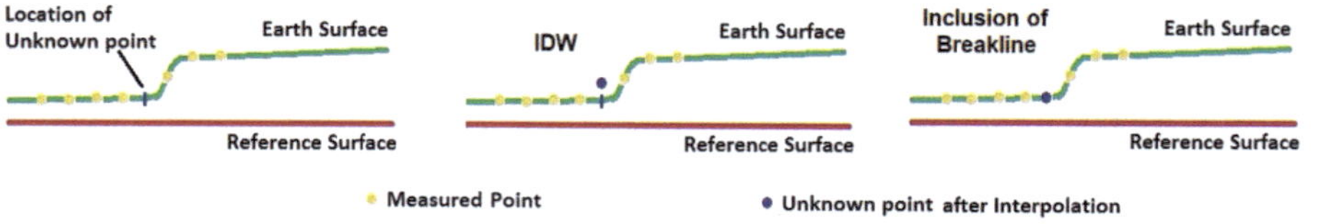

Figure 9.14 Abrupt terrain changes (discontinuities) cause wrong interpolation results (middle) and require incorporation of terrain shape features such as breaklines

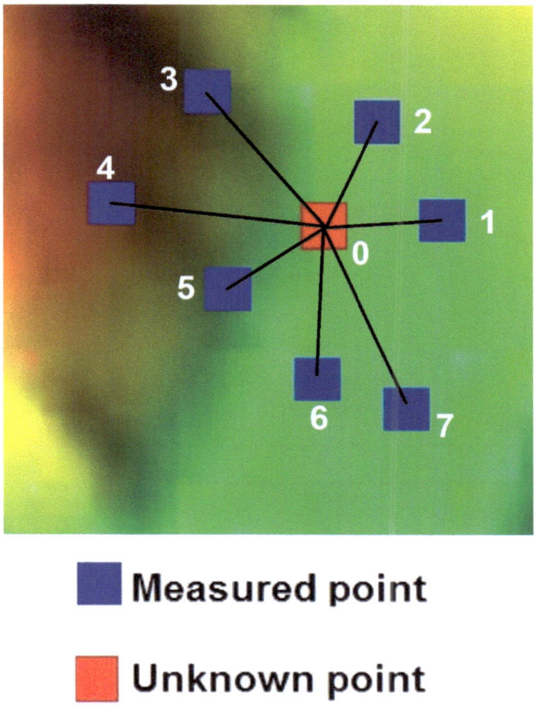

Figure 9.15 Hilly terrain with abrupt elevation changes

Because points in the vicinity are involved which are located on the slope of the hill, the interpolated height using IDW becomes 151.41 m (Table 9.1).

To perform well on terrain surface which contains abrupt changes, interpolation should be preceded by the inclusion of terrain knowledge preferably in the form of breaklines. When a breakline cuts an interpolation neighbourhood, the part of the neighbourhood in which the unknown point is located should be used (Figure 9.16). If breaklines are not available from additional sources, an alternative approach is to carry out a region-growing technique to segment the area into regions with similar height values or slope characteristics. Region growing is a well-known segmentation method used in the analysis of digital images (see e.g. Ballard and Brown, 1982; Haralick and Shapiro, 1985; Besl and Jain, 1988; Adams and Bischof, 1994; Pratt, 2007). The term growing refers to the fact that neighbouring raster cells are selected as initial pixels at the start of the procedure. These initial pixels, called seed points, are joined with neighbouring pixels if these fulfil one or more similarity criteria. This means that by joining adjacent pixels based on similarity, the seeds grow to regions. Within the framework of DEMs, such properties may include elevation, slope and aspect. Region-growing techniques are implemented as modules in programming languages such as Python and Matlab (matrix laboratory). Since breaklines are characterised by sudden changes in slope, the use of an edge detection algorithm would also be a feasible approach. Like region growing, edge detection has also been developed in the realm of computer vision (see Section 11.4). Probably the best-suited methods are based on the integration of region growing

Table 9.1 IDW interpolation with power 2 in a neighbourhood with abrupt terrain changes, shown in Figure 9.14

Point	Z (m)	r (m)	$1/r^2$	Z/r^2
1	141.65	32.1	0.0010	0.13747
2	140.71	45.5	0.0005	0.067968
3	178.37	58.9	0.0003	0.051415
4	207.56	72.7	0.0002	0.039271
5	156.09	29.6	0.0011	0.178152
6	135.21	51.2	0.0004	0.051579
7	131.14	83.4	0.0001	0.018854
	Average: **151.41**		$\sum 1/r^2$: **0.0036**	$\sum Z/r^2$: **0.544709**

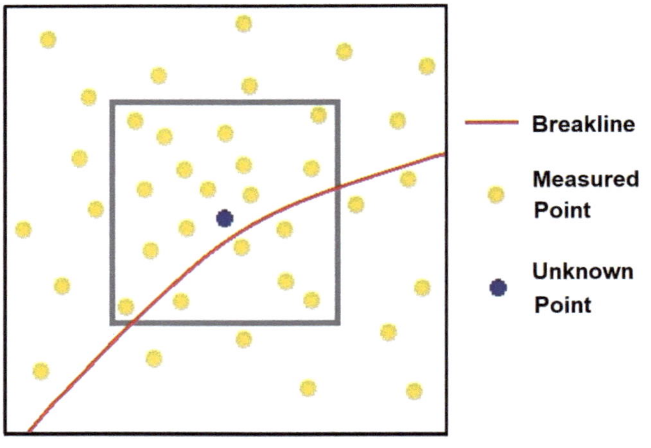

Figure 9.16 Inclusion of breaklines enables the exclusion of points, which do not have the same terrain characteristics, from interpolation

techniques and edge detection algorithm. Methods based on a combination of regions and edge detection are described amongst others in Pavlidis and Liow (1990), Chu and Aggarwal (1993), Gevers and Smeulders (1997) and Cufí et al. (2003).

9.11 Choosing an Interpolation Method

Often the interpolation method of choice depends on availability within a GIS software package. The principles underlying the algorithms of the software tools may have nothing to do with a point cloud captured by laser scanning or derived from image matching applied on overlapping images. Therefore, the point cloud analyst should examine the performance of existing interpolation techniques and seek approaches which reflect the basic characteristics of the point cloud. These characteristics are primarily related to the point cloud data acquisition technology, the type of scene and the information one wants to extract from the point cloud.

No interpolation method can be applied to all situations. Some are more exact and useful than others but take longer to calculate. They all have advantages and drawbacks. In practice, the choice of a particular interpolation method should depend on the criteria mentioned earlier. Choosing from different interpolation methods present within GIS software requires some effort and may proceed as follows:

- Examine the distribution of points over the area and the characteristics of the terrain.
- Select an interpolation method based on this examination, the task at hand and your craftsmanship.
- Probably the ultimate method is not obvious in advance, so several methods have to be scrutinised to arrive at the most optimal choice.
- Compare the results, keeping the task at hand in mind, and select the most proper method.

The above-mentioned workflow may seem time consuming and a waste of effort but is necessary to gain experience and knowledge of the peculiarities of the various interpolation methods.

Developers of an interpolation method may take a number of constraints and practical considerations into account. For example, in developing an interpolation method for daily precipitation data based on the observed data at 753 stations for the period 1951–2005 in China, Chen et al. (2010) went in search of a practical method which creates daily high-quality data on a grid covering the entire territory of China and is simple enough to be implemented in operational use. They compared five interpolation methods and found OK gives the best performance, closely followed by IDW with a power of 2. The interpolated daily precipitation data can be accessed by researchers via the internet.

9.12 Concluding Remarks

Interpolation is a topic of great attention to geostatisticians, a research topic in itself and still evolving. The main aim of this chapter has been to give the reader a flavour of this complex topic and to create awareness that craftmanship is required to apply the abundant techniques developed in course of time for a plethora of different applications. Selecting and applying the proper interpolation technique to a geo-related analysis problem is elusive and the rules of thumb are not available, and if they are available they focus on a limited set of applications. In the meantime, it is recommendable to remember the words of Peter Burrough who wrote in 1986 in his seminal book *Principles of Geographical Information Systems for Land Resources Assessment*:

> It is unwise to throw one's data into the first available interpolation technique without carefully considering how the results will be affected by the assumptions inherent to the method. A good GIS should include a range of interpolation techniques that allow the user to choose the most appropriate method for the job at hand.

References

Adams, R., Bischof, L. (1994) Seeded region growing. *IEEE Transactions on Pattern Analysis and Machine Intelligence*, 16(6), pp. 641–647.

Baarda, W. (1968) *A Testing Procedure for Use in Geodetic Networks*. Publications on Geodesy 9, 2(5), Netherlands Geodetic Commission, Delft, The Netherlands.

Ballard, D.H., Brown, C.M. (1982) *Computer Vision*. Englewood Cliffs, Prentice-Hall.

Bater, C.W., Coops, N.C. (2009) Evaluating Error Associated with Lidar-derived DEM Interpolation. *Computers & Geosciences*, 35(2), pp. 289–300.

Besl, P.J., Jain, R.C. (1988) Segmentation through Variable-order Surface Fitting. *IEEE Transactions on Pattern Analysis and Machine Intelligence*, 10(2), pp. 167–192.

Burrough, P.A. (1986) *Principles of Geographical Information Systems for Land Resources Assessment*. Clarendon Press, Oxford.

Burrough, P.A., McDonnell, R.A. (1998) *Principles of Geographical Information Systems*. 2nd Edition, Oxford University Press, New York.

Burrough, P.A., McDonnell, R.A., Lloyd, C.D. (2015) *Principles of Geographical Information Systems*. 3rd Edition, Oxford University Press, New York.

Chen, D.L., Ou, T.H., Gong, L., Xu, C.-Y., Li, W.J., Ho, C.-H., Qian, W.H. (2010) Spatial Interpolation of Daily Precipitation in China: 1951–2005. *Advances in Atmospheric Sciences*, 27(6), pp. 1221–1232.

Chilès, J-P., Delfiner, P. (2012) *Geostatistics: Modeling Spatial Uncertainty*. 2nd Edition, Wiley, New York.

Chu, C.C., Aggarwal, J.K. (1993) The Integration of Image Segmentation Maps Using Region and Edge Information. *IEEE Transactions on Pattern Analysis and Machine Intelligence*, 15(12), pp. 1241–1252.

Cressie, N. (1991) *Statistics for Spatial Data*. Wiley, New York.

Cufí, X., Muñoz, X., Freixenet, J., Martí, J. (2003) A Review of Image Segmentation Techniques Integrating Region and Boundary Information. *Advances in Imaging and Electron Physics*, 120, pp. 1–39.

Daya, A.A. (2015) Ordinary Kriging for the Estimation of Vein Type Copper Deposit: A Case Study of the Chelkureh, Iran. *Journal of Mining and Metallurgy*, 51A(1), pp. 1–14.

Fisher, P.F., Tate, N.J. (2006) Causes and Consequences of Error in Digital Elevation Models. *Progress in Physical Geography*, 30, pp. 467–489.

Garnero, G., Godoneb, D. (2013) Comparisons between Different Interpolation Techniques. *International Archives of the Photogrammetry, Remote Sensing and Spatial Information Sciences*, XL-5/W3, pp. 139–144.

Gevers, T., Smeulders, A.W.M. (1997) Combining Region Growing and Edge Detection through Guided Delaunay Image Subdivision. In *IEEE Computer Society Conference on Computer Vision and Pattern Recognition*, Puerto Rico, pp. 1021–1026.

Goovaerts, P. (1997) *Geostatistics for Natural Resources Evaluation*. Oxford University Press, Oxford.

Guo, Q.H., Li, W.K., Yu, H., Alvarez, O. (2010) Effects of Topographic Variability and Lidar Sampling Density on Several DEM Interpolation Methods. *Photogrammetric Engineering and Remote Sensing*, 76, pp. 701–712.

Haralick, R.M., Shapiro, L.G. (1985) Image Segmentation Techniques. *Computer Vision Graphics and Image Processing*, 29, pp. 100–132.

Hardy, R.L. (1990) Theory and Applications of the Multiquadrics—Biharmonic Method. *Computers and Mathematics with Applications*, 19, pp. 163–208.

Huff, D. (1954) *How to Lie with Statistics*. Norton, New York; reprinted in 1991, Penguin.

Hutchinson, M.F.A. (1989) New Procedure for Gridding Elevation and Stream Line Data with Automatic Removal of Spurious Pits. *Journal of Hydrology*, 106, pp. 211–232.

Hutchinson, M.F., Xu, T., Stein, J.A. (2011) Recent Progress in the ANUDEM Elevation Gridding Procedure. In: Hengel, T., Evans, I.S., Wilson, J.P., Gould, M. (Eds), *Geomorphometry*, Redlands, pp. 19–22.

Isaaks, E.H., Srivastava, R.M. (1989) *Applied Geostatistics*. Oxford University Press, New York.

Johnston, K., Verhoef, J.M., Krivoruchko, K., Lucas, N. (2001) *Using ArcGIS Geostatistical Analyst*. ESRI, Redlands.

Journel, A.G., Huijbregts, C.J. (1978) *Mining Geostatistics*. Academic Press, Harcourt Brace & Company, London, San Diego.

Krige, D.G. (1951) A Statistical Approach to Some Basic Mine Valuation Problems on the Witwatersrand. *Journal of the Chemical, Metallurgical and Mining Society of South Africa*, 52, 119–139.

Lemmens, M.J.P.M. (1999) Quality Description Problems of Blindly Sampled DEMs. In Shi, W., Goodchild, M.F., Fisher, P.F. (Eds.), *Proceedings of the International Symposium on Spatial Data Quality '99*. Hong Kong Polytechnic University, Hong Kong, pp. 210–218.

Li, J., Heap, A.D. (2008) A Review of Spatial Interpolation Methods for Environmental Scientists. *Geoscience Australia*, 23, p. 137.

Lloyd, C.D. (2010) *Spatial Data Analysis: An Introduction for GIS Users*. Oxford University Press, New York.

Matheron, G. (1963) Principles of Geostatistics. *Economic Geology*, 58, pp. 1246–1266.

Meyer, T.H. (2004) The Discontinuous Nature of Kriging Interpolation for Digital Terrain Modeling. *Cartography and Geographic Information Science*, 31(4), pp. 209–216.

Mitas, L., Mitasova, H. (1999) Spatial Interpolation. Chapter 34 in *Geographical Information Systems: Principles, Techniques, Management and Applications*, Vol. 1. Wiley, New York, pp. 481–492.

Monmonier, M. (2018) *How to Lie with Maps*. 3rd Edition, University of Chicago Press, Chicago and London.

Pavlidis, T., Liow, Y. (1990) Integrating Region Growing and Edge Detection. *IEEE Transactions on Pattern Analysis and Machine Intelligence*, 12(3), pp. 225–233.

Pavlova, A.I. (2017) Analysis of Elevation Interpolation Methods for Creating Digital Elevation Models. *Optoelectronics, Instrumentation and Data Processing*, 53(2), pp. 171–177.

Pratt, W.K. (2007) *Digital Image Processing*. 4th Edition, Wiley, Los Altos, CA.

Robeson, S.M. (1997) Spherical Methods for Spatial Interpolation: Review and Evaluation. *Cartography and Geographic Information Systems*, 24(1), pp. 3–20.

Shepard, D. (1968) A Two-dimensional Interpolation Function for Irregularly-spaced Data. In *Proceedings of the 23rd ACM National Conference*, New York, pp. 517–524.

Sibson, R. (1981) A Brief Description of Natural Neighbour Interpolation. In Barnett, V. (Ed.), *Interpreting Multivariate Data*. Wiley, Chichester, West Sussex, New York, pp. 21–35.

Šiljeg, A. (2013) Digital Terrain Model in the Analysis of Geomorphometrical Parameters: The Example of Nature Park Lake Vrana). PhD Thesis, University of Zagreb (written in Croatian).

Wackernagel, H. (1995) *Multivariate Geostatistics*. Springer Verlag, Berlin.

Walden, A.T., Guttorp, P. (1992) *Statistics in the Environmental and Earth Sciences.* Halstead Press, Sydney and Canberra.

Wang, J.G., Liu, G.R. (2002) A Point Interpolation Meshless Method Based on Radial Basis Functions. *International Journal for Numerical Methods in Engineering*, 54, pp. 1623–1648.

Watson, D.F. (1992) *Contouring: A Guide to the Analysis and Display of Spatial Data.* Oxford, Pergamon.

Ground Filtering

CHAPTER

10

Since the emergence of airborne Lidar in the mid-1990s, ground filtering has been an active research area. This demonstrates the importance of digital terrain models (DTMs) for studying many earth-related phenomena. As a result, DTMs belong to the main products created from aerial photogrammetry and airborne Lidar. As discussed in Chapter 8, DTMs consist of points which represent the bare ground surface. This is the terrain surface without buildings, vegetation and other above-ground objects. Often a digital surface model (DSM), which is the raw point cloud source for DTM generation, contains billions of points, which makes manual removal of non-ground points infeasible and impracticable. To automate the ground filtering process, researchers and developers have spent a lot of effort in the development of algorithms, which can conduct the task automatically once implemented in software running on a (fast) computer. Over time, particularly since the mid-1990s, many algorithms have been developed, based on a variety of mathematical principles and assumptions about the characteristics of the landscape. All these research and development efforts have yielded a plethora of ground filtering algorithms. This is a clear symptom and hints that distinguishing ground from non-ground points is a difficult classification task especially when the terrain surface shows high variability in elevation and object types. Completely automated generation of DTMs from raw point clouds is still in the realm of inventions to be done. Manual editing of the output of automated filtering approaches is still necessary to obtain high-quality DTMs. The manual editing efforts, which are strongly related to labour and time and thus associated with costs, depend for a major part on how correct automated classification of ground points and removal of non-ground points has been conducted. Many software packages contain advanced tools to support the human operator in editing point clouds.

The fitness for use of a DTM depends on point precision, point density, homogeneity of point distribution and also – probably for the major part – how well non-ground points have been removed. If the rate of non-ground points is high, the quality of the analysis carried out on the DTM, such as calculating the water drainage capacity of an area, will be low. So, ground filtering is an essential and crucial pre-processing task in the creation of reliable DTMs on which GIS analysts can rely to obtain the information necessary for their field of work.

Currently, ground filtering approaches, which appear to be successful for certain application domains, have been implemented in commercial and open-source software packages. This chapter doesn't aim at scrutinising the sheer amount of algorithms and their variations and extensions to these approaches proposed, designed, studied and tested by researchers. Rather, this chapter aims at providing insight into the concepts and ideas underlying the design of ground filtering techniques and to learn from this which concepts appear to be feasible given certain types of landscape. Particularly, the methods considered here are the reconstruction of the ground surface using TIN-based approaches (Section 10.2), weighted linear least squares local surface fitting (Section 10.3), hierarchical robust interpolation (Section 10.3), multi-scale curvature classification (MCC) (Section 10.4) and morphological filters (Section 10.5). These algorithms are often used and well known. Before considering the ground filtering methods mentioned above in greater detail, this chapter starts with a brief overview of the basics and design principles of ground filtering algorithms (Section 10.1). After outlining the basics of the ground filtering methods mentioned above, we will have a look at airborne Lidar returns which are best suited for carrying out ground filtering (Section 10.6), the exploration of ancillary data sources (Section 10.7), considerations on the development of a ground filtering procedure (Section 10.8) and quality measures (Section 10.9). This chapter ends with some concluding remarks (Section 10.10).

10.1 Basics and Design Principles

Basically, ground filtering is a binary classification process. Ground filtering aims at separating points reflected on the ground surface from those reflected on above-ground objects. Ground points and above-ground points are also called terrain points and off-terrain points, respectively. Above-ground or non-ground objects include immobile objects such as buildings, bridges, trees and shrubs but also mobile or dynamic objects such as cyclists, cars, cows, horses and other livestock. In most point clouds captured from airborne platforms, buildings and vegetation, including trees, are the major non-ground objects.

10.1.1 Need for Automation

The primary geodata sources, based on taking measurements above the terrain, for DTM production include aerial and satellite images, airborne Lidar data and interferometric synthetic aperture Radar (IFSAR) data. The vast majority of the ground filtering methods have been developed for removing non-ground points in airborne Lidar point clouds. Before the time when airborne Lidar became a preferred technique for acquiring (large-scale) DTMs, the main source was photogrammetry. In this established technology, a human operator collected elevations at predefined intervals using an analytical plotter (see Chapter 4). In contrast, airborne Lidar, which is an active remote sensing system, was able to capture DTM points highly automatically. Since the blind sampling process collected both ground and non-ground points without discrimination, a compelling need was felt to exploit the methods for dividing the massive amount of points into two classes – ground and non-ground – highly automatically. This need has given a boost in research. Ground filtering methods developed for Lidar point clouds are also suited for point clouds created through dense image matching (DIM), although there are some restrictions and limitations. For example, DIM point clouds see through the canopy only rarely. In areas covered by trees and shrubs significantly fewer ground points will be captured compared to Lidar point clouds. This impedes ground filtering of photogrammetric point clouds.

10.1.2 Naïve Approach

Sometimes researchers start from the idea that designing a ground filtering approach is a rather simple exercise. One overlooks that ground filtering is a classification problem, admittedly a binary one, but this reduces complexity only seemingly. This is because the non-ground class is actually a bulk class of many different object types of different sizes and shapes. Intuitively it seems reasonable to detect individual non-ground points by fitting a surface through the points in the neighbourhood of the point under consideration. If the distance between the point and the surface exceeds a threshold, the point is classified as a non-ground point (Figure 10.1). It is as simple as what the researcher who has never been confronted with the pertinacity of the problem might think. However, the point itself and its neighbours may all be part of a roof, resulting in assigning the class ground to a roof point. The only thing this approach can achieve is to detect whether the individual point belongs to the same type of object as its neighbouring points. In addition, the group of neighbouring points may consist of a mixture of points belonging to the ground and above-ground objects resulting in a surface which doesn't accurately represent the ground surface.

10.1.3 Observations Demonstrating Complexity

In general, ground surfaces will show low elevations. Often the lowest elevation in a neighbourhood whose the extent is not too big will belong to the ground surface (Meng et al., 2010). Of course, this rule assumes that multipath errors have been successfully removed at forehand. This is because multipath may result in points with an elevation below the ground surface. This is the first observation underlying the design of ground filtering algorithms. A second observation is that in a neighbourhood consisting of a mixture of ground and above-ground objects, the elevation differences will be large compared to neighbourhoods covered by ground surface alone. A third observation is that the intra-object variations in elevation of trees and buildings will be larger than will be the case for ground surfaces of similar size. This means the local neighbourhoods of ground surfaces are relatively smooth. A fourth observation is that the surface slope will be lower between neighbouring bare ground points than between bare ground

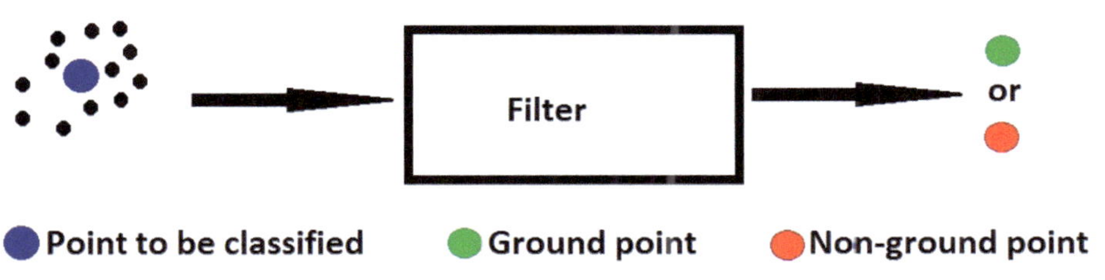

Figure 10.1 A naïve approach to ground filtering is to check whether individual points belong to the same group as their neighbours

points and non-ground points (Vosselman, 2000; Zhang and Whitman, 2005). So, if the slope or gradient is larger than a prespecified threshold, the point should be classified as a non-ground point. The setting of the threshold will depend on the landscape type. Mountains or wooded hills will have steep slopes and require a higher threshold than urban areas with flat surfaces. In general, the four observations mentioned earlier and summarised in Table 10.1 are valid for the majority of landscapes and underpin the design of the majority of ground filtering algorithms.

10.2 Reconstruction of Ground Surface

The ground surface or DTM can be reconstructed from the point cloud by starting with a coarse set of points well distributed over the area. The selection of these points, which are usually called seed points, is done such that it is pretty sure that they belong to the DTM. Next, this initial surface, which consists of a few seed points, is densified by iteratively detecting ground points and adding these to the surface. This densification process can be done along several lines as discussed in the sequel.

10.2.1 Seed Points

One of the most widely used surface reconstruction methods based on iterative, also called progressive, densification assumes that the terrain surface can be divided into a set of piecewise planar surfaces, in which elevation variations are small. The assumption originates in *Observation 2*: In a neighbourhood consisting of a mixture of ground and above-ground objects, the elevation differences will be large compared to neighbourhoods covered by ground surface alone. A convenient and well-established way to obtain piecewise planar surfaces is by creating a triangular irregular network (TIN) (see Section 8.3). In the progressive TIN densification approach, the terrain surface is described by a concatenation of triangles, which do not overlap nor do they have gaps. The process starts with the reconstruction of an initial surface from a sparse set of ground points. Next, the coarse initial TIN surface is iteratively densified by adding ground points. The selection of new ground points is done by looking for one point in each TIN triangle which fulfils a prespecified criterion. In the first few iterations, each TIN triangle will likely contribute one new point, but when the iterations progress, the

Table 10.1 Four observations underlying the design of ground filtering algorithms

	Observation
1	Lowest elevation in a small neighbourhood usually belongs to the ground surface
2	In a mixed neighbourhood elevation differences are large compared to ground surface neighbourhoods
3	Local neighbourhoods of ground surfaces are relatively smooth
4	Surface slope will be lower between neighbouring bare ground points than between bare ground points and non-ground points

number of added ground points will decrease. After each iteration, the DTM becomes denser. This process goes on until no ground points can be found anymore. The various TIN approaches differ from each other with respect to the selection of the following aspects:

- Selection of the seed points
- Decision criterion on which a point should be added to the reconstructed surface

The sparse start subset of seed points can be created by simply selecting the lowest point in each of the four corners of the area (Sohn and Dowman, 2002). Another method is cutting the area covered by the DTM into square or rectangular blocks of relatively large sizes and selecting the lowest point in each block as the seed point (Axelsson, 2000). Other methods can also be used (Hansen and Vögtle, 1999). Since one starts with a coarse set of seed points, which is iteratively densified by adding ground points, these approaches have received the tag: progressive TIN (PTIN) densification.

10.2.2 Decision Criteria

The second criterion concerns the decision of whether a point within the TIN triangle should be added or not. A feasible approach is to consider initially only points that lie below the triangle and to select the lowest point. This criterion assumes that the three vertices spanning up the triangle are indeed ground points. If no point can be found below the triangle anymore, the points lying above the triangle are examined and the point which is closest to the triangle is selected as the ground point. However, caution should be exercised here. Intuitively, the assumption that a distance criterion would be suitable seems feasible. This criterion means that if the distance of the point to the triangle is smaller than a threshold defined by the user, it will be accepted as a ground point and added to the TIN surface. The issue is that such a point might violate the smoothness criterion (Figure 10.2). To avoid this drawback of solely using a distance criterion, an angle criterion can be introduced. This can be done by calculating the three angles between the point under consideration and the TIN triangle (Figure 10.3). If the three

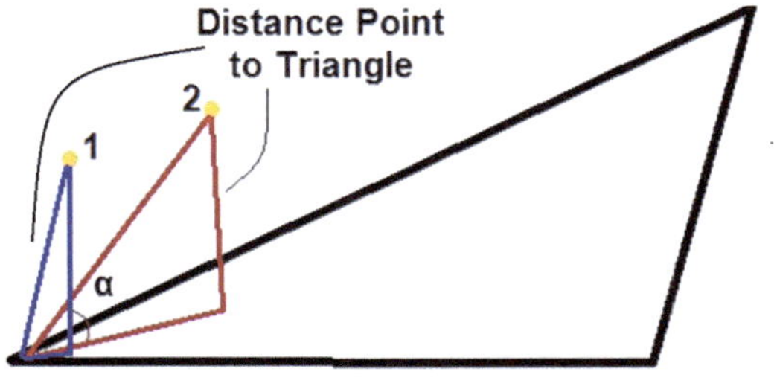

Figure 10.2 Two points are close to the TIN triangle; point 1 is unlikely to be a ground point since it violates the smoothness constraint

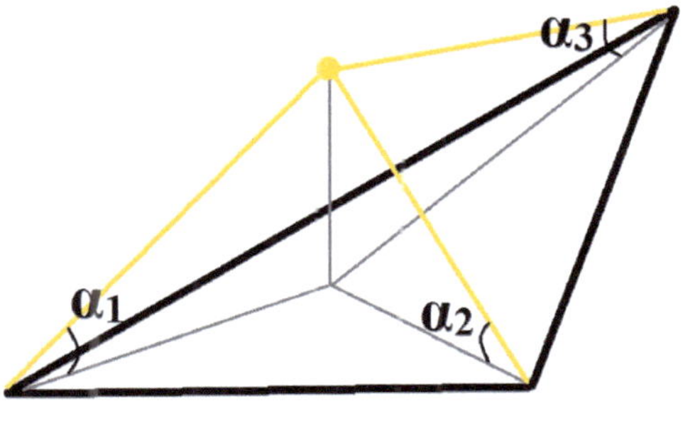

Figure 10.3 Whether a point above the TIN triangles belongs to the ground surface is checked by examining the three angles, which should be smaller than a given threshold

angles are smaller than a prespecified threshold, the point is added as a ground point (Axelsson, 2000).

10.2.3 Software Implementation

The progressive TIN densification approach is well investigated by multiple academic researchers. The results brought the researchers to praise its reliability and accuracy. However, PTIN might meet difficulties when the terrain surface contains abrupt elevation changes, such as break lines and step edges. In these places, ground points are often classified as non-ground points, while the reverse is true for points of objects which are close to and attached to the ground surface (Lin and Zhang, 2014). These non-ground points are often classified as ground points. Added to this, the processing of dense point clouds covering larger areas is time consuming (Chen et al., 2017). The commercial Lidar processing software TerraScan, a product of TerraSolid based in Helsinki, Finland, has been designed based on PTIN. The algorithm is also available in LASground, one of the multiple tools of LASTools, which has been specifically developed for Lidar data processing (Isenburg, 2015). As a result of its ready implementation in software, PTIN is widely used in engineering applications.

10.3 Local Surface Fitting

The design of local surface fitting approaches starts with exploiting *Observation 3: Local neighbourhoods of ground surfaces are relatively smooth.* A smooth surface can be mathematically described by, for example, a second- or third-order polynomial. The non-ground points are treated as outliers, which are also known as gross errors. Outliers are points which do not fit within the underlying model, here, for example, mathematised as a second-order polynomial. This allows exploitation of the apparatus of data snooping (for explanation of data snooping, see text box on Non-Linear Least Squares Adjustment, Chapter 4). The criterion to decide whether a point is an outlier, which means a non-ground point, is based on the distance to the computed surface, which approximately describes the ground surfaces. The surfaces should be computed using more points than there are coefficients in the surface function.

10.3.1 Weighted Linear Least Squares (WLS)

This mathematical surface is fitted through the points within the local neighbourhood using a least-squares methodology. Some points will lie close to the surface others deviate a little and some a lot. This means that points which do have a distance to the surface larger than a threshold are likely non-ground points and should be removed. However, these outliers have also influenced the computation of the surface function.

The remedy is well known in land surveying, photogrammetry and geodesy and is called data snooping. This approach of removing outliers proceeds as follows. First, the largest outlier, which means the point with the largest distance to the surface, is removed. Next, the surface is computed again using all remaining points. This procedure is repeated till all points fall within the distance criterion. This sounds easy, and in many circumstances it is. However, the straightforward implementation of this approach is associated with snags. First, data snooping is feasible if the number of outliers is modest. The method has been developed for data sets in which less than 1 percent of the measurements are outliers. However, within a local neighbourhood of a point cloud, many non-ground points may be present. The local neighbourhood may contain a building or a tree. So, the number of outliers may be 50% or even 100% in case the local surface is fitted through roof points only. This makes the straightforward implementation of data snooping unfeasible. Second, data snooping has been designed for removing outliers independent of their location with respect to the surface. However, points which lie (high) above the modelled surface are more likely to be outliers than points below the surface. This is a consequence of *Observation 1: Lowest elevation in a small neighbourhood usually belongs to the ground surface.* Third, the approach is computationally time consuming. Fourth, because of the many outliers which may be present in a neighbourhood, the removal of outliers may be unreliable.

To adapt the data snooping approach to the ground/non-ground classification problem, several adjustments can be introduced. One may question why use all the points in a local neighbourhood? Why not take a subset? This can be done by superimposing a raster of square cells on the point cloud, as discussed in Section 8.4. The question arises as to which point to select to assign to each raster cell of the tessellation. There are several possibilities to assign an elevation value to each raster cell: The mean and median are well-accepted statistical measures to characterise a set of observations. Especially the mean can be computed easily and rapidly since it is a linear function, while the median is a non-linear function, which consumes computation time. Introducing *Observation 1: Ground points have low elevation*, it is feasible to select the lowest point in each cell. Alternatively one may select the point which is close to the centre of the cell. Pfeifer et al. (2001) discuss some of the pros and cons of these choices. Taking the lowest point in each cell accelerates the process, but the lowest point may also be an outlier for example caused by multipath. Hence, this approach should be used with caution. Choosing the point closest to the centre of each raster cell gives a more regular distribution of the points used in the process.

The original design of this ground filtering approach (Kraus and Pfeiffer, 1997; Pfeiffer et al., 1998) is known as weighted linear least squares (WLS). This approach was extended by Pfeifer et al. (2001) to a hierarchical approach, called hierarchical robust interpolation.

10.3.2 Hierarchical Robust Interpolation

The main line of the hierarchical robust interpolation approach proceeds as follows. Superimpose a coarse raster over the point cloud followed by selecting for each cell the point with the lowest elevation and fit a local surface through the neighbourhood of raster cells. The number of points and thus the number of raster cells within a local neighbourhood should be larger than the number of coefficients of the surface function. In this first step, the surface is computed with equal weights for the Lidar points, resulting in a computed surface which is a rough approximation of the actual ground surface and lies in between the actual ground and the non-ground objects. Ground points are more likely to be located below the surface and buildings, vegetation and other points above the surface.

Next, the distances, also called residuals, to this surface are computed. The points closer to the computed surface will likely belong to the ground surface compared to points lying farther away. This likelihood or importance criterion can be implemented in the procedure by assigning weights to the points. These weights should depend on whether points lie above or below the computed surface and should be a function of the distance to this surface: The farther a point is from the local surface, the more unlikely it will be that it belongs to that surface and thus the lower the weight should be. If the distance is large, a point receives a zero weight; it is classified as a non-ground point at forehand and removed. After the first iteration, the coarse raster is refined. Per cell, the point with the lowest elevation is selected, and a new surface is computed, which will lie closer to the actual ground surface. This hierarchical interpolation approach is repeated until a stable situation has been reached, which means no outliers, which are associated with non-ground points, are detected anymore. To avoid endless loops, a maximum on the number of iterations should be set.

The procedure described above leaves us with three questions:

- Which mathematical description is best suited for approximation of the ground surface, meaning which function should be used as a fitting function?
- How should the weights be computed depending on distance and above or below ground location?
- How should elevations be assigned to the raster cells?

The fitting function should replicate a 2D smooth curved surface in 3D space. Different fitting functions can be used. For example, Deng and Shi (2013), who combined the filtering results of two popular filter algorithms, PTIN and hierarchical robust interpolation, adopted a quartic function to approximate the ground surface:

$$\begin{aligned}(x,y) = a_0 + a_1 x + a_2 y + a_3 xy + a_4 x^2 + a_5 y^2 \\ + a_6 x^2 y + a_7 xy^2 + a_8 x^2 y^2\end{aligned} \tag{10.1}$$

10.3.3 Weights

Points which lie close to the computed surface should have more influence in the next iteration step than the points which have a large distance. Furthermore, points lying below the surface should receive higher weights than the points above the surface, even if their distances to the computed surface are the same. The dependency on the distance to the surface and its location above or below the surface can be implemented by an asymmetric weight function, in which weights w_i are determined as:

$$w_i = \begin{array}{ll} 1 & \Delta h \le \tau_{\text{below}} \\ 1/\left(a\left(\Delta h - \tau_{\text{below}}\right)^b\right) & \tau_{\text{below}} < \Delta h \le \tau_{\text{above}} \\ 0 & \Delta h > \tau_{\text{above}} \end{array} \tag{10.2}$$

In Equation 10.2, Δh is the distance to the surface. If the point lies above the surface, the sign of Δh is positive, and if lying below the surface, the sign is negative; τ_{below}, the

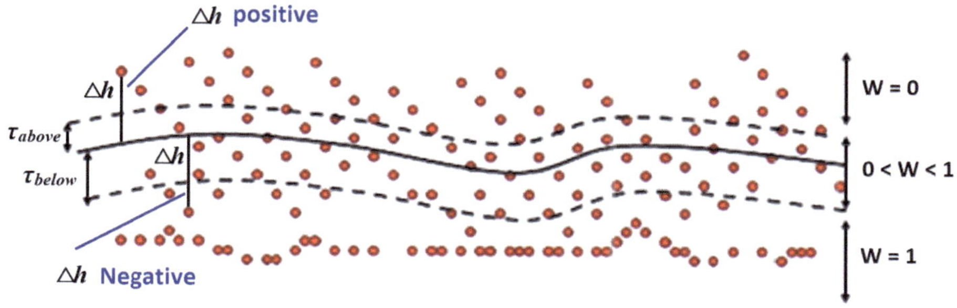

Figure 10.4 Assignments of weights in the hierarchical robust interpolation approach (source: SCOP++ manual; modified)

below surface threshold, has a negative sign, and τ_{above}, the above surface threshold, has a positive sign (Figure 10.4). The parameters a and b determine the steepness of the weight function. For most applications, values of 1.0 and 4.0 for a and b, respectively, have produced adequate results (Mc Gaughey, 2018). Points which lie below the surface by more than τ_{below} receive a weight of 1. Points which lie above the surface by more than τ_{above} receive a weight 0. It is possible to change the values of τ_{below} and τ_{above} after each iteration (Kraus and Pfeifer, 1998). After the final iteration, all points with elevations that satisfy the first two conditions of the weight function are considered to belong to the ground surface.

In the above approach, the assumption is that, hypothetically, all points within the neighbourhood are part of the ground surface. Points which violate this assumption are removed as outliers. As stated above, data snooping is not well suited if the number of outliers is high. The solution discussed above was to carry out the surface fitting in a hierarchical manner. Another approach would be to use the random sample consensus method, which works by selecting a few ground points in a neighbourhood. The algorithm was first published by Fischler and Bolles (1981). The number of points is at least sufficient to compute the coefficients of the surface fitting function. Next points which are closest to the computed surface are incorporated in the readjustment of the surface and so on till a stable situation has been reached. This also leads to an iterative approach.

10.3.4 Software Implementation

The iterative robust interpolation algorithm has been embedded in SCOP Lidar, which is one of the SCOP++ software packages (https://photo.geo.tuwien.ac.at/software/scop/). The SCOP++ suite has been designed for the management, interpolation and visualisation of digital elevation models (DEMs). Developed and continuously extended in a joint effort between the Institute of Photogrammetry and Remote Sensing, Vienna, Austria, and INPHO GmbH, Stuttgart, Germany (presently a subsidiary of Trimble and called Trimble Imaging), SCOP++ has a reputation for scientific soundness among professionals. Under the name GroundFilter, the WLS algorithm is implemented in Fusion/LDV software. This software has been developed by the United States Department of Agriculture (USDA) to help researchers understand, explore and analyse Lidar data (McGaughey, 2018). By the time airborne Lidar became a major survey technique, the use of large Lidar data sets required extensive pre-processing in commercial GIS or image processing software. Due to their massive volume and the variety of ASCII text formats used by the various vendors, seemingly simple tasks, such as extracting a subset of the point cloud lying within a field plot, were complicated. The software is regularly updated. GroundFilter does not filter out all non-ground points. Particularly, points on large, flat and slanted surfaces such as roofs are not completely removed (McGaughey, 2018). Most vegetation returns are removed with the appropriate coefficients for the weight function and sufficient iterations. The default settings for the weight function generate good results in point clouds with a point density of 4 returns/m^2 or higher. Point clouds with lower density require trial-and-error tests on the setting of the coefficients of the weight function. The filtered output is suited for calculating vegetation heights. The experiences of the USDA demonstrate the forestry origin of the filter.

10.4 Multi-scale Curvature Classification (MCC)

Another approach based on fitting a surface through points is the multi-scale curvature classification (MCC) method. As with the WLS approach, MCC has been specifically developed for separating the ground from non-ground airborne Lidar returns in high biomass forested areas. The developers, Evans and Hudak (2007), were with the Science Laboratory of the USDA Forest Service and got confronted with airborne Lidar data over extensive forested environments with complex terrain surfaces.

MCC takes account of the fact that objects, such as trees, can have various sizes. At a coarse point density, adjacent objects may even fuse or objects may be undetectable or absent in the point cloud. To cope with different object sizes, Evans and Hudak (2007) introduced a multi-scale approach consisting of three levels based on point spacing. The approach follows a fine to coarse strategy by starting at the highest resolution. In the resulting iterative approach, non-ground points are traced by computing the curvature of the surface fitted through the points. If the curvature is positive and exceeds a prespecified threshold, points are classified as non-ground points. The surface, which is fitted through the points, is a thin-plate spline of which the spline tension parameter enables the tuning of the influence of points which are lying far away from the surface. Before checking whether points obey the curvature threshold, the computed surface is smoothed using a low-pass filter. Usually, the low-pass filtering is done by a moving 3 × 3 unweighted smoothing kernel.

Since the location of the surface will be affected by non-ground points, in the next iteration step, the surface is computed again after the removal of returns which have been identified as non-ground points. After the iteration, again non-ground points will be detected and removed. This process repeats until the number of non-ground points is low compared to the previous iteration. For example, in their paper Evans and Hudak (2007) introduced a convergence threshold of 0.10%, meaning that if just one in every thousand returns, which remained from the previous iteration, is classified as a non-ground point, the iteration will terminate and the algorithms go to the next scale level. If the last scale level has been reached as well as convergence, the process ends.

In addition to the setting of the spline tension parameter and the window size of the smoothing kernel, the MCC iterative algorithm requires the setting of the following tuning parameters:

- Number of scale levels (N_s)
- Scale level or cell size ($\alpha_i \cdot$NPS; $i = 1, \ldots, N_s$)
- Curvature threshold or tolerance (τ_{cur})
- Adjustment parameter(s) of the curvature threshold at subsequent scale levels ($\Delta\tau_{cur}$)
- Convergence threshold (τ_{con})

The optimal scale levels are a function of the size of the objects and the nominal point spacing (NPS).

Usually, the number of scale levels is limited to three, and the scale parameter or cell size is defined as a function of nominal point density. For example, if the nominal point density is four points per square metre, NPS will be 50 cm. Scale levels defined by the developers are 0.5 NPS, 1.0 NPS and 1.5 NPS. In this way, the cells become progressively larger. The lower the scale level, which means the lower the point density, the more generalised the surface will be. Therefore it is necessary to change the curvature threshold when going to the next scale level. The algorithm consists of two iteration levels. The major iteration level consists of the fine to coarse strategy. Within each scale level, a further iteration takes place until convergence has been reached. A flowchart of the MCC algorithm, presented in a generalised way, is depicted in Figure 10.5. Note that in the software implementation the number of scale levels has been fixed to three.

In available software the user may give the curvature threshold a value between 0.1 and 1, which does not mean that the user may not deviate from the given range. For example, for a 38-ha mountainous area in southwestern Idaho, USA, with heights above sea level varying from 2,023 to 2,139 m and slopes reaching 35°, Tinkham et al. (2011) found a curvature threshold of 0.05 to be the best value. The area, covered by a patchwork of shrub-steppe, meadow and bare ground with large stands of coniferous and deciduous forests, was captured through airborne Lidar with a mean point density of 5.7 points/m^2.

Since there is a functional relationship between scale level and curvature, the setting at subsequent scale levels of the curvature adjustment value is done automatically by adding a prespecified value. In the original design of the algorithm, this adjustment value was set at 0.1. A small curvature threshold may prevent points on low vegetation to be classified as ground, but may also remove ground points in terrain with a high elevation variability.

Parts of each airborne Lidar pulse may reflect on vegetation and parts on the ground surface. If this is the case, airborne Lidar point clouds will contain multiple returns. Only points which are single returns or the last return of multiple returns form the input of the MCC algorithm. First or intermediate returns will usually stem from trees, shrubs or other vegetation. The input point cloud can be further thinned by excluding the last returns of which the elevations are significantly higher than the adjacent last returns.

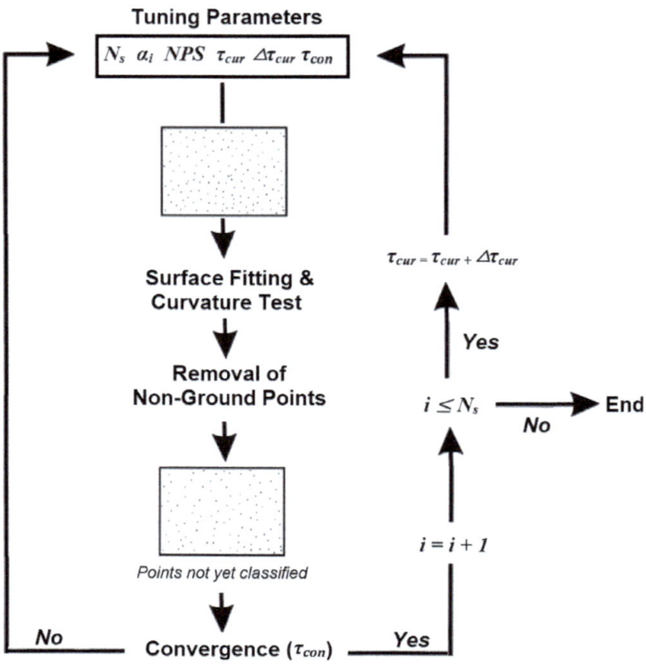

Figure 10.5 Flow diagram of the MCC ground filtering approach. There are two levels of iteration: one on scale level and the other depending on convergence per scale level

10.5 Morphological Filtering

One of the first ground filtering methods developed for airborne Lidar point clouds was based on mathematical morphology (Lindenberger, 1993). Mathematical morphology has originally been developed for removing noise in binary digital images, and the approach has a sound basis in digital image processing. When applying morphological filtering on a point cloud, the point cloud is first transferred to a raster representation. The basic idea of morphological filtering is to geometrically describe the above-ground objects as structure elements. It is like checking whether a hat is fitting a head. Of course, the crux is in the translation of the idea of structure elements into a mathematical description. The biggest issue with this type of filtering is how to cope with the fact that buildings and vegetation may appear in many different shapes and sizes. So the main effort in the design of morphological filtering is the handling of the different shapes and sizes.

To outline the concept of morphological filtering, we will use a binary image in which raster cells, which are covered by an object, are indicated by one, while empty raster cells receive the value zero. To detect objects in such a binary representation, a model of the object is created and transferred to a raster representation, called structure element. The cells in the structure element which represent the object are indicated by one. The structure element is superimposed on the raster and moved stepwise over the raster as a moving window, examining the one raster cell after the other. The moving window is shifted along the columns and rows of the raster. The basic operations are dilation and erosion. The dilation operation checks if one of the cells covered by the structure element contains an object, which means one of the cells has the value one. The centre cell covered by the structure element receives the value one if one or more cells covered by the structure element contains the value one also when the centre cell itself contains a zero. The erosion operation checks if one of the cells covered by the structure element contains a zero. The centre cell covered by the structure element receives the value zero if one or more cells covered by the structure element contain the value zero also when the centre cell itself contains a one.

By alternating dilation and erosion, other operations can be derived from these two basic operations. The most important derived operations are opening and closing. The opening operation first conducts an erosion and next a dilation. This operation results in the removal of small objects and the reduction of the size of objects in the binary images. The closing operation first conducts a dilation and next an erosion. This operation results in the removal of small holes where no objects are present or an enlargement of the objects.

The several operators discussed above are demonstrated in Figures 10.6–10.8 using as structure element a 3 × 3 window, in which an object is represented by the value 1. The figures show that the use of morphological operations requires the storage of the point cloud in a raster. So, one of the first issues one has to deal with is the transfer of two or more elevations in one cell to one elevation and the filling of an empty cell with an interpolated elevation value (see Section 8.4). One of the solutions is based on the consideration that if two or more elevations fall within the same cell, the point with the lowest elevation is most likely a ground point. The elevation of this point is assigned to the cell. If no elevation is present in the cell, the cell can be filled by nearest neighbourhood interpolation (Zhang et al., 2003). In the transformation from point cloud to raster, the cell size usually depends on the point density, which means average point spacing. This is done to avoid the loss of information present in the point cloud and thus not fully exploited.

For an airborne Lidar point cloud, dilation means assigning the maximum elevation value to the centre cell. Erosion means taking the smallest elevation and assigning this elevation to the centre cell. An example of an opening operation is shown in Figure 10.9, which shows a profile through an airborne Lidar point cloud. The unfiltered points are sampled every 1 × 1 m along the profile. The window size of the structure element is 11 m. Returns from trees with a size smaller than 11 m are removed, but the buildings are preserved because their footprint is larger than 11 m. First, the erosion reduces the size of the footprint, but the dilation restores the shapes of buildings.

The performance of morphological filters is sensitive to the setting of the size of the opening and closing windows (Chen et al., 2016; Zhao et al., 2016). Buildings and other above-ground objects may have many sizes. Small windows may detect small objects but may overlook large objects. As a result, large buildings are classified as being part of the terrain surface. Large windows can detect these objects but will also remove dikes, dunes, peaks and other salient terrain features. A remedy that most morphological approaches apply is the iterative use of a sequence of window sizes, a process called progressive

1	1	1
1	1	1
1	1	1

Structure element

	1	2	3	4	5	6	7	8	9	10	11
1	1	1	1	1	1	1	1	1	1	1	1
2	1	0	1	1	1	1	1	1	1	1	1
3	1	1	1	1	1	1	1	1	1	1	1
4	1	1	0	1	1	1	1	1	1	1	1
5	1	1	1	1	1	1	1	1	1	1	1
6	1	1	1	1	1	1	0	0	0	0	1
7	1	0	0	0	1	1	0	1	0	1	1
8	1	0	0	0	1	1	0	0	0	1	1
9	1	0	0	0	1	1	1	1	1	1	1
10	1	1	0	1	1	1	1	1	1	1	1
11	1	1	1	1	1	1	1	1	1	1	1

Binary raster representation

Figure 10.6 Binary raster representation of a small part of a DSM (right) and structure element (left)

	2	3	4	5	6	7	8	9	10
2	1	1	1	1	1	1	1	1	1
3	1	1	1	1	1	1	1	1	1
4	1	1	1	1	1	1	1	1	1
5	1	1	1	1	1	1	1	1	1
6	1	1	1	1	1	1	1	1	1
7	1	1	1	1	1	1	1	1	1
8	1	0	1	1	1	1	1	1	1
9	1	1	1	1	1	1	1	1	1
10	1	1	1	1	1	1	1	1	1

Dilation

	2	3	4	5	6	7	8	9	10
2	0	0	1	1	1	1	1	1	1
3	0	0	0	1	1	1	1	1	1
4	0	0	0	1	1	1	1	1	1
5	0	0	0	1	0	0	0	0	0
6	0	0	0	0	0	0	0	0	0
7	0	0	0	0	0	0	0	0	0
8	0	0	0	0	0	0	0	0	0
9	0	0	0	0	0	0	0	0	0
10	0	0	0	0	1	1	1	1	1

Erosion

Figure 10.7 Dilation and erosion conducted on the binary raster shown in Figure 10.6 using the 3 × 3 structure element

	3	4	5	6	7	8	9
3	1	1	1	1	1	1	1
4	1	1	1	1	1	1	1
5	1	1	1	1	1	1	1
6	1	1	1	1	1	1	1
7	1	1	1	1	1	1	1
8	1	1	1	1	1	0	0
9	0	1	1	1	1	1	1

Closing: dilation next erosion

	3	4	5	6	7	8	9
3	1	1	1	1	1	1	1
4	0	1	1	1	1	1	1
5	0	1	1	1	1	1	1
6	0	1	1	1	0	0	0
7	0	0	0	0	0	0	0
8	0	0	0	0	0	0	0
9	0	0	1	1	1	1	1

Opening: Erosion next dilation

Figure 10.8 Closing and opening conducted on the binary image shown in Figure 10.6 using the 3 × 3 structure element

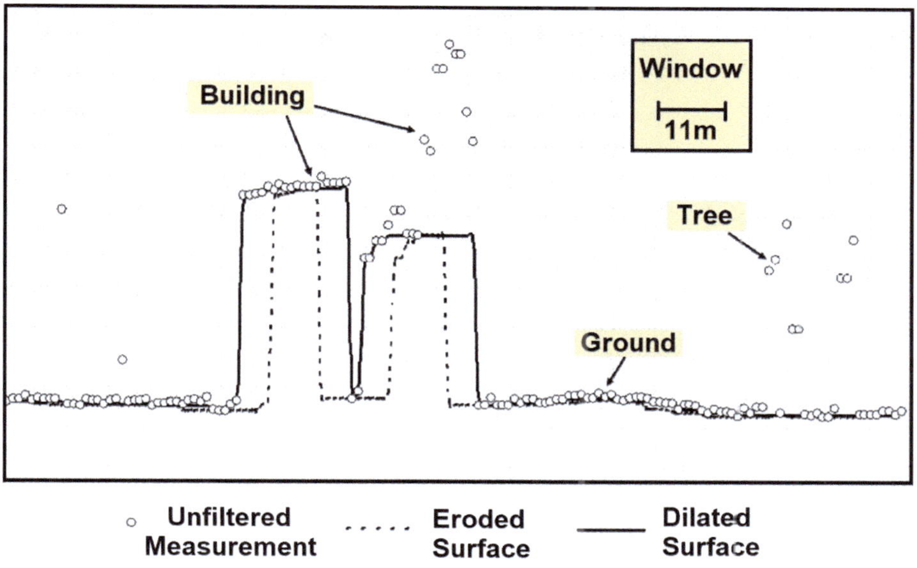

Figure 10.9 Opening operation on an airborne Lidar point cloud (source: Zhang et al., 2003; modified)

morphological filtering (PMF). Nevertheless, this progressive method entails the settings of minimum and maximum window sizes, which requires prior knowledge of the smallest and largest objects present in the point cloud (Zhang et al., 2003). To tackle this issue, Li et al. (2017) developed a filter based on geodesic transformations of mathematical morphology, which eliminates the need of defining window sizes. The method requires the removal of multipath returns, which are also called low outliers, at the forehand. Geodesic transformations apply an iterative process until stability has been reached, eliminating the need of defining window sizes. Thus, this algorithm adopts the elementary filtering window without considering how to select different window sizes or determine the maximum window size, which enhances the robustness and automation over a variety of landscapes. Figure 10.10 shows the results of applying a geodesic morphological filter on two patches of the National Elevation Model of different landscape types in the Netherlands. Morphological methods are simple in concept and easy to implement and detect non-ground points effectively (Zhang et al., 2003; Arefi and Hahn, 2005). The progressive morphological filter algorithm is available in the free software ALDPAT, which was developed by the National Center for Airborne Laser Mapping (Zhang and Zheng, 2007).

10.6 Selection of Returns

All airborne Lidar systems are able to record the first and the last return of each emitted pulse. In addition, many systems are able to record more than two returns, and they register multiple returns. Some systems are also able to digitise the full waveform of the

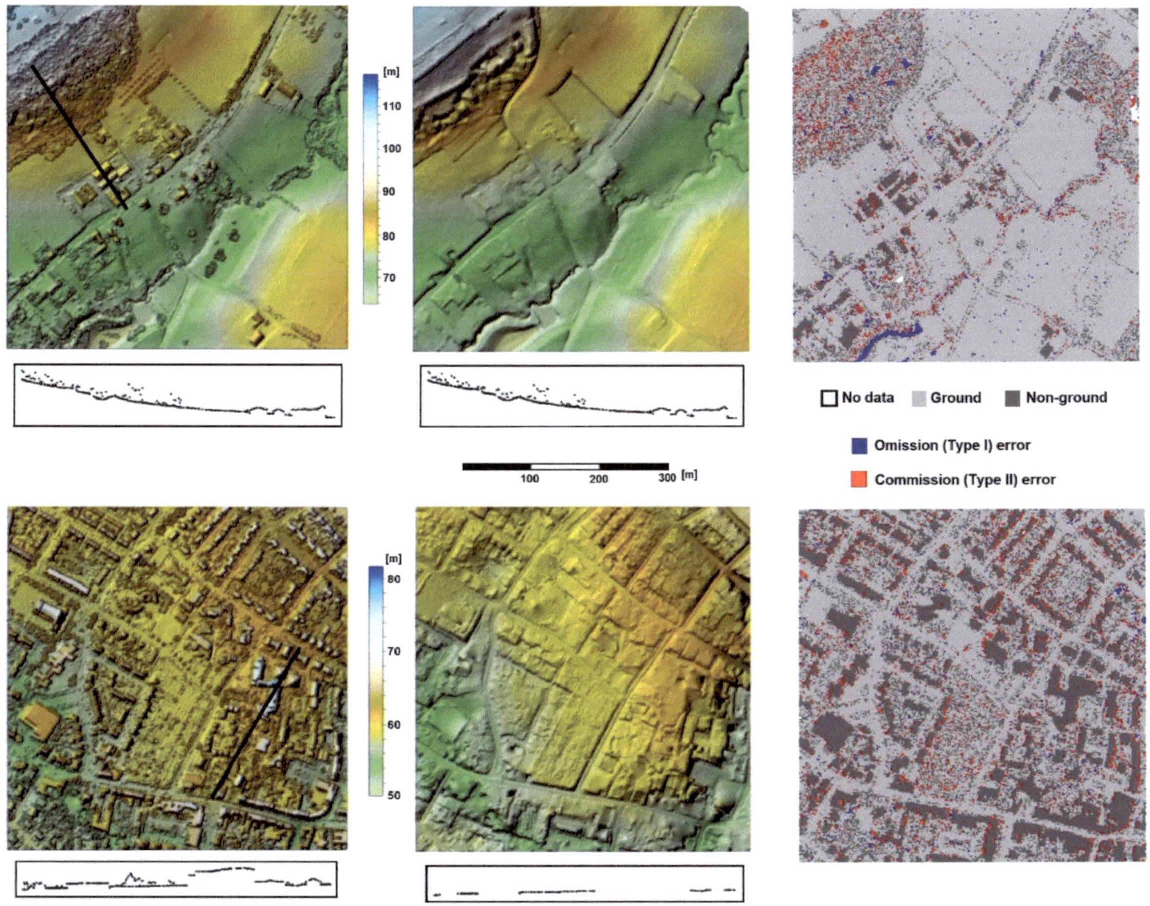

Figure 10.10 Application of a geodesic morphological filter on the National Elevation Model of two landscape types in the Netherlands, the black lines indicate the position of profiles. The left part of the figure depicts the two DSMs, the middle the two DTMs after filtering and the right part the errors of omission and errors of commission (source: Li et al., 2017)

return (see Section 5.7.1). These technological advancements induce two questions with respect to ground filtering:

- Which return should be chosen in the generation of DTMs and thus in the ground filtering approach: the first return, the last return or a mixture of multiple returns?
- How can full-waveform digitisation contribute to the improvement of ground filtering?

This section briefly considers these two questions.

Before starting to apply a ground filtering algorithm, the Lidar return(s), which are best suited for filtering, have to be selected (Meng et al., 2010). Should it be the first return, the last return or a mixture of multiple returns? The number of returns per pulse depends on the characteristics of the terrain, particularly on the smoothness of elevation variations within the footprint of the pulse. For example, a road or a building roof will backscatter one return, while a building edge or vegetation will backscatter multiple returns. In urban landscapes, the majority of the first and last returns will coincide because urban scenes contain many impervious surfaces, which are hit on top and thus result in only one return. Therefore, the use of the first return is preferred for urban scenes. For ground filtering of forested environments, the last returns are preferred. This is because the last returns stem from reflections of parts of the pulse which have penetrated vegetation. They are reflected on the ground surface or surfaces close to the ground. This doesn't mean that the last returns are always backscattered by the ground surface. This would be a misinterpretation of the interaction of Lidar pulses with object surfaces. The selection of last returns is an aid in the ground filtering process, but no more than an aid and can't replace ground filtering.

Airborne Lidar systems equipped with a full-wavelength digitisation facility sample the intensity of the backscattered pulse energy. Full-waveform digitisation provides more detailed information about objects along the path of the pulse than discrete-return digitisation, which may deliver multiple returns depending on the type of object hit by the footprint. The full waveform may contribute to improving ground filtering – for some ground filtering methods waveform information is an aid in the selection of seed points (Xing et al., 2017). An elongated waveform is caused by the same terrain characteristics as multiple returns. This means that an elongated waveform is likely the response of a pulse, which is reflected on vegetation or edges of buildings. This information can be fruitfully explored to screen initial seed points on the likelihood of being a ground point. The waveform can also give a measure of the weight of points in surface fitting approaches. Since full-waveform digitisation captures many useful features, which are not provided by multiple return Lidar, it provides good opportunities for ground filtering of scenes covered by a large variety of object types and sizes which are often characterised by sudden jumps in elevations such as those present in urban areas (Mallet et al., 2011).

10.7 Ancillary Data

In the classification of remotely sensed multispectral imagery, ancillary data is often used. This is data stemming from other sources, such as topographic, soil or geological maps as well as DTMs. To improve the classification results, ancillary data can be exploited before, during and after the classification (Hutchinson, 1982). The use of ancillary data prior to classification involves a division of the image in sub-images covering areas of similar characteristics, i.e. they are homogeneous (Ricchetti, 2000). Exploiting ancillary data during classification can be done in several ways. One of these approaches is incorporating the data as one or more extra features in addition to the multispectral bands. If available, Lidar point clouds are also often used as ancillary data in the classification of multispectral images. It is generally recognised by remote sensing experts that combining airborne Lidar point clouds with multi-spectral and hyperspectral imagery improves land cover classification (Singh et al., 2012; Luo et al., 2016).

10.7.1 Multispectral Imagery

Up to around 2015, the use of ancillary data, such as multispectral images, in DTM generation from airborne Lidar data was limited. The reason for this oversight may be that the classification of multispectral images is based on exploring the radiometric characteristics of images in the form of spectral features, while ground filtering has been primarily approached as a geometric problem. However, the view that ground filtering methods benefit from ancillary data is gaining ground. Including spectral information from images is probably a better solution for DTM generation in complex landscapes than spending a lot of effort on developing new ground filtering methods using Lidar data only (Chen et al., 2017). Indeed, ancillary data can be fruitfully exploited in the binary classification problem of separating ground points from non-ground points. Which ancillary data can be used depends on availability, suitability, operator experience and price. See Sections 11.7.1 and 11.7.2 for explanation of the basics on how to obtain object classes from multispectral imagery.

10.7.2 Building Outlines from Topographic Maps

In addition to starting to investigate the benefits gained from multispectral images, researchers are also attempting to explore information on building footprints depicted in topographic maps. The idea of using building outlines in the ground filtering process of airborne Lidar data is not new (Lemmens, et al., 1997), but since 2015 a revitalisation is ongoing. For example, building outlines have been used for generating urban DTMs from airborne Lidar data by Nordbo et al. (2015). The idea of using digital topographic maps is actually quite straightforward since today the topographic mapping agencies of most countries have digital topographic maps with outlines of building footprints. The usefulness of these building outlines depends on the positional precision, among others. These outlines can be projected in the point cloud, provided that the two geodata sets have been accurately georeferenced in the same coordinate system. When one may assume that returns of which the X,Y coordinates fall within the building outline captured on the digital topographic map are backscattered from roofs and facades, these points can be easily classified as non-ground points and removed.

However, as with so many approaches the snag lies in the detail. It is highly unlikely that the point cloud and map would capture the situation at the same moment in time. Usually, time differences between capturing the point cloud and topographic maps have to be counted in years. In some countries, the time difference may even span several decades. In the time gap between capturing the one geodata set and the other geodata set, the situation on the ground may have changed, sometimes drastically. This is, for example, the case when suburbs have been constructed at the outskirts of a city, or an office area has been erected in the vicinity of a railway station or along the main road. The recently recorded DSM contains the newly built dwellings or office buildings, but they are absent in the topographic map, of which the last update was conducted several years back.

Discrepancies may not only occur due to time gaps but also because of the positional inaccuracy of the building outlines. For example, the difference between the location of buildings on a 1:10,000 topographic map and their actual location on the ground may reach several metres. The third source of discrepancies is induced by the generalisation of maps. The outlining of buildings does not exactly follow the border of the objects. Bay windows, front doors protruding from the façade and other bulges may be left out in the representation of terraced houses. Indeed mapping is based on idealisation and generalisation. Temporary protrusions, such as sun blinds or awnings, may reflect Lidar pulses but are not indicated on topographic maps. Another difference between the map and reality is the result of the mapping specifications. For example, the mapping specification may state that tiny cottages and sheds, which are smaller than 10 m^2, should be left out. Another rule may instruct that adjacent buildings, of which the walls are less than 3 m apart, should be represented as one building with a class assignment: built-up area. As a result of the latter specification, detached houses lying close together in a row along the street may be represented on the map as one big building. The small ground strips in between the houses are represented as buildings, while in reality, they may be lawns or ground covered by shrubs. Some of the deviations between map representation and the situation on the ground are demonstrated in Figure 10.11. The left part of this figure shows the building outlines as represented in the topographic map. The right part shows the building outlines in reality at the time of the airborne Lidar survey. The Lidar points which deviate from the situation indicated on the map are shown in colour and specified in the legend.

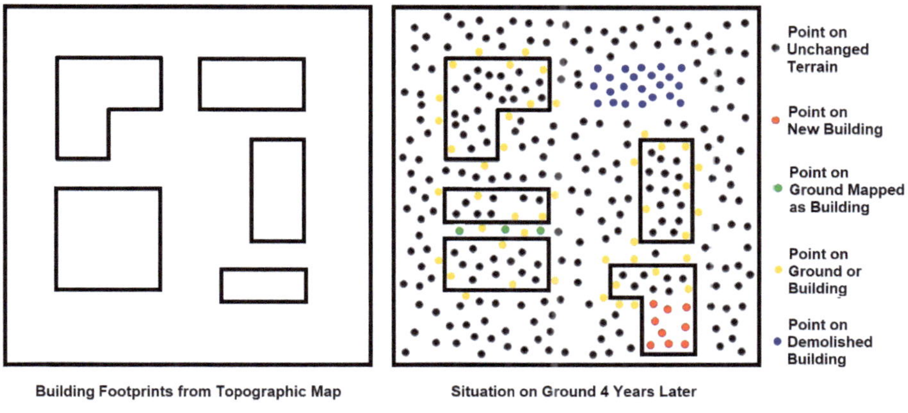

Figure 10.11 Deviations between a building's footprints in a topographic map and the actual situation on the ground at the time of capturing point cloud

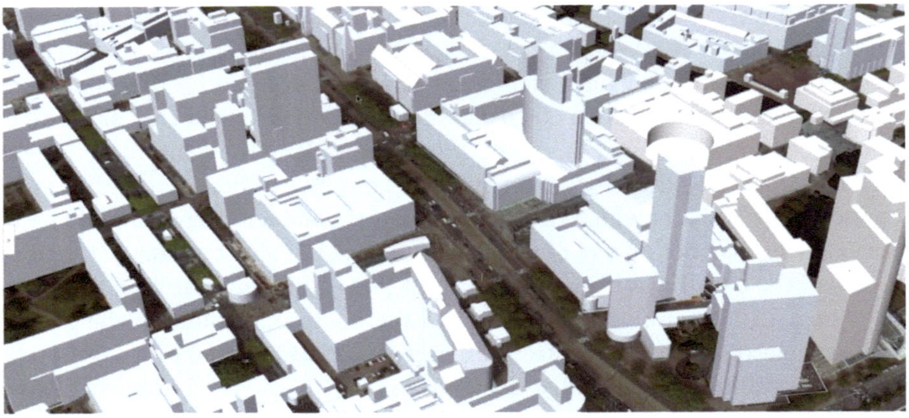

Figure 10.12 3D map at LoD1 of a part of Rotterdam automatically created by combining buildings' footprints with airborne Lidar point clouds (courtesy: Municipality of Rotterdam)

The discrepancies sketched above are of the same type as those one gets confronted with when creating Level of Detail 1 (LoD 1) 3D representations of cities or parts of cities by extruding building footprints outlined on existing maps using the height component of airborne Lidar point clouds. For example, airborne Lidar enabled Rotterdam to generate a 3D map of buildings, which was made openly available in 2011. The basic data source was a detailed map with footprints of buildings. The airborne Lidar point cloud was acquired in 2010. The accuracy and point density of 30 points/m^2 enabled the automatic extrusion of the footprints resulting in a Level of Detail 1 (LoD1) representation of Rotterdam in CityGML (Figure 10.12). However, manual editing was required to eliminate anomalies.

10.7.3 Clipping

The separation of ground points from non-ground points can't be solely and exclusively based on examining the absolute terrain elevations. If the shape of the terrain surface is not incorporated in ground filtering, wrong classifications are likely to occur, especially in areas where the terrain slopes are steep (Kraus and Pfeifer, 2001). This is because ground points often do have a higher elevation than non-ground points in terrain with large elevation variations (Figure 10.13). This doesn't mean that examining points on absolute elevation would be useless in the ground filtering process. If ancillary data is available, flat, horizontal areas can be masked out. It will be very likely that in flat, horizontal terrain the higher points will be non-ground points. In the pre-laser scan era, national governmental agencies may have established course elevation models captured by levelling or photogrammetric means. Such a course elevation model may have a point density of one point per hectare, for example. If adjacent points in this coarse model do have similar elevation values, this is an indication that the area is flat. This information can be used for clipping elevations which are higher than a threshold and to classify these as non-ground points prior to entering the actual ground filtering process. Also, a human operator can outline flat, horizontal areas using his visual interpretation skills possibly combined with aerial images simultaneously acquired during the Lidar survey.

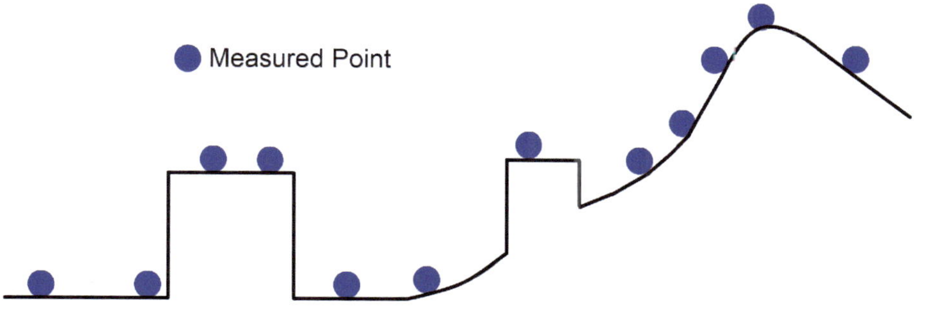

Figure 10.13 Height alone is not suitable for distinguishing ground from ground points; at best, it can be an aid

Many national mapping agencies have created elevation contours which are superimposed on their topographic maps. Often they are already digitised for the creation of a national DTM for supporting GIS analysts. Also, this DTM can provide approximate elevation values for clipping the point cloud.

10.8 Developing a Ground Filtering Procedure

In an ideal world one could dump raw point clouds, consisting of billions of points, on a mass computer memory by pushing the ground filtering button, and after a few hours or maybe a night of intensive computational flapping, the point cloud would be cleaned and stripped from all the dirt, as fresh from the laundry. However, such an ideal world doesn't exist. Ground filtering is essentially a binary classification problem and as all classification of remote sensing data is a challenging task. Ground filtering is only seemingly a simple task, and its complexity may be easily underestimated. This is because only the ground surface can be fairly well defined. However, the non-ground surfaces consist of a plethora of objects with far different characteristics. All these different object classes are collected in the grab bag of just one superclass called non-ground. From the field of remote sensing, one may learn that analysts are satisfied when they achieve an overall accuracy of 85% or a kappa coefficient of 80%. Fully automated classification is still in the realm of inventions yet to be done.

Ground filtering is usually conducted by the survey company that captured the point cloud commissioned by a governmental agency or another organisation. To control the expenses, ground filtering will be usually conducted with standardised procedures in which human involvement, such as editing, is limited to the minimum. The development of standardised procedures will be based on the experience of geomatics experts and may include a concatenation of several ground filtering methods, which may be implemented in the form of a cascade – one method is applied after the other. A standardised procedure may also exploit just one ground filtering method but applied in a sequence using different settings of the tuning parameters.

Since the mid-1990s, a plethora of ground filtering approaches have been developed by photogrammetric and computer vision researchers. If the landscape is flat and covered by smooth surfaces, most ground filtering approaches separate non-ground points from ground points fairly well, but in urban areas, regions with abrupt terrain discontinuities or terrain overgrown with complex vegetation their performance may quickly deteriorate (Sithole and Vosselman, 2004). The extensive efforts in the development of ground filtering have not yet resulted in an approach which operates faultlessly on a broad variety of terrain types. Since the quest for an ideal ground filtering method has not yielded the ultimate supreme approach, the mindset of researchers has gradually changed. Instead of going in search of another new algorithm, researchers are increasingly looking for the integration of two or more filtering techniques to arrive at satisfactory results. Combining the merits of several ground filtering methods and (partly) eliminating their deficiencies can significantly decrease misclassification of points (Chen et al., 2017). The integration of different ground filtering methods can be done along two lines. The first line is to apply different approaches in sequence, which means the output of the one approach is the input for the other filtering approach. For example, Mongus and Zalik (2014) combined three ground filtering methods including morphological operators resulting in a high success rate.

The development of a ground filtering procedure may also consist of combining the outputs of two or more filtering approaches and cleaning the resulting point cloud from contradictions. Assuming that nearly all non-ground Lidar points have been removed and the results between the two algorithms are quite similar, Deng and Shi (2013) developed a method to combine the outputs of the PTIN algorithm and the hierarchical robust interpolation method – both ground filtering approaches are implemented in commercial software as discussed above. From each of the outputs, G_1 and G_2, regular grid DTMs are created: DTM_1 and DTM_2. The two grid DTMs act as an intermediate storage layer to compare the corresponding ground points in G_1 and G_2. This is done by calculating the elevation differences between each ground point in G_1 and the corresponding cell in DTM_2. If this difference is smaller than a prespecified threshold, the G_1 point is accepted as a ground point. The same is done for G_2 by confronting the ground points with the corresponding cells in DTM_1. Next, the accepted G_1 and G_2 ground points are combined to create a new grid DTM_0. If the differences between both a G_1 ground point and a G_2 ground point with the corresponding point in the grid DTM_0 is smaller than a threshold, the point is definitely accepted as a ground point.

10.9 Quality Measures

Since ground filtering is a binary classification problem, the classification quality is commonly expressed in three error descriptors: omission, commission and total errors. Omission, or Type I, error is the portion of ground points which are wrongly rejected as ground points. Commission, or Type II, error is the portion of non-ground points which are wrongly classified as ground points. The total error is the portion of all misclassified points. The portion is usually expressed in the form of a percentage. To compute the three types of errors, ground truth is necessary. Ground truth consists of points with which it is known whether they belong to the ground or non-ground. Ground truth is acquired by another independent observation method and usually consists of a small fraction of the total number of points in the filtered point could. The locations of ground truth collection should be well distributed over the survey area. Ground truth can be gathered visually by a human operator. The counting of correctly classified points and

Table 10.2 Cross-matrix (confusion matrix) confronting a small fraction of the points in the filtered point cloud with ground truth

	Filtered point cloud		
Ground truth	Ground points	Non-ground points	
Ground points	Ng	Ngn	Ng + Ngn
Non-ground points	Nng	Nn	Nn + Nng
	Ng + Nng	Nn + Ngn	N = Ng+Nn+Nng+Ngn

wrongly classified points is tabulated in a confusion matrix. Since only two classes are present, the confusion matrix is often called cross-matrix (Table 10.2).

The portion Type I error can be computed from the cross-matrix as: $N_{gn}/\left(N_g + N_{gn}\right)$
The portion Type II error can be computed as: $N_{ng}/\left(N_n + N_{ng}\right)$
And the total error as: $\left(N_{gn} + N_{ng}\right)/N$
With: $N = N_g + N_n + N_{gn} + N_{ng}$

The right part of Figure 10.10 shows errors of omission and commission after the application of a geodesic morphological filter on two small parts of complex landscapes in the Netherlands. Instead of total error, which has a negative connotation, a more positive measure is often used in the field of remote sensing: the overall accuracy also called agreement. This is the ratio between the points which received the right label – observed agreement – and the total number of ground truth points. In a binary classification scheme the overall accuracy (OA), using the notation above, is:

$$\mathrm{OA} = \left(N_g + N_n\right)/N \tag{10.3}$$

Total error (TE) can be expressed in terms of the overall accuracy as follows:

$$\mathrm{TE} = \left(N_{gn} + N_{ng}\right)/N = 1 - \left(N_g + N_n\right)/N \tag{10.4}$$

Another measure which is often used for quantifying the classification accuracy is the kappa coefficient of agreement (Cohen, 1960; Bishop et al.,1975), which has been introduced in the field of remote sensing by Congalton et al. (1983) (Hudson and Ramm, 1987). The kappa coefficient measures the agreement between classification results and ground truth by taking into account the possibility of the agreement occurring by chance. For a binary classification problem as is the case for ground filtering, the kappa coefficient is defined as:

$$K = \frac{OA - Q}{1 - Q} \tag{10.5}$$

With *OA* the Overall Accuracy (Equation 10.3) and $Q = \left(\left(N_g + N_{ng}\right)\left(N_g + N_{gn}\right) + \left(N_n + N_{gn}\right)\left(N_n + N_{ng}\right)\right)/N^2$.

Table 10.3 Binary classification in terms of true and false positives and negatives

	Filtered point cloud	
Ground truth	Ground points	Non-ground points
Ground points	*True positives*	*False negatives*
Non-ground points	*False positives*	*True negatives*

A perfect, faultless classification results in $OA = 1$ and thus gives a kappa equal to 1. In all other cases kappa will be smaller than OA.

In the fields of pattern recognition and binary classifications, other measures are in use to evaluate classification algorithms. These measures are Recall (R), Precision (P) and F_1 score. To elucidate these measures, let's list Table 10.2 in more general statistical terms (see Table 10.3).

Recall is the capability of a ground filtering algorithm to find all the ground points and is defined as the number of ground points which have been labelled as ground points (true positives) divided by the total number of ground points, which means true positives plus false negatives: Recall = true positives/(true positives + false negatives), or in terms of Table 10.2:

$$R = N_g / \left(N_g + N_{gn} \right) \tag{10.6}$$

The values of the Recall measure lie between zero and one. A problem with using only the Recall measure is that a completely wrong classification could result in a high Recall score. To demonstrate this, suppose we would assign to all points the label ground point; this could be done without any classification effort. Then there would be no false negatives, but only false positives, resulting in a Recall score of one. Obviously, very wrong indeed; the classification would be very imprecise because of the presence of many false positives. To avoid this statistical bias, a second measure has been developed: Precision. Precision takes into account the false positives and thus expresses the capability of a ground filtering algorithm to avoid false positives. Precision is defined as the number of ground points which have been labelled as ground points (true positives) divided by the total number of points classified as ground point, which means true positives pulse false positives: Precision = true positives/(true positives plus false positives), or in terms of Table 10.2:

$$P = N_g / \left(N_g + N_{ng} \right) \tag{10.7}$$

In the Recall-Precision performance metric, an overall performance measure can be determined by combining the two measures. Most commonly the two measures are combined in the so-called F_1 score, which is defined as:

$$F_1 = 2 \frac{R \cdot P}{R + P} \tag{10.8}$$

Both Recall and Precision can take values in the range zero to one, which means they are normalised measures. In the case, there are no false positives and false negatives, both R and P will take the value one, which means the denominator in Equation 10.7 can take values up to two. To ensure that also the F_1 score is normalised, the fraction is multiplied by 2.

Now, the question arises: Which error does affect the quality of a DTM more – wrongly classified ground points or wrongly classified non-ground points? Wrongly classified ground points are removed from the data set because they have been assigned to the non-ground point class. The effect is that the DTM will be less dense. Maybe some parts of the terrain surface are affected more than other parts. But in general, one may state that wrongly removing ground points will have a modest influence on the quality of subsequent analyses or derived products. In contrast, wrongly accepting non-ground points as ground points will influence processing in later stages more, although the extent to which this happens will depend on the application. It may occur that the overall error will be low due to low errors with respect to wrongly assigning the non-ground class to ground points. But many non-ground points may be classified as ground points. This is because the overall error is an average error measure. So, it would be unjustified to claim that a ground filtering approach would be superior to other methods just based on comparing the overall error.

Confronting the above discussion with the Recall and Precision performance measure reveals that a high Recall score is less important than a high Precision score. When the Recall score is high, only a few ground points have been classified as non-ground points. As a result, they have been wrongly removed from the point cloud, which has – in most cases – little effect on the final quality of the DTM as the discussion above has shown. A low Precision means that many non-ground points are classified as ground points, which has a significant effect on the quality of the final DTM. A fair comparison of the performance of ground filtering methods should, therefore, be based on the Precision measure given in Equation 10.7.

The error types mentioned above focus on classification accuracy. Classification is done on a nominal scale, and on this scale only qualitative error metrics can be used, basically counting rightly and wrongly classified items. However, a DTM consists of elevation values which are expressed in an interval scale, which allows a quantitative description of quality. The most feasible measures are root mean square error (RMSE) and systematic error. See Section 7.7 for an explanation of the definition and use of these measures for quantifying the height accuracy of DTMs.

10.10 Concluding Remarks

The ground filtering methods, outlined within this chapter, have been specifically developed for airborne Lidar point clouds. Since 2010, very dense photogrammetric point clouds can be created using semi-global matching techniques, called dense image matching (DIM). The ground filtering methods discussed in this chapter are also suited for being applied at photogrammetric point clouds, especially when the images cover urban areas. The penetration rate of DIM points over forested environments is low, and for these areas the ground filtering methods will be less successful. The primary DTM data source for capturing forests will continue to be airborne Lidar. In the meantime, Lidar technology continues to advance. The point density is still increasing. The performance of ground filtering algorithms depends partly on the point density. As a result, analysts who generate DTMs from raw point clouds will have to adjust their ground filtering procedures regularly. Also, the type of landscape greatly affects the performance of ground filtering algorithms. No general-purpose ground filtering technique exists, which performs equally well on the large diversity of landscape types.

References

Arefi, H., Hahn, M.A. (2005) Morphological Reconstruction Algorithm for Separating Off-terrain Points from Terrain Points in Laser Scanning Data. *The International Archives of the Photogrammetry, Remote Sensing and Spatial Information Sciences*, 36, pp. 120–125.

Axelsson, P. (2000) DEM Generation from Laser Scanner Data Using Adaptive TIN Models. *International Archives of Photogrammetry and Remote Sensing*, 17–22 July, Amsterdam, Vol. 33, Part B4, pp. 110–117.

Bishop, Y.M.M., Feinberg, S.E., Holland, P.W. (1975) *Discrete Multivariate Analysis - Theory and Practice*. MIT Press, Cambridge, MA.

Chen, C., Li, Y., Yan, C., Dai, H., Liu, G., Guo, J. (2016) An Improved Multi-resolution Hierarchical Classification Method Based on Robust Segmentation for Filtering ALS Point Clouds. *International Journal of Remote Sensing*, 37, pp. 950–968.

Chen, Z., Gao, B., Devereux, B. (2017) State-of-the-Art: DTM Generation Using Airborne LIDAR Data. *Sensors*, 17(1), doi:10.3390/s17010150.

Cohen, J. (1960) A Coefficient of Agreement of Nominal Scales. *Educational and Psychological Measurement*, 20(1), pp. 37–46.

Congalton, R.G., Oderwald, R.G., Mead, R.A. (1983) Assessing Landsat Classification Accuracy Using Discrete Multivariate Analysis Statistical Techniques. *Photogrammetric Engineering and Remote Sensing*, 49(12), pp. 1671–1678.

Deng, S.S., Shi, W.Z. (2013) Integration of Different Filter Algorithms for Improving the Ground Surface Extraction from Airborne Lidar Data. *International Archives of the Photogrammetry, Remote Sensing and Spatial Information Sciences*, XL-2/W1, pp. 105–110.

Evans, J.S., Hudak, A.T. (2007) A Multiscale Curvature Algorithm for Classifying Discrete Return LiDAR in Forested Environments. *Geoscience and Remote Sensing*, 45(4), pp. 1029–1038.

Fischler, M.A., Bolles, R.C. (1981) Random Sample Consensus: A Paradigm for Model Fitting with Application to Image Analysis and Automated Cartography. *Communications of the ACM*, 24(6), pp. 381–395.

Hansen, W. von, Vögtle, T. (1999) Extraktion der Geländeoberfläche aus flugzeuggetragenen LaserscannerAufnahmen. *Photogrammetrie Fernerkundung Geoinformation*, 4, pp. 229–236.

Hudson, W.D., Ram, C.W. (1987) Correct Formulation of the Kappa Coefficient of Agreement. *Photogrammetric Engineering and Remote Sensing*, 53(4), pp. 421–422.

Hutchinson, C.F. (1982) Techniques for Combining Landsat and Ancillary Data for Digital Classification Improvement. *Photogrammetric Engineering & Remote Sensing*, 48, pp. 123–130.

Isenburg, M. (2015) LAStools: Efficient Tools for LiDAR Processing. lastools.org

Kraus, K., Pfeifer, N. (1997) A New Method for Surface Reconstruction from Laser Scanner Data. *International Archives of Photogrammetry and Remote Sensing*, 32(3/2W4), pp. 80–86.

Kraus, K., Pfeifer, N. (1998) Determination of Terrain Models in Wooded Areas with Airborne Laser Scanner Data. *ISPRS Journal of Photogrammetry and Remote Sensing*, 53, pp. 193–203.

Kraus, K., Pfeifer, N. (2001) Advanced DTM Generation from Lidar Data. In *International Archives of Photogrammetry and Remote Sensing*, Vol. XXXIV-3/W4, Annapolis, MD, pp. 23–30.

Lemmens, M.J.P.M., Deijkers, H., Looman, P.A.M. (1997) Building Detection by Fusing Airborne Laser-altimeter DEMs and 2D Digital Maps. *IAPRS*, 32(Part 3-4W2), pp. 42–49.

Li, Y., Yong, B., van Oosterom, P., Lemmens, M., Wu, H., Ren, L., Zheng, M., Zhou, J. (2017) Airborne LiDAR Data Filtering Based on Geodesic Transformations of Mathematical Morphology. *Remote Sensing*, 9(11), doi:10.3390/rs9111104.

Lin, X., Zhang, J. (2014) Segmentation-Based Filtering of Airborne LiDAR Point Clouds by Progressive Densification of Terrain Segments. *Remote Sensing*, 6, pp. 1294–1326.

Lindenberger, J. (1993) Laser-Profilmessungen zur topographischen Geländeaufnahme. PhD Thesis, Deutsche Geodätische Kommission, Series C, No. 400, Munich.

Luo, S., Wang, C., Xi, X., Zeng, H., Li, D., Xia, S., Wang, P. (2016) Fusion of Airborne Discrete-Return LiDAR and Hyperspectral Data for Land Cover Classification. *Remote Sensing*, 8(1), p. 3.

Mallet, C., Bretar, F., Roux, M., Soergel, U., Heipke, C. (2011) Relevance Assessment of Full-waveform Lidar Data for Urban Area Classification. *ISPRS Journal of Photogrammetry and Remote Sensing*, 66(6 supplement), pp. 71–84.

McGaughey, R.J. (2018) *FUSION/LDV: Software for LIDAR Data Analysis and Visualization August 2018: FUSION Version 3.80*. United States Department of Agriculture, Forest Service.

Meng, X., Currit, N., Zhao, K. (2010) Ground Filtering Algorithms for Airborne LiDAR Data: A Review of Critical Issues. *Remote Sensing*, 2(3), pp. 833–860.

Mongus, D., Zalik, B. (2014) Computationally Efficient Method for the Generation of a Digital Terrain Model from Airborne LiDAR Data Using Connected Operators. *IEEE Journal of Selected Topics in Applied Earth Observations and Remote Sensing*, 7(1), pp. 340–351.

Nordbo, A., Karsisto, P., Matikainen, L., Wood, C.R., Järvi, L. (2015) Urban Surface Cover Determined with Airborne Lidar at 2 m Resolution: Implications for Surface Energy Balance Modelling. *Urban Climate*, 13, pp. 52–72.

Pfeifer, N., Köstli, A., Kraus, K. (1998) Interpolation and Filtering of Laser Scanner Data: Implementation and First Results. *International Archives of Photogrammetry and Remote Sensing*, Vol. XXXII, Part 3/1, Columbus, pp. 153–159.

Pfeifer, N., Stadler, P., Briese, C. (2001) Derivation of Digital Terrain Models in the SCOP++ Environment. In Torlegard, K., Nelson, J. (Eds.), *OEEPE Workshop on Airborne Laserscanning and Interferometric SAR for Detailed Digital Terrain Models*, Stockholm, Sweden.

Ricchetti, E. (2000) Multispectral Satellite Image and Ancillary Data Integration for Geological Classification. *Photogrammetric Engineering & Remote Sensing*, 66(4), pp. 429–435.

Singh, K., Vogler, J.B., Shoemaker, D.A., Meentemeyer, R.K. (2012) LiDAR-Landsat Data Fusion for Large-area Assessment of Urban Land Cover: Balancing Spatial Resolution, Data Volume and Mapping Accuracy. *ISPRS Journal of Photogrammetry and Remote Sensing*, 74, pp. 110–121.

Sithole, G., Vosselman, G. (2004) Experimental Comparison of Filter Algorithms for Bare Earth Extraction from Airborne Laser Scanning Point Clouds. *ISPRS Journal of Photogrammetry and Remote Sensing*, 59, pp. 85–101.

Sohn, G., Dowman, I. (2002) Terrain Surface Reconstruction by the Use of Tetrahedron Model with the mdl Criterion. *The International Archives of the Photogrammetry, Remote Sensing and Spatial Information Sciences*, 34, pp. 336–344.

Tinkham, W.T., Huang, H., Smith, A.M.S., Shrestha, R., Falkowski, M.J., Hudak, A.T., Link, T.E., Glenn, N.F., Marks, D.G. (2011) A Comparison of Two Open Source LiDAR Surface Classification Algorithms. *Remote Sensing*, 3, pp. 638–649.

Vosselman, G. (2000) Slope based Filtering of Laser Altimetry Data. *International Archives of the Photogrammetry, Remote Sensing and Spatial Information Sciences*, 33, pp. 935–942.

Xing, S., Lia, P., Xua, Q., Wang, D., Lia, P. (2017) Surface Fitting Filtering of LiDAR Point Cloud with Waveform Information. *ISPRS Annals of the Photogrammetry, Remote Sensing and Spatial Information Sciences*, IV-2/W4, pp. 179–184.

Zhang, K., Whitman, D. (2005) Comparison of Three Algorithms for Filtering Airborne LiDAR Data. *Photogrammetric Engineering and Remote Sensing*, 71(3), pp. 313–324.

Zhang, K., Zheng, C. (2007) *Airborne LIDAR Data Processing and Analysis Tools*. International Hurricane Research Center, Department of Environmental Studies, Florida International University, Miami, FL.

Zhang, K., Chen, S.C., Whitman, D., Shyu, M.L., Yan, J., Zhang, C. (2003) A Progressive Morphological Filter for Removing Nonground Measurements from Airborne LIDAR Data. *IEEE Transactions on Geoscience and Remote Sensing*, 41(4), pp. 872–882.

Zhao, X., Guo, Q., Su, Y., Xue, B. (2016) Improved Progressive TIN Densification Filtering Algorithm for Airborne Lidar Data in Forested Areas. *ISPRS Journal of Photogrammetry and Remote Sensing*, 117, pp. 79–91.

Part IV
3D Mapping of Point Clouds

Feature Detection from Images and Point Clouds

CHAPTER 11

Many processing and analysis tasks in the field of pointcloudmetry require the comparison of the characteristics of an object with the characteristics of a prototype of that object. An object in an image or point cloud doesn't have the meaning that human perceives in everyday language: house, table, lamppost, tree, etc. The object may be a boundary or edge between two objects, a corner or a T-junction in an image, or even the individual pixel of a multispectral image or a single point in a point cloud. The image parts which are valuable for a certain task have been given many names, such as image primitives, tokens, characteristic points, salient points, significant points, edges, interest points, key points and multispectral pixels. These image parts which are valuable within the next steps of an image processing pipeline are called features. The value of a feature may differ per task. For example, detecting pixels lying on a boundary – called edge detection – is valuable for outlining the footprints of buildings, or the boundaries of agricultural parcels, which may often correspond to property boundaries and thus beneficial for land administration purposes. On the other hand, for the determination of the land cover of parcels using a pixel-based classification strategy, all pixels have high relevance, because the reflectance values of the diverse spectral bands are used as attributes within the classification task. The assignment of land cover classes to pixels in a multispectral image is called multispectral classification, a term which is increasingly replaced by the term machine learning.

Exploiting features is also necessary for the (co-)registration of overlapping images and laser scans for detecting tie-points, also called key points. It is also essential for image matching aimed at the generation of digital elevation models; segmentation of a point cloud into planes, cylinders and spheres; generation of image mosaics from a set of overlapping images, such as photogrammetric orthomosaics; motion detection; and 3D mapping.

Feature extraction consists of two fundamental steps: (1) feature detection and (2) feature description. The first step – feature detection – is usually the hard part and will receive major attention in this chapter. In the feature description part, one or more measures are derived from the image patch or a neighbourhood of points in a point cloud in which the detected feature lies. Feature measures may consist of the original measurements. These are, for example, the reflectance values of a laser point cloud or the RGB or other spectral bands of multispectral images. Often, although not always, these measurements will have been corrected for distortions, such as atmospheric effects or a decrease in the signal strength of laser beams with the distance travelled through the atmosphere. The feature measures are usually combined in a *feature vector* in the case of two or more attributes. Commonly features are extracted to compare the values of their attributes with the attributes of candidate prototypes and to decide, based on the similarity between the values, to which prototype or class the feature belongs. In the case of image matching or co-registration of laser scans, the feature detected in the one image or scan is compared to candidate features in the other image or scan. The candidate features are identified by defining a search area.

The comparison between a given feature and relevant prototypes or candidate features is conducted by computing a similarity measure. This means that the attributes or feature descriptors have to be expressed in numerical measures. These feature descriptors are called *feature measures*. Depending on the task as well as the discipline, the determination of the similarities between the features of a given object and the features of a set of prototype or candidate objects is called classification, matching, 3D mapping, segmentation, object recognition, pattern recognition, structure from motion, or solving the correspondence problem. This list is certainly not limitative. In the task of pixel-based classification of multispectral images, the feature is simply the individual pixel. In the task of image matching, the features may be the grey values in image patches in the case of area-based matching, or interest points in the case of feature-based matching. Interest points may have different shapes. In the image matching task most popular are L-junctions, Y-junctions, T-junctions, Arrow-junctions and X-junctions (Figure 11.1). If the task concerns edge detection in an image, there are only two prototypes: edge and non-edge and the decision whether a pixel is lying on

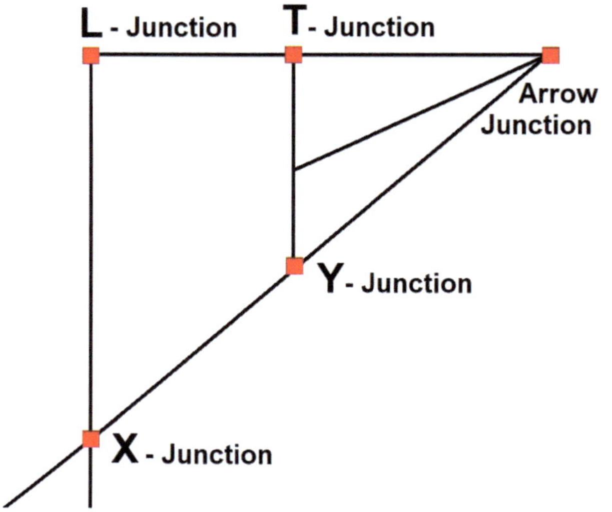

Figure 11.1 L-junctions, Y-junctions, T-junctions, Arrow-junctions and X-junctions are important interest points for feature-based image matching

an edge or not is made by thresholding: If the response is larger than the threshold, the pixel is accepted as an edge pixel.

In general, the process of finding the corresponding feature of a given feature consists of four stages:

1. Detection of the feature in the data set (in the case of multispectral classification, the pixel is the feature and detection can be omitted).
2. Assigning feature measures – also called attributes – to the features. These feature measures should be the same for the detected feature and the prototype features. Usually, the feature measures are arranged in a vector, called a feature vector.
3. Finding the level of similarity between the feature measures of the given feature and the feature measures of prototypes.
4. Applying a decision rule to decide to which prototype the feature belongs, or whether it doesn't show sufficient similarity with any of the prototypes.

This chapter primarily focuses on the automated detection of particular feature types in images, called interest points. As such this chapter should be read as a prelude to Chapter 4, which outlines photogrammetry and structure from motion (SfM) also called image matching. We will also briefly outline the assignment of geometric feature measures to small neighbourhoods of points in a point cloud to support (semi-)automatic segmentation, classification and 3D mapping.

Section 11.1 focuses on the modelling of a digital image. Section 11.2 briefly considers point operations carried out on a raster. Point operations carry out computations on two or more points which capture the same point in object space. Since the extraction and description of features mainly rely on applying statistical concepts, Section 11.3 recaps statistical measures including measures of correlation between two data sets, the covariance matrix and its eigenspace, and the principles of least-squares adjustment, which aims at estimating the coefficients of the parameters of a known mathematical relationship between two data sets. Section 11.4 outlines the basics of differentiation on a discrete grid as a prelude to Section 11.5, which covers the detection of interest points in images. Section 11.6 considers the extraction of geometric features from point clouds. Section 11.7 turns to the application of feature measures for the classification of multispectral images as well as feature-based matching. In this section, we introduce the Mahalanobis distance, named after the Indian scientist and statistician P.C. Mahalanobis (1893–1972), which provides weights to the values in two or more data sets. The weights are inversely proportional to the correlation. This chapter ends with concluding remarks.

In this chapter, feature detection will be mainly treated with digital images in mind, particularly because image features play a fundamental role in image matching, which is an important topic in photogrammetry and computer vision. Nevertheless, the approaches discussed can also fruitfully be applied to point clouds. This chapter builds upon and extends the work of Lemmens (1988). This publication brought the issues and solutions of digital photogrammetry and computer vision together in one survey for the first time. Feature extraction from images is a broad subject. An in-depth treatment covering an entire book is provided in Nixon and Aguado (2012).

11.1 Image Modelling

An image is a collection of grey values (also each band of an RGB image consists of a single digital number per pixel, which we will call grey value) arranged on a square or rectangular raster. The cells of the raster are usually squares and same-sized. Four main scene factors are responsible for the grey values in an image (Marr, 1982):

1. Illumination source (strength and position with respect to the scene)
2. Position and orientation of the camera
3. Surface reflectivity
4. Surface geometry

Figure 11.2 sketches this dependency among the four scene factors. Although the variations in grey values are partly induced by the surface geometry and thus the objects present in the scene, it is often convenient to look at an image as a 2D discrete signal. This view of an image is the classical one and frequently applied. It has the advantage of being fully compatible with the well-defined and thoroughly studied concepts of digital signal processing. Especially radiometric restoration, enhancement by smoothing, and edge detection gain profit from a signal approach. One of the methods for generating photogrammetric point clouds, which means point clouds from overlapping images, exploits correlation techniques applied to image patches, which approximately capture the same part of the scene. This method is called area-based matching. A second main method is feature-based matching. The first factor – illumination source – demonstrates that the smaller the time span between the recordings of overlapping images, the better the image matching process will succeed.

A second way to look at an image is as a representation of different objects. The vast majority of the pixels covering an object will show not much difference in grey values with neighbouring pixels or the differences are rather chaotic; they do not show a clear

Figure 11.2 Four scene factors responsible for the visibility of features (source: Lemmens, 1988; modified)

pattern and don't contribute much to the image content. The fluctuations of grey values are usually due to noise and texture. On the other hand, significant or salient points may be identified such as pixels lying on the boundary between two or more objects, particularly corner-like junctions. These salient points and their surrounding pixels, which show a high level of contrast, are of interest for use for the reliable finding of corresponding tie-points. They are also used for feature-based matching. The detection of salient points is also indispensable for automated boundary outlining of objects.

To summarise the above discussion. Over time, two main approaches for modelling an image have been developed. An image is either considered to be:

- A signal, which is a 2D spatial discrete distribution of intensity values, or
- A set of L-shaped corners and other types of junctions, edges, blobs and other salient points, called features

Looking at an image as a set of features is the most natural way to do so, because features are a representation of objects or parts of objects, and extracting objects is in most cases the ultimate aim of capturing a scene. The treatment of an image as a 2D signal is an intermediate attempt to get there.

Often it is convenient to model the grey values of an image $g(i,j)$ to consist of two parts: a deterministic part, $f(i,j)$, and a part consisting of random noise. Even when we know the noise model, it is not possible to reconstruct the deterministic part, because the noise model describes the group behaviour of the noise over the entire image, but not how much it affects the grey values of individual pixels. It is convenient and often not too far from reality to model the noise as an additive, space invariant, zero-mean Gaussian distributed noise field (Pratt, 1991):

$$g(i,j) = f(i,j) + n(i,j); i = 1,\ldots n; j = 1,\ldots,m \tag{11.1}$$

The image consists of n columns and m rows; $n(i,j)$ is a number stemming from a zero-mean Gaussian distribution with standard deviation σ_n. The value of $n(i,j)$ is independent of the position of the pixel in the image, which means its value is space invariant. This modelling allows the computation of the effect of, for example, applying a 3 × 3 unweighted smoothing filter on an image contaminated by a noise field. Such a filter reduces the standard deviation σ_n of the noise field to 1/3 σ_n. A 3 × 3 unweighted smoothing filter consists of a 3 × 3 window or kernel, in which each of the nine elements contains the number one. The outcome is divided by nine. Using the same error propagation method, it can be shown that the application of differential operators increases the noise level. This exercise is left to the reader.

11.2 Point Operations

The operations carried out on pixels in a digital image or a point in a laser point cloud can be distinguished into two broad groups: (1) point operations and (2) neighbourhood operations (Figure 11.3). This section focuses on point operations. The operators

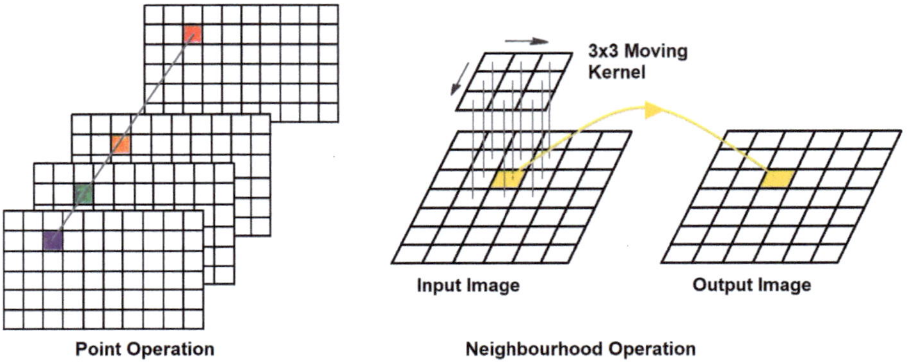

Figure 11.3 Sketch showing the difference between a point operation (left), which is applied to two or more (multispectral) images of the same scene, and neighbourhood operation, which involves the moving of a window, e.g. a 3 × 3 kernel, over the image; the response is stored in an output image (right)

discussed in other sections of this chapter focus on neighbourhood operations, which involve a group of adjacent points. Point operations may be further divided into (1) point operations on a single point and (2) point operations on multiple points, which capture the same spot in object space.

Point operations on a single point focuses on modifying the value of a single point according to some mathematical expression. Within image processing, point operations are used to manipulate the image contrast and to correct for atmospheric or other distortions. Contrast stretching aims at exploiting the entire bit depth of an image to improve the visibility of the image content. A well-known point operation is histogram equalisation, which enhances the contrast in low contrast shadow image patches or in bright, overexposed image patches. The contrast can also be rigorously changed by dividing the grey values into two groups. This operation is called grey-level thresholding. This operation converts an image containing one object against a sharply contrasting background into a binary image in which the pixels covering the object receive the value one and the background pixels the value zero. Also, several thresholds may be used, which divide the histogram into a number of slices; each slice receives one value. This operation is called level slicing. Another phenomenon, which frequently occurs in multispectral satellite images, is the corruption of the intensity values due to atmospheric effects or other degrading sources. If a correction model and its coefficients are available, the grey values of each and every pixel can be adjusted. The parameters of such a model may be obtained by placing reference reflectance panels in the terrain, while the scene is recorded or from published models, developed by others.

Another often applied point operation is band rationing, which is the computation of the ratio between the grey value of each pixel in one band and the grey value of the corresponding pixel in a second band. Band rationing reveals the differences between the spectral reflectance values, which is beneficial for remote sensing applications. Band rationing also reduces the effect of shadows since the ratio of the illuminated and non-illuminated areas of the same type of object surface will show much resemblance in different bands. This is beneficial for image matching and the classification of multispectral images and laser points enriched with RGB values. Point operations are also often applied for the pixel-based classification of multispectral images into land cover classes using machine learning approaches (see Section 11.7).

11.3 Statistical Preliminaries

Processing of images and raw point clouds often requires statistical inference. The measurements are contaminated by noise and degrading effects. By taking samples, a quantification of the dispersion of the measurements can be obtained by calculating the variance. Two or more data sets may have a correlation which can be quantified by calculating the covariances and a measure derived from the covariances, called the normalised correlation coefficient. Insight into the kind of correlation between two data sets is also obtained by transforming the covariance matrix to its eigenspace, leading to eigenvectors and eigenvalues. The relationship between two data sets may be known in the form of a mathematical model. Such models may take the form of a linear relationship: $x = \alpha y + \beta$, with x and y values of corresponding points in the two data sets and α and β the coefficients. Frequently, these coefficients are unknown in advance and have to be computed by identifying and using the values of corresponding points in the two data sets in a learning stage. To improve precision and reliability of the estimation of α and β, usually more than two points, which are minimally necessary to exactly determine α and β, are taken. When more measurements are available than unknowns to be solved, a general apparatus to arrive at an optimal solution is least-squares adjustment. These are the topics considered in this section.

11.3.1 Statistical Measures for Neighbourhoods

In image processing two basic types of neighbourhood operators can be distinguished. One type of operation is based on taking a statistical measure of a patch of adjacent points. The statistical measures include: mean, weighted mean, median, variance, most common values (majority), values that occur least (minority), number of different values (multiplicity), maximum, minimum or a value derived from subtracting the minimum from the maximum (extent). They are usually calculated over rectangular image patches, for example, 3×3 or 5×5 windows. The response of the operator is assigned to the centre pixel of the window in the output image (see Figure 11.3, right). The mean and weighted mean operations are aimed at reducing the effect of random noise; this process is called smoothing. The median, majority, minority, maximum and minimum operations are used for removing strongly deviating values, also called outliers. The median does this by ranking the grey values from small to large and next taking the middle value. Extreme values are filtered out in this way. The variance and extent operations provide the measures of dispersion or variation of the grey values within the image patch. Calculating the mean over an image patch is called unweighted smoothing. Calculating the weighted mean is usually done by a kernel which is derived from the 2D Gaussian bell-shaped curve and accordingly is called Gaussian smoothing. The values in the smoothing kernels are always positive, and the basic mathematical operation of these kernels is addition. These operations may therefore also be called integrating filters.

In contrast to (weighted) means, which are integrating operators – often called filters – which aim at reducing noise, differential operators aim at detecting locations in the image with sudden jumps in grey values. These grey value jumps may indicate the location of boundaries between objects, for example, agricultural parcel boundaries or boundaries between buildings and paved surfaces. The kernels of differential filters contain always positive and negative numbers, which are distributed in a structured way over the kernel. Well-known differential kernels are the Laplace operator, the Prewitt operator and the Sobel operator. In contrast to the last two operators, which are named after their designers, the Laplace operator is named after the French mathematician and physicist Pierre-Simon Laplace (1749–1827) who formulated this type of differential operator, which is widely used in mathematics and physics, for the first time. Before discussing differential operators in greater detail in Section 11.4, this section continues with variances, covariances and least-squares adjustment.

11.3.2 Variance

The computation of the variance of grey values within an image patch provides powerful means to characterise the local image function. Usually, the image patch is defined as a square window of a certain size containing n pixels and thus n grey values. To avoid the mathematical root operation, the variance instead of the standard deviation can be better used. The variance of the grey values in an image patch P consisting of n pixels is defined as:

$$\sigma_g^2 = \frac{1}{n-1} \sum_{i \in P} \left(g_i - \bar{g}\right)^2 \tag{11.2}$$

where

$$\bar{g} = \frac{1}{n} \sum_{i \in P} g_i \tag{11.3}$$

To save computation time, σ_g^2 can be rewritten in a form which can be derived from Equations 11.2 and 11.3:

$$\sigma_g^2 = \frac{1}{n-1} \left[\sum_{i \in P} g_i^2 - \frac{1}{n} \left(\sum_{i \in P} g_i \right)^2 \right] \tag{11.4}$$

The above variance measure describes the dispersion of grey values around a mean value, which is usually computed from the values within the image patch.

11.3.3 Covariance

Often one has two or more data sets for which one wants to get insight into the degree of agreement or correlation. Area-based matching between image patches in two overlapping images is an example in which the degree of correlation is a measure of correspondence. A measure of the degree of correlation is the covariance. Suppose we do have two image patches extracted from overlapping images of which we may expect that they cover the same area in object space. If the two images – $g(i,j)$ and $h(i,j)$, $i = 1, \ldots, n$; $j = 1, \ldots m$ – are taken quickly after each other, the differences in illumination will be small and it makes sense to compute the variance to validate whether our assumption that the two image patches correspondent is true. Defining the two image patches as above the covariance is computed as follows:

$$\sigma_{gh} = \sigma_{hg} = \frac{1}{n-1} \sum_{i \in P} \left(g_i - \bar{g}\right)\left(h_i - \bar{h}\right) \tag{11.5}$$

It is convenient to arrange the two variances of the data sets and their covariance into one, so-called covariance matrix, also called variance-covariance matrix:

$$\boldsymbol{C} = \begin{pmatrix} \sigma_g^2 & \sigma_{gh} \\ \sigma_{gh} & \sigma_h^2 \end{pmatrix} \tag{11.6}$$

Transformation of the covariance matrix to the eigenspace and computing the eigenvalues, λ_1 and λ_2, provide insightful information about the data sets. The eigenvector with the largest eigenvalue points in the direction of the highest variance. In the general case of an n-dimensional covariance matrix the next largest eigenvector points in the direction with the next highest variance and so on. The eigenvectors are orthogonal to each other. A multispectral image is taken as an example of an application. The reflectance values of some multispectral bands may be highly correlated, which may degrade the quality of classification. In order to decorrelate the spectral correlation, a principal component transformation can be carried out prior to classification. The transformation is based on computing the eigenvalues and eigenvectors of the covariance matrix of the bands. A multispectral image consisting of four bands results in a 4 × 4 covariance matrix. The eigenspace analysis may reveal that the correlation between bands is such that the information content can be put into two bands. These two bands are obtained by a principal component transformation, which significantly simplifies the classification burden.

The eigenvalues of matrix **C** are those λ for which determinant of the covariance matrix minus λ times the identity matrix is zero. For the detection of interest points, the eigenvalues of a 2 × 2 covariance matrix are of particular interest. Denoting the covariance matrix for convenience as:

$$\boldsymbol{C} = \begin{pmatrix} C_{11} & C_{12} \\ C_{12} & C_{22} \end{pmatrix}$$

The eigenvalues are computed from:

$$\det\begin{bmatrix} C_{11} - \lambda & C_{12} \\ C_{12} & C_{22} - \lambda \end{bmatrix} = \left(C_{11} - \lambda\right)\left(C_{22} - \lambda\right) - C_{12}^2 = 0 \tag{11.7}$$

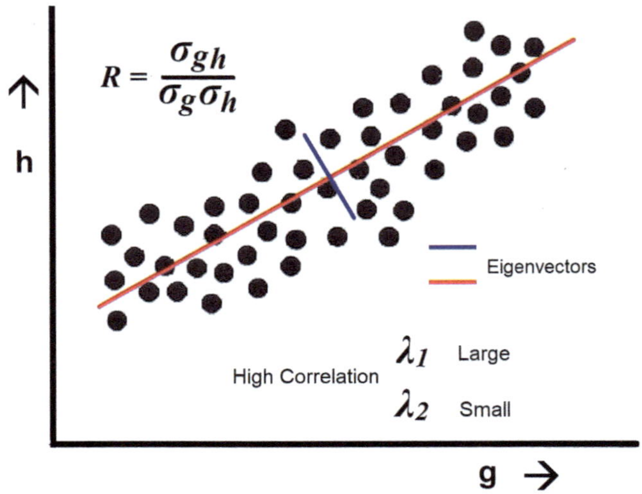

Figure 11.4 Two values (g and h) of a sample of objects plotted in a scatter diagram show high correlation, which can be quantified by the normalised cross correlation (R) and the eigenvalues of the covariance matrix

The solution reads:

$$\begin{aligned}\lambda_{1,2} &= \frac{C_{11}+C_{22}}{2} \pm \frac{\sqrt{(C_{11}+C_{22})^2 - 4(C_{11}C_{22} - C_{12}{}^2)}}{2} \\ &= \frac{1}{2}(C_{11}+C_{22}) \pm \sqrt{\left(\frac{C_{11}-C_{22}}{2}\right)^2 + C_{12}{}^2}\end{aligned} \tag{11.8}$$

The eigenvector with the largest eigenvalue points in the direction in which the variation is largest. The other eigenvector is in the direction of the smallest variation (Figure 11.4). In Section 11.5, we will meet eigenspace solutions for the detection of interest points in images.

11.3.4 Cross Correlation

The covariance of two or more data sets provides powerful information about the relationship between these data sets. Within the field of pointcloudmetry, the covariance is exploited for detecting features in images, among others. A modified form of the covariance matrix is used to quantify similarity in two corresponding image patches for image matching purposes. This measure is normalised by dividing the covariance by the product of the standard deviations of the two data sets. This operation results in a normalised, dimensionless number in the range −1 to +1. Values close to 1 indicate good similarity, values around 0 indicate no or very little similarity and when the values reach −1 the correlation is negative. With respect to an image, the latter would mean that a positive image is correlated with its negative. The normalised cross correlation (R) of two image patches is defined as:

$$R = \frac{\sigma_{gh}}{\sigma_g \sigma_h} = \frac{\sum_{i \in P} (g_i - \bar{g})(h_i - \bar{h})}{\sqrt{\sum_{i \in P} (g_i - \bar{g})^2} \cdot \sqrt{\sum_{i \in P} (h_i - \bar{h})^2}} \tag{11.9}$$

Note that the division operation eliminates the factor $1/(n - 1)$.

11.3.5 Least-Squares Adjustment

In the field of pointcloudmetry, it frequently happens that a model of the relationship between two or more variables is known and that the coefficients of that model have to be estimated. An example is the fitting of a plane or a second-order polynomial to a neighbourhood of points. A model description of the relationship of the 3D coordinates of points lying on a plane reads as $\alpha X + \beta Y + \gamma Z + \delta = 0$. The task is now to determine the coefficients α, β, γ and δ, from points lying on that plane. In vector notation a linear relationship reads as:

$$\boldsymbol{x} = \boldsymbol{A}\boldsymbol{y} \tag{11.10}$$

where **x** is a vector containing the observations, **y** the vector of unknown coefficients which have to be determined and **A** a matrix which connects the unknown coefficients to the measurements. This matrix is usually called a design matrix. Often there will be more observations than unknown parameters. Throwing away redundant data is no option. By doing so, we would give out of our hands the possibility to detect outliers, improve the accuracy of the fit and also obtain a quality assessment of the goodness of the fit. The solution is to apply least-squares adjustment (LSA), which is an optimisation technique used for solving many engineering and statistical problems. Since there are more observations than unknowns, the system of equations will be overdetermined. Without taking provisions, a unique solution is difficult to obtain.

LSA obtains a unique solution by introducing the fundamental consideration that the measurements are done by physical instruments operating in a physical environment. It is a general experience that such types of measurement are contaminated by small deviations, called random noise. One may experience this oneself through a group experiment. One person draws a straight line on a piece of paper with a length of 12 cm. The other group members measure this distance one by one using their own rulers. The outcomes will differ, and some may even produce a strongly deviating length, an outlier. LSA copes with this problem by allowing the measurements to deviate a little bit from their actual values as measured. Mathematically we do this by introducing an error vector (**v**) of the same size as the measurement vector:

$$\boldsymbol{x} + \boldsymbol{v} = \boldsymbol{A}\boldsymbol{y} \tag{11.11}$$

To arrive at a solution, a criterion has to be put on the error vector. A lot of research has shown that the criterion introduced by Gauss provides an optimal solution, provided that the random noise field is additive and has zero mean. The criterion is that the sum of squared errors:

$$v^T v = (Ay - x)^T (Ay - x) \tag{11.12}$$

should be minimum. A function reaches an extremum where the first derivative is zero. This consideration results in the solution:

$$y = (A^T A)^{-1} A^T x \tag{11.13}$$

Some measurements may have a higher quality than other measurements. To incorporate this empirical observation, a weight matrix (**W**) may be introduced. A convenient and theoretical sound approach is to use the inverse of the covariance matrix as a weight matrix, if available, resulting in the following least-squares solution:

$$y = (A^T W A)^{-1} A^T W x = (A^T C^{-1} A)^{-1} A^T C^{-1} x \tag{11.14}$$

The nice property of LSA is that after computation of the unknown parameters, adjusted measurements can be computed by substituting the computed unknown parameters according to Equation 11.14 into Equation 11.11. From this, the error vector can be computed. In this stage the errors are often called residuals:

$$v = Ay - \tilde{x} \tag{11.15}$$

where $\tilde{x}$ are the adjusted measurements. If a residual value is large and far exceeds the range which may be expected from the known covariance matrix of the observations, the corresponding measurement is probably an outlier, which means the residual can't be explained by the characteristics of the random noise. Just removal of all residuals of which the absolute values lie above a threshold determined from the covariance matrix would neglect that the adjusted values of the unknown parameters are affected by the outliers. In order to prevent good observations from being detected as outliers or outliers not being detected, the measurements with large residuals are removed one by one starting with the measurement with the largest residual. After removal of the measurement with the largest residual, the unknown parameters are computed again but now without the outlier. This process is repeated until no outliers are detected anymore. This process is called data snooping. The application of LSA for solving the Geomatics problem is a topic in itself. A recent textbook on this subject is Ogundare (2018).

11.4 Differential Operators

In the previous section, we discussed smoothing operators in which the grey values of pixels with an image patch are averaged in some way. Another type of neighbourhood operation concerns differential operators of which the kernels contain positive and negative numbers. Now the question pops up: How to calculate the first derivative or gradient from grey values arranged in a raster, which is not a continuous representation but a discrete one. There are many ways to approach this issue. Since an image is a two-dimensional scalar field, gradients should be computed along two orthogonal directions. The most natural way to select the orthogonal directions are the rows and columns, but also the two diagonals can be chosen as orthogonal directions, which is the case for the Roberts edge detector.

11.4.1 Discrete Differentiation

At each pixel, two gradients can be computed: g_x and g_y. The most simple one and commonly accepted is to define the gradient as the difference in grey values between two adjacent pixels, i.e. along the row and column directions the discrete first derivatives become:

$$
\begin{aligned}
g_x(i,j) &= g(i,j) - g(i+1,j) \\
g_y(i,j) &= g(i,j) - g(i,j+1)
\end{aligned}
\tag{11.16}
$$

which can be described as convolving the image with the kernels:

$$
G_x = G * \begin{bmatrix} 1 & -1 \end{bmatrix}; \quad G_y = G * \begin{bmatrix} 1 \\ -1 \end{bmatrix}
\tag{11.17}
$$

By fitting two perpendicular planes through the local grey value function and discretising the least-squares solution in a 3 × 3 kernel or mask, Prewitt (1970) designed the edge detector named after her. The two 3 × 3 kernels are convolved with the image (G) resulting in two discrete derivatives G_x and G_y, where G_x indicates the vertical intensity changes and G_y the horizontal intensity changes.

$$
\begin{aligned}
G_x &= G * \begin{bmatrix} +1 & 0 & -1 \\ +1 & 0 & -1 \\ +1 & 0 & -1 \end{bmatrix} \\
G_y &= G * \begin{bmatrix} +1 & +1 & +1 \\ 0 & 0 & 0 \\ -1 & -1 & -1 \end{bmatrix}
\end{aligned}
\tag{11.18}
$$

where * denotes convolution. The working of the Prewitt operator can also be interpreted as first smoothing the image with 3 × 2 and 2 × 3 asymmetric unweighted smoothing kernels and next applying the elementary differential operators defined in Equation 11.17:

$$
\begin{bmatrix} +1 & 0 & -1 \\ +1 & 0 & -1 \\ +1 & 0 & -1 \end{bmatrix} = \begin{bmatrix} 1 & 1 \\ 1 & 1 \\ 1 & 1 \end{bmatrix} * \begin{bmatrix} 1 & -1 \end{bmatrix} \qquad \begin{bmatrix} +1 & +1 & +1 \\ 0 & 0 & 0 \\ -1 & -1 & -1 \end{bmatrix} = \begin{bmatrix} 1 & 1 & 1 \\ 1 & 1 & 1 \end{bmatrix} * \begin{bmatrix} 1 \\ -1 \end{bmatrix}
\tag{11.19}
$$

Edge detection, using for example the Prewitt or Sobel operator, is a well-known image processing operator. Neighbourhood operations are usually implemented as moving one or more convolution kernels over the points.

The gradient magnitude or edge strength ($M(i,j)$) can be computed by combining the two responses, $g_x(i,j)$ and $g_y(i,j)$, using the Pythagorean distance computation in Euclidian space:

$$M(i,j)=\sqrt{g_x^2(i,j)+g_y^2(i,j)} \tag{11.20}$$

The two responses G_x and G_y provide also information about the direction of the gradient with respect to the underlying raster:

$$\theta(i,j)=\mathrm{a}\tan\frac{g_y(i,j)}{g_x(i,j)} \tag{11.21}$$

The two kernels of the Sobel operator deviate only slightly from those of the Prewitt operators. The two 3 × 3 kernels are convolved with the image (G) resulting in two discrete derivatives G_x and G_y, where G_x indicates the horizontal intensity changes and G_y the vertical intensity changes.

$$\begin{aligned} G_x &= G * \begin{bmatrix} +1 & 0 & -1 \\ +2 & 0 & -2 \\ +1 & 0 & -1 \end{bmatrix} \\ G_y &= G * \begin{bmatrix} +1 & +2 & +1 \\ 0 & 0 & 0 \\ -1 & -2 & -1 \end{bmatrix} \end{aligned} \tag{11.22}$$

In terms of the discrete differential kernels defined in Equation 11.17, the Sobel operator can be described as first carrying out a smoothing in both column and row directions as follows:

$$\begin{bmatrix} +1 & 0 & -1 \\ +2 & 0 & -2 \\ +1 & 0 & -1 \end{bmatrix} = \begin{bmatrix} 1 & 1 \\ 2 & 2 \\ 1 & 1 \end{bmatrix} * \begin{bmatrix} 1 & -1 \end{bmatrix} = \begin{bmatrix} 1 \\ 1 \end{bmatrix} * \begin{bmatrix} 1 \\ 1 \end{bmatrix} * \begin{bmatrix} 1 & 1 \end{bmatrix} * \begin{bmatrix} 1 & -1 \end{bmatrix} \tag{11.23}$$

The Sobel kernel for calculating G_y has a similar convolution scheme. This scheme also shows that the image is smoothed by a 3 × 2 weighted smoothing function. The outcome should be divided by eight because the sum of the values within the smoothing kernel, which is basically an averaging, is eight. This would give the actual gradient value. However, in practice, this division by eight is rarely done, because of computational reasons. When using this operator for edge detection purposes, the threshold to determine whether the pixel lies on an edge or not can be adjusted accordingly. To obtain the exact gradient value of the Prewitt operator, the response to each of the two kernels should be divided by six. But this also is never done, because of the reason mentioned above.

Other orthogonal differential operators are the Roberts cross with kernels:

$$G_x = G * \begin{bmatrix} 0 & -1 \\ 1 & 0 \end{bmatrix} \quad G_y = G * \begin{bmatrix} -1 & 0 \\ 0 & 1 \end{bmatrix} \tag{11.24}$$

And Frei–Chen operator, with kernels:

$$G_x = G * \begin{bmatrix} +1 & 0 & -1 \\ +\sqrt{2} & 0 & -\sqrt{2} \\ +1 & 0 & -1 \end{bmatrix} \quad G_y = G * \begin{bmatrix} +1 & +\sqrt{2} & +1 \\ 0 & 0 & 0 \\ -1 & -\sqrt{2} & -1 \end{bmatrix} \tag{11.25}$$

11.4.2 Edge Detection

The differential operators, discussed earlier, are primarily applied for edge detection purposes. Edges are sudden discontinuities in the grey value distribution which may be caused by the presence of different adjacent objects in the scene. Edge detection of images is carried out by moving two differential kernels over the image in row and column directions and combining the two responses according to Equations 11.20 and 11.21 (Figure 11.5). The Prewitt operator and Sobel operator provide similar responses, but the latter has slightly better noise reduction abilities and therefore is often preferred over the Prewitt or Frei–Chen operator. These differential operators are often used for edge detection purposes by thresholding the magnitude (edge strength) of each pixel, resulting in a binary image, the edges represented by one and the non-edges by zero.

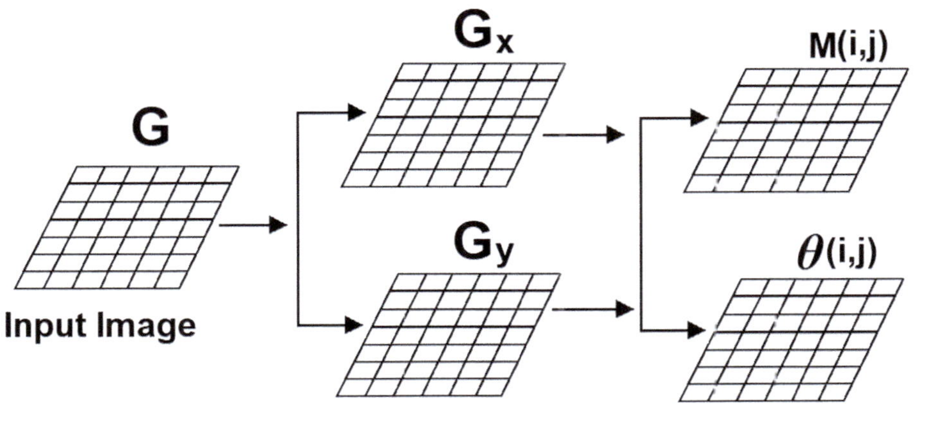

Figure 11.5 Diagram demonstrating the pipeline of using differential operators acting as edge detectors

The two responses of the orthogonal differential operators are typically developed for a boundary detection problem consisting of four stages (Lemmens, 1996a):

1. Smoothing to reduce the effects of noise
2. Local edge detection
3. Thresholding and thinning
4. Edge linking

Thresholding is a decision process in which the label edge or non-edge is assigned to each pixel based on the response of the local edge detector. The choice of the threshold affects errors of omission and errors of commission. This means the setting of the threshold determines the number of spurious edges, i.e. pixels which receive the label edge wrongly and the number of actual edge pixels which are rejected as lying on an edge. Thinning is necessary to obtain one-pixel-thick edges. Finally, the edges are linked to form a boundary of connected pixels that may be generalised and vectorised. This approach works well in highly conditioned scenes such as factories, where products lying on a conveyor belt have to be checked and the illumination conditions can be optimally controlled. However, for images of complex scenes with unconditioned illumination, such as aerial and satellite images, the approach fails to find closed boundaries. Often the results consist of many fragmented line elements.

The basic assumption of this type of local edge detector is that abrupt grey value changes in the image correspond to meaningful object boundaries in the scene. However, intensity changes may also occur where no boundaries are present, because of, for example, noise, texture and shadow. Furthermore, relevant object boundaries may produce no or little intensity changes. As a consequence, the results are poor and unreliable. To tackle this problem, Lemmens (1996a) introduced the concept of structure-based edge detection. This concept is based on exploiting local radiometric and spatial structures, based on non-parametric statistics. Within the research, a local edge detector has been developed to explore the local radiometric structure. The basic idea underlying the design of this edge detector is that the homogeneity of the areas at both sides of a hypothesised edge is, besides contrast, a measure of the presence of an edge: The higher the homogeneity, the lower the contrast has to be. In this way, low contrast boundaries can be traced, while incorrect detection of edges, due to noise and texture, remains limited. The edge detector developed is based on non-parametric statistics and performs uniformly well on a wide range of image characteristics. Experiments have shown that this operator has strong capabilities and desirable properties for carrying out the first step of the boundary outlining problem of outdoor scenes. The advantage of this operator is that it automatically adapts to the contrast differences within a window. The operator has never found wide acceptance probably due to the computational burden non-parametric statistics require. The most widely used edge detector has been developed by Canny (1986). A survey of boundary delineation methods is provided in Lemmens (1996b).

11.5 Interest Operators

The naming of operators, such as edge detection operators and interest operators, is not primarily induced by the mathematics underlying the design of the operator but by the aim they have been designed for. Edge detectors can also detect corners (L-junctions), T-junctions, blobs and so on, but they are unable to discriminate among these features. Detection of corresponding points in images taken from a moving platform or photogrammetric images should be based on reliable and discriminative features. Edges do not belong to these features, because correspondence can be found along many places of the boundary (Figure 11.6). To cope with this problem, operators have been designed which are able to discriminate edges from corner-like features. Operators which are able to detect features with preferred properties are called interest operators, key point operators, salient point detectors, significant point detectors or corner detectors. For obtaining reliable matches in overlapping images, the salient points should be preferably located at T-junctions, corners, blobs and the like features (Figure 11.7).

Within the field of robotics, researchers aim at designing vehicles, which move through an unknown environment autonomously. This research area started in the 1960s and resulted half a century later in the hype of self-driving cars. Equipped with cameras – laser scanners were not yet available in the shop selves at that time – the vehicles moved and "saw" the road and its environment through one or more cameras producing a sequence of images. To predict the path to follow, the images, which were taken while the robot moved along the path, had to be stitched together. It was early recognised that the best way to do this was by extracting characteristic points or landmarks in the images recorded while the vehicle was moving. This approach not only tremendously reduced the data content of the image but also allowed for quick matching

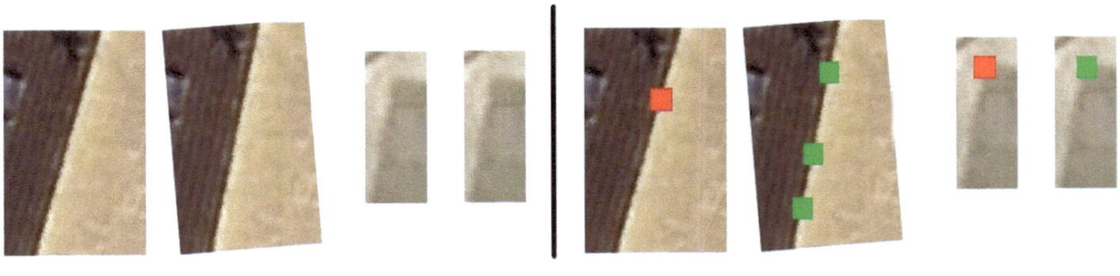

Figure 11.6 Edges are unsuited for matching; the corresponding point can lie all along the edge; corners are well-suited interest points for detecting corresponding points

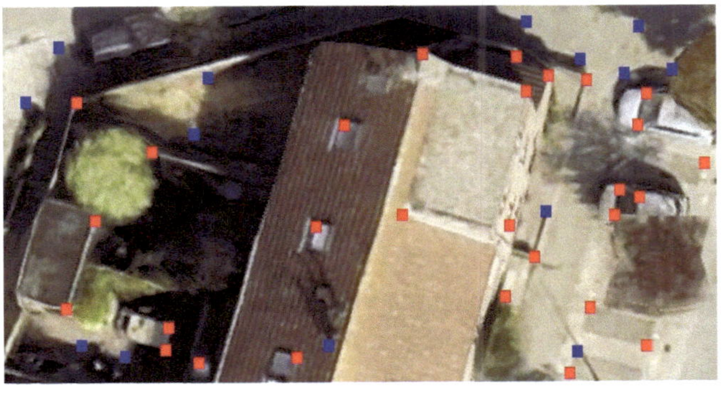

Figure 11.7 Sample of points detected with an interest operator; also shadow corners are detected (blue squares)

of the images. At that time, data reduction was very important because of the limited computation speed and storage capabilities of computers. Here the term structure from motion, usually abbreviated to SfM, comes from a term which has been adopted by the photogrammetric community, although in this discipline the aim is not to guide a vehicle in motion autonomously through an environment but to create 3D maps of the real world. Since the images were taken from a vehicle in motion, the orientation of the images will differ, sometimes significantly. The interest points should be detected even when the viewing perspective changes, which means they should be rotation invariant. So, it appeared to be important to design operators which could not only discriminate among different feature types but also had the property to be rotation invariant. Next, descriptors or attributes have to be assigned to each interest point, which is exploited to determine the similarity of a point in the one image and candidate points in the overlap. The descriptors are computed from the grey values of the image patch of the interest point and should be distinct or discriminative, robust to noise and invariant to geometric and radiometric transformations.

11.5.1 Moravec Operator

Starting from this point of beginning discussed above, Moravec (1977, 1981) developed an operator to detect interest points. The basic idea of this operator can be formulated as follows. Slide a strong magnifying glass over the image so that only small patches are clearly visible. If an edge is present within the patch, moving the lens across the edge will pop up as high-intensity variation. Sliding along the edge no intensity variation will be visible. At a corner or blob, the intensity variation is large in any direction. The Moravec operator implements this idea by evaluating the variances, defined as the sum of the squared grey value differences between adjacent pixels, in the four main directions (rows, columns and two diagonals) in an image patch, usually a 5×5 window. The operator produces four responses within a 5×5 window:

$$
\begin{aligned}
M_1(i,j) &= \frac{1}{20}\sum_{k=-2}^{2}\sum_{l=-2}^{1}\left[g(i+k,j+l)-g(i+k,j+l+1)\right]^2 \\
M_2(i,j) &= \frac{1}{20}\sum_{k=-2}^{1}\sum_{l=-2}^{2}\left[g(i+k,j+l)-g(i+k+1,j+l)\right]^2 \\
M_3(i,j) &= \frac{1}{16}\sum_{k=-2}^{1}\sum_{l=-2}^{1}\left[g(i+k,j+l)-g(i+k+1,j+l+1)\right]^2 \\
M_4(i,j) &= \frac{1}{16}\sum_{k=-2}^{1}\sum_{l=-2}^{1}\left[g(i+k,j+l+1)-g(i+k+1,j+l)\right]^2
\end{aligned}
\tag{11.26}
$$

The minimum of the four responses should exceed a threshold to receive the label: interest point. When applying this operator to each and every pixel (i,j) of the image, at the neighbourhood of an interest point, a clutter of interest points will appear which is undesirable because of the computational burden it introduces in the follow-up processing steps as well as it may result in unreliable matches due to ambiguity. During his experiments, Moravec recognised that the interest operator was not entirely rotation invariant as he had hoped. This was due to discretisation effects, which means the evaluation of grey value differences on a raster. An edge which is oriented in between the four principal directions (horizontal, vertical and two diagonals) will produce high responses. Since the minimum of these responses will be large the edge will be incorrectly detected as an interest point. The Moravec operator is easy to implement but has several shortcomings. In addition to being non-rotational invariant its ad hoc character has two other drawbacks (Dreschler, 1981):

- Not the real corner is found, but a shift is introduced, the larger the size of the operator, the larger the shift.
- The operator is sensitive to low-resolution features, i.e. small points cause an extended operator response.

11.5.2 Marr-Hildreth Operator

Based on the human visual system, Marr and Hildreth (1980) developed an operator for detecting abrupt changes in grey values in images. This edge detection operator is still widely used in computer vision applications, although often in a modified form. This operator has not been specifically developed for detecting corners and other interest points. However, this operator lies on the basis of other interest operators, such as the Scale-Invariant Feature Transform (SIFT) operator which will be outlined in Section 11.5.5. The basic idea underlying this operator designed for the detection of edges in an image is to accurately detect the location of edges. This is done by the use of second derivatives, in particular the rotation-invariant Laplace operator.

The Laplacian applied to a discrete function f with n variables $x_i = 1, \ldots, n$ is defined as:

$$\Delta f = \sum_{i=1}^{n} \frac{\partial^2 f}{\partial x_i^2} \tag{11.27}$$

For a two-dimensional image, in which differentiation takes place along the row and column directions which are orthogonal, the Laplacian becomes:

$$\Delta G(x, y) = \frac{\partial^2 G}{\partial x^2} + \frac{\partial^2 G}{\partial y^2} \tag{11.28}$$

The first derivatives taken along the row and column directions have been defined by Equation 11.16 and can be described as convolving the image with the kernels:

$$G_x = G * \begin{bmatrix} 1 & -1 \end{bmatrix} \qquad G_y = G * \begin{bmatrix} 1 \\ -1 \end{bmatrix}$$

The second derivative is obtained by applying the above two convolution kernels a second time resulting in:

$$G_{xx} = G * \begin{bmatrix} 1 & -2 & 1 \end{bmatrix} \qquad G_{yy} = G * \begin{bmatrix} 1 \\ -2 \\ 1 \end{bmatrix} \tag{11.29}$$

Inserting Equation 11.29 into Equation 11.28 results in the kernel of the Laplace operator:

$$\Delta G = G * \begin{bmatrix} 0 & 1 & 0 \\ 1 & -4 & 1 \\ 0 & 1 & 0 \end{bmatrix} \tag{11.30}$$

In contrast to the first derivatives which are able to detect peaks at the location of edges, the Laplacian produces a zero response located in between a positive and a negative peak at the location of the edge (Figure 11.8). By interpolation between the positive and negative peaks of the response, the location of the zero crossing can be found.

The detection of features based on second derivatives is sensitive to noise. To reduce the effect of noise, the computation is usually combined with a Gaussian smoothing kernel. Formally, the Gaussian smoothing kernel is derived from the 2D Gaussian function:

$$G(x, y) = \frac{1}{2\pi\sigma^2} e^{-\frac{x^2+y^2}{2\sigma^2}} \tag{11.31}$$

The size of the kernel is determined by the parameter σ. Applying Gaussian smoothing – some may prefer to call it blurring – to a digital image corresponds to convolving the image with a kernel derived from the 2D Gaussian function. For $\sigma = 0.8$ and a kernel size of 3×3, the Gaussian smoothing kernel becomes:

$$\frac{1}{16} \begin{bmatrix} 1 & 2 & 1 \\ 2 & 4 & 2 \\ 1 & 2 & 1 \end{bmatrix} = \frac{1}{16} \begin{bmatrix} 1 \\ 2 \\ 1 \end{bmatrix} * \begin{bmatrix} 1 & 2 & 1 \end{bmatrix} \tag{11.32}$$

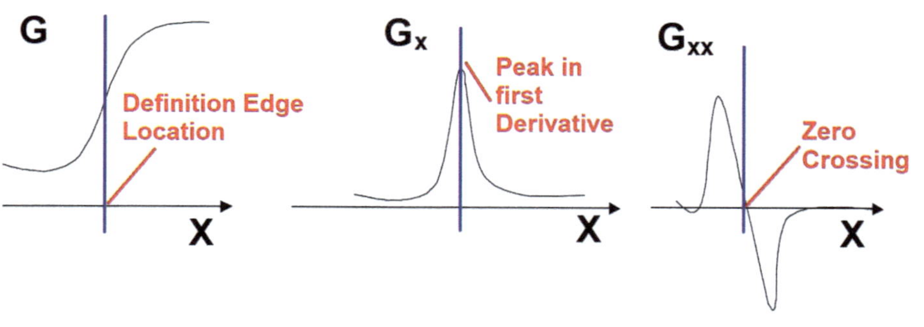

Figure 11.8 Response of the first derivative on an image edge is a peak, and a second derivative operator determines zero crossings

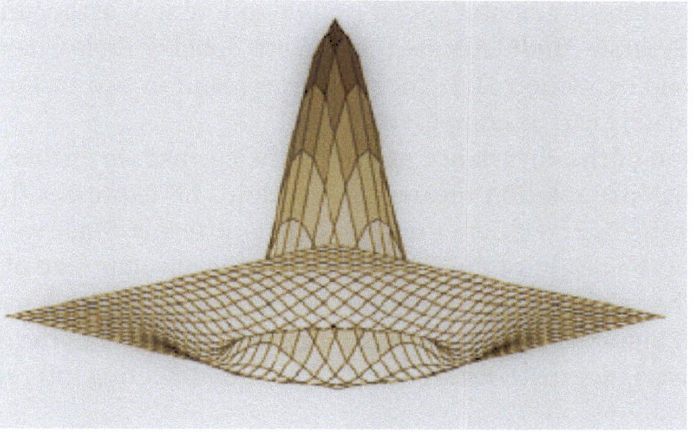

Figure 11.9 Marr-Hildreth operator or Laplacian of Gaussian (LoG), turned upside-down, resembles a Mexican hat

To apply the Gaussian function on a digital image, a discretisation has to be carried out to produce a kernel which can be convolved with the image. The values within a kernel will not sum up to one and therefore the outcome has to be divided by the sum of the individual values within the kernel. The use of a Gaussian kernel will also weigh the influence of the first or second derivatives with distance. The larger away from the pixel under consideration, the lower the influence. To save computation time, the implementation of Gaussian smoothing can be conducted by convolving the image with two 1D Gaussian functions in the vertical and horizontal directions, the one after the other, as shown in Equation 11.32. The larger the 2D kernel, the more computational time efficiency will be gained.

Since the operator is based on convolving a Laplace operator over a Gaussian smoothed image, the Marr-Hildreth operator is also often called Laplacian of Gaussian (LoG). Because of its shape, the LoG operator is also called the Mexican hat operator (Figure 11.9).

The operator is computationally expensive and therefore often replaced by the difference between two Gaussian smoothed versions of the same image. One Gaussian kernel is smaller than the other kernel. This operation is called Difference of Gaussians (DoG).

The Laplacian is isotropic, which means the output of convolving this operator with an image is independent of the orientation of the features such as edges with respect to the raster. As a result, the output is insensitive to the orientation of the camera with respect to the scene. In image processing, the Laplace operator is often used for enhancing edges. The SIFT interest operator, developed within computer vision, is based on the LoG properties but is implemented as DoG for computational reasons.

11.5.3 Förstner–Harris Operator

To improve the rotation invariance of the Moravec operator, an interest operator has been developed based on computing the gradients along the two orthogonal directions defined by the raster, which means along the column direction and row direction. For the computation of the gradients (g_x and g_y), two differential orthogonal operators have to be exploited. Feasible candidates are the Prewitt, Sobel or Frei–Chen operators, which have been outlined in Section 11.4. The responses result in two raster outputs of which the covariance matrix can be computed.

Transformation of the covariance matrix to the eigenspace enables the computation of measures which are rotation invariant. Equation 11.8 expresses the computation of the eigenvalues of a 2 × 2 covariance matrix in generic terms. Since we are not interested in the location of the origin of the eigenspace nor the absolute size of the eigenvectors, but only in the shape of the eigenspace and the strength of the response, the subtraction of the means in Equations 11.2 and 11.5 can be omitted as well as the division by $(n - 1)$. The modified covariance matrix suited as the basis for detecting interest points becomes accordingly:

$$\boldsymbol{C} = \begin{pmatrix} \Sigma g_x^2 & \Sigma g_x g_y \\ \Sigma g_x g_y & \Sigma g_y^2 \end{pmatrix} \tag{11.33}$$

The eigenvalues of this covariance matrix can be computed according to Equation 11.8 resulting in:

$$\lambda_{1,2} = \frac{1}{2}\left(\Sigma g_x^2 + \Sigma g_y^2\right) \pm \sqrt{\left(\frac{\Sigma g_x^2 - \Sigma g_y^2}{2}\right)^2 + \left(\Sigma g_x g_y\right)^2} \tag{11.34}$$

The above approach has been independently developed in the field of digital photogrammetry by Förstner (1986) and in computer vision by Chris Harris (Harris and Stephens, 1988) more or less at the same time. This is not astonishing since, in the late 1970s and the early 1980s, researchers got confronted with the problem to identify corresponding points in the overlaps of digital images. A number of researchers have turned to this problem, resulting in among others the interest operators outlined in this section. As stated earlier, in computer vision the aim was primarily to let a robot move autonomously through an environment using image sequences. In photogrammetry the aim was the accurate georeferencing of images and the creation of digital elevation models.

It should be noted that the formulation of the covariance matrix or structure tensor in Equation 11.33 is according to the formulation of Wolfgang Förstner. Harris proposed a slightly different computation. The squares of the two differential operators and their cross products lying within a window are not summed up but are combined by using a Gaussian smoothing operator. This may be interpreted as a weighted adding within a neighbourhood: The farther away from the centre of the point under consideration, the smaller the influence of the responses will be on the further processing. From this vision, the Förstner operator may be regarded as combining the squares of the two differential operators and their cross products within a window using an unweighted smoothing operator.

The computation of the eigenvalues can be interpreted as fitting an ellipse through the scatter plot of the responses of the two orthogonal differential filters (Figure 11.10).

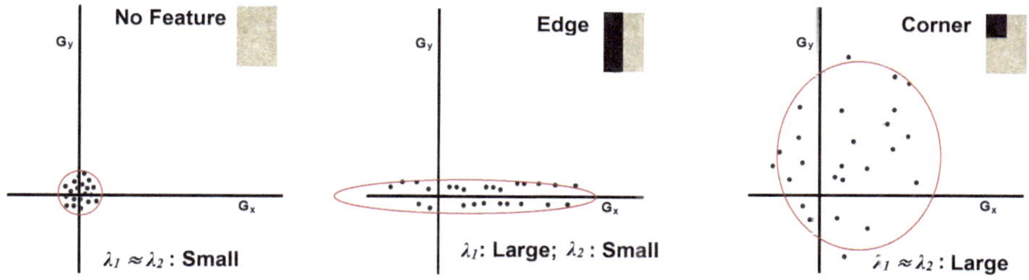

Figure 11.10 Computation of eigenvalues corresponds to fitting an ellipse through the scatter plot of the responses of the two orthogonal differential filters

Examining the eigenvalues, arranged from large to small, i.e. $\lambda_1 \geq \lambda_2$, provides a clue about the presence of a corner. This will be the case when the values of λ_1 and λ_2 lie close together. It is possible to normalise the response by taking the ratio between the difference of the two eigenvalues and their sum:

$$Q = \frac{\lambda_1 - \lambda_2}{\lambda_1 + \lambda_2} \tag{11.35}$$

A corner or other junction type suited to act as an interest operator will be present if Q is close to 0. To avoid the acceptance of weak feature strength responses, the response has to be checked against a threshold. Both Förstner and Harris use the following feature strength response measure:

$$w = \frac{\lambda_1 \lambda_2}{\lambda_1 + \lambda_2} = \frac{\sum g_x^2 \sum g_y^2 - \left(\sum g_x g_y\right)^2}{\sum g_x^2 + \sum g_y^2} \tag{11.36}$$

The feature strength value should exceed a predefined threshold. The explicit computation of the eigenvalues can be avoided by introducing domain-specific measures. To distinguish corners from edges, Förstner and Gülch (1987) introduced the following roundness measure:

$$q = 1 - \left[\frac{\lambda_1 - \lambda_2}{\lambda_1 + \lambda_2}\right]^2 = 4\frac{\lambda_1 \lambda_2}{\left(\lambda_1 + \lambda_2\right)^2} = 4\frac{\sum g_x^2 \sum g_y^2 - \left(\sum g_x g_y\right)^2}{\left[\sum g_x^2 + \sum g_y^2\right]^2} \tag{11.37}$$

Inserting Equation 11.36 into Equation 11.37 results in:

$$q = \frac{4w}{\sum g_x^2 + \sum g_y^2} \tag{11.38}$$

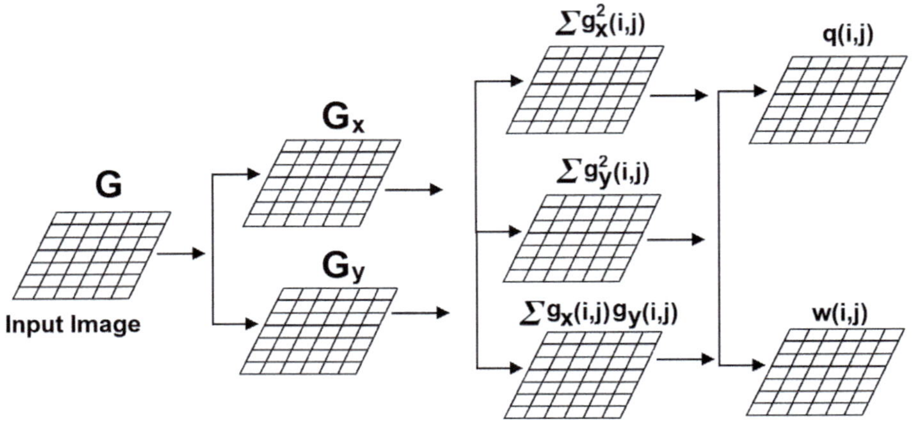

Figure 11.11 Diagram demonstrating the pipeline of the Förstner–Harris interest operator

The values of q range from 0 to 1. If q lies close to 1, a corner or other junction type has been detected. If q lies close to 0, the operator has detected an edge. Figure 11.11 outlines the pipeline of the Förstner–Harris operator.

11.5.4 Dreschler–Nagel Operator

Another way to look at an intensity distribution over 2D space is to treat it as a curved 2D surface in 3D space. Such a view of the grey value function of an image allows exploration of the basics of differential geometry. This view has been explored and investigated by Dreschler (1981) and Dreschler and Nagel (1982, 1983). From differential geometry, it is known that the Gaussian curvature ($K = k_1 \times k_2$) and thus the principal curvatures (k_1 and k_2) are invariant against rotation, and such a measure is very appropriate for interest points as we have seen above. The curvature can be computed by computing the second derivatives of a function. For an image, the second derivatives are G_{xx}, G_{yy} and G_{xy}. Since just the kind of curvature and not its absolute value is of interest, the principal curvatures can be found from an eigenvalue analysis of the Hessian, which is a symmetric matrix:

$$\boldsymbol{H} = \begin{bmatrix} G_{xx} & G_{xy} \\ G_{xy} & G_{yy} \end{bmatrix} \tag{11.39}$$

The second derivatives are obtained by convolving the image with the three kernels as follows:

$$G_{xx} = G * \begin{bmatrix} 1 & -2 & 1 \end{bmatrix} \quad G_{yy} = G * \begin{bmatrix} 1 \\ -2 \\ 1 \end{bmatrix} \quad G_{xy} = G * \begin{bmatrix} 1 & -1 \\ -1 & 1 \end{bmatrix} \tag{11.40}$$

The above kernels can be extended to 3 × 3 kernels:

$$\begin{bmatrix} +1 & -2 & +1 \\ +1 & -2 & +1 \\ +1 & -2 & +1 \end{bmatrix} \begin{bmatrix} +1 & +1 & +1 \\ -2 & -2 & -2 \\ +1 & +1 & +1 \end{bmatrix} \begin{bmatrix} +1 & 0 & -1 \\ 0 & 0 & 0 \\ -1 & 0 & +1 \end{bmatrix}$$

Second derivatives are not only sensitive to grey value jumps in the image but also to noise. Therefore, it is appropriate to apply a Gaussian smoothing on the image prior to calculating the second derivatives. The principal curvatures k_1 and k_2 of a curved surface in 3D space can be calculated by solving the quadratic form:

$$\kappa^2 - \kappa\left(G_{xx} + G_{yy}\right) + G_{xx}G_{yy} - G_{xy}^2 = 0 \tag{11.41}$$

resulting in the solution:

$$\begin{aligned} \kappa_{1,2} &= \frac{1}{2}\left[G_{xx} + G_{yy} \pm \sqrt{\left(G_{xx} + G_{yy}\right)^2 - 4\left(G_{xx}G_{yy} - G_{xy}^2\right)}\right] \\ &= \frac{G_{xx} + G_{yy}}{2} \pm \sqrt{\left(\frac{G_{xx} - G_{yy}}{2}\right)^2 + G_{xy}^2} \end{aligned} \tag{11.42}$$

- If k_1 and k_2 are both small, only little grey value differences or random texture will be present.
- If k_1 is large and k_2 small, the type of curvature indicates the presence of an edge.
- If $k_1 \approx k_2$ and both are large, the type of curvature indicates the presence of a corner-like junction or blob type of feature.

Note that the computation of the two principal curvatures exactly corresponds to the computation of the eigenvalues as outlined in Equation 11.8. Figure 11.12 outlines the pipeline of the Dreschler–Nagel operator.

11.5.5 SIFT Operator

Probably the most widely used interest operator is the Förstner–Harris operator. This operator and Dreschler–Nagel operator are invariant to the orientation of the image with respect to the scene (rotation-invariant), which is a considerable improvement compared to the Moravec interest operator. Under different rotations of the camera, the same features are detected. However, these operators are not scale invariant. In one image, a feature may be visible in a corner, but at a larger scale, it may appear as a curved edge within the image patch (Figure 11.13). In conventional airborne photogrammetry, scale invariance is not a big issue since the overlapping images are usually recorded with the same (metric) camera at approximately the same flight height and more or less parallel to each other. The latter means that the optical axis of the camera approximately points in the vertical direction. However, scale invariance is becoming

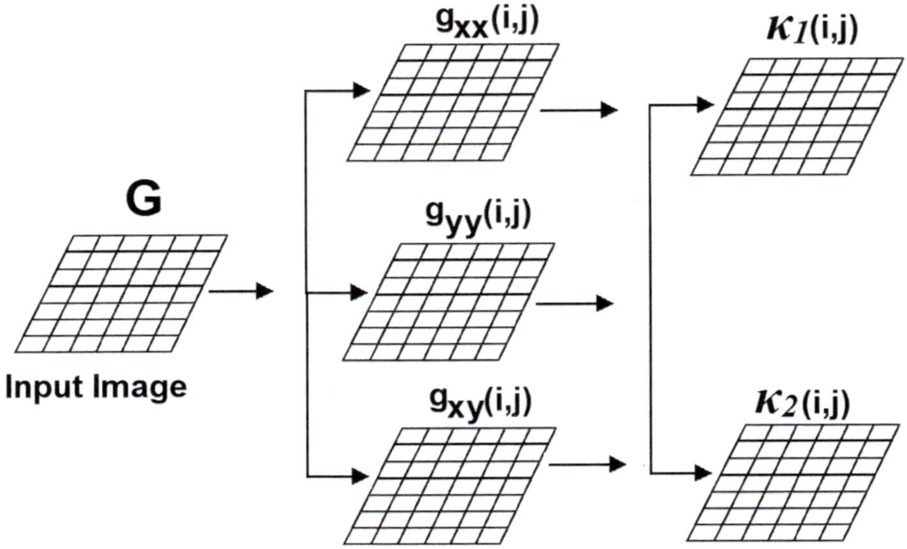

Figure 11.12 Diagram demonstrating the pipeline of the Dreschler–Nagel interest operator

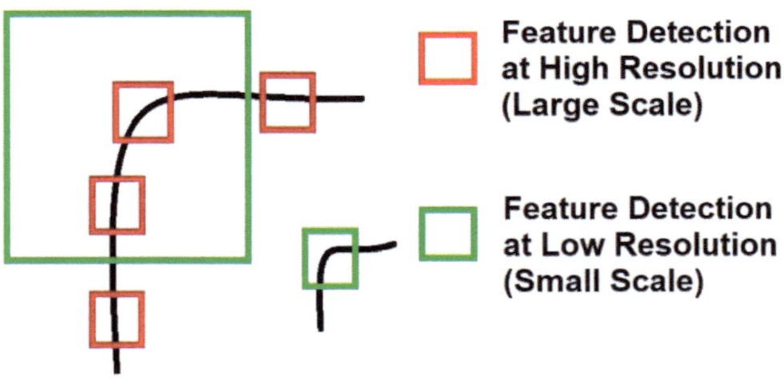

Figure 11.13 Corner detection at high resolution (large scale) and low resolution (small scale)

increasingly important in photogrammetry because of the use of oblique aerial cameras and the use of cameras mounted on unmanned airborne systems (UASs), which capture objects from a diversity of orientations.

To achieve scale invariance, Lowe (1999, 2004) developed an empirical approach, called Scale-Invariant Feature Transform (SIFT). It is called a transformation because the operator transforms the image into a set of features, which are invariant to image translation, rotation and scaling. The operator can also partially handle illumination changes and projective distortions. The basic idea for solving the scale issue is the processing of the image with interest operators of different sizes. Because of its rotation-invariant property, the LoG is best suited for this purpose. Instead of changing the size of the LoG operator, the size of the image is reduced by constructing an image pyramid with at its base the original, full-size image and at subsequent higher levels images generated from uniting $n \times n$ pixels, where n is, for example, 2 (Figure 11.14). This may be repeated until an image of just one pixel remains. A general scheme to generate an

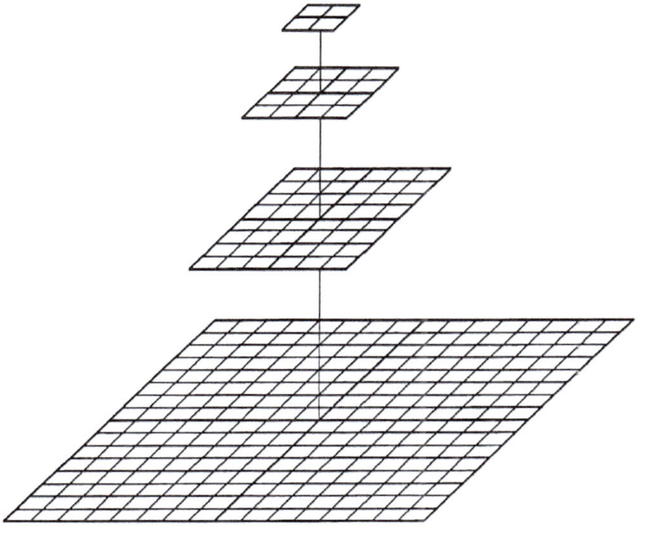

Figure 11.14 Schematic representation of the structure of a base-two image pyramid; the scale is reduced successively by a factor of two

image pyramid is by convolving the image with a Gaussian smoothing kernel. Each next lower-resolution level of the pyramid is obtained by a weighted average of the previous level followed by a resampling. The sequence of increasingly blurred images at a lower resolution is called scale space, of which the concept has been introduced by Marr and Hildreth (1980) for edge detection purposes. They state:

> A major difficulty with natural images is that changes can and do occur over a wide range of scales. No single filter can be optimal simultaneously at all scales, so it follows that one should seek a way of dealing separately with the changes occurring at different scales. This leads to the basic idea, in which one first takes local averages of the image at various resolutions and then detects the changes in intensity that occur at each one. To realise this idea, we need to determine (a) the nature of the optimal smoothing filter, and (b) how to detect intensity changes at a given scale.

Scale-space properties have been intensively investigated by Koenderink (1984) and Lindeberg (1994). The convolution of an image with the Marr-Hildreth or LoG operator is computationally costly. As a feasible alternative, the image is pre-processed by Gaussian smoothing with different sigma, resulting in a sequence of images which are increasingly blurred. In other words, instead of using an LoG operation an approximation of the LoG is used based on subtracting the output of differing Gaussian smoothed versions of the image. This Difference of Gaussians (DoG) operation is used to detect potential interest points, which are invariant to the orientation of the image plane with respect to the scene. This is the first stage of the four stages of the SIFT operator. The output of the first stage is a set of candidate interest points – called key points – which have to be further scrutinised. In order to improve accuracy, the precise locations of the interest points are determined in the second stage. Up to the second stage, no difference has been made between edges and corners, the latter being the actual points of interest. The detection of corners on other relevant junctions is carried out by the Dreschler–Nagel operator, which in turn is based on the Hessian matrix, from which the curvature can be determined. The explicit calculation of the two principal curvatures (see Equation 11.41) is avoided by introducing an alternative measure:

$$\frac{\left(G_{xx}+G_{yy}\right)^2}{G_{xx}G_{yy}-\left(G_{xy}\right)^2}<\frac{\left(r+1\right)^2}{r} \tag{11.43}$$

where r is the ratio between the largest principal curvature and the smallest one. The point is accepted as an interest point if the above criterion is met. This measure allows defining the decision criterion by the explicit definition of the ratio between the two principal curvatures. Lowe (2004) suggests a ratio of 10. Next, the orientations of the gradients of the pixels within the image patch of the interest point are computed using Equation 11.21. This is followed by plotting the orientations in a compass card histogram uniformly divided into segments, for example, 36 segments. The orientations are weighted according to their gradient magnitude, which is computed according to Equation 11.20 and the distance of the concerning pixel to the interest point. The latter is done by a Gaussian decaying weighting. The maxima detected in the histogram are identified as interest points suited for further processing. The last stage consists of the generation of a feature descriptor necessary for finding the corresponding points in other images. This is done by defining a 16 × 16 image patch around the interest point containing the weighted orientations calculated in the previous stage. This 16 × 16 image patch is subdivided into 16 4 × 4 windows, each window containing 16 pixels. Per 4 × 4 window, the orientations are mapped into a compass card subdivided into eight segments, i.e. each segment covers 45°. Figure 11.15 sketches this process. The subdivision of the 16 × 16 image patch into 4 × 4 windows has been empirically determined. In total there are 4 × 4 × 8 orientation segments, which are grouped into a vector of length 128 in a fixed order.

The SIFT operator and its modifications have been extensively applied for feature-based image matching applications (see Chapter 4). The choices made in the design of the SIFT operator heavily originate from computational efficiency considerations. The operator has been widely used in image mosaicking, tie-point matching of overlapping images and object recognition. The computation of the 128 descriptors is costly, which impedes its real-time use in, for example, simultaneous localisation and mapping or real-time object tracking (Miksik and Mikolajczyk, 2012). In the course of time, many modifications of the SIFT operator have been developed to cope with issues including differences in contrast among multi-source images, the computational burden of large images and real-time applications, heterogeneous distribution of interest points and matching accuracy. Especially the size of the vector of descriptors causes a computational burden. A modification which resulted in a near real-time computation of matches has been developed by Bay et al. (2008). In the processing of satellite images, one faces the challenging problem of georeferencing for which ground control points

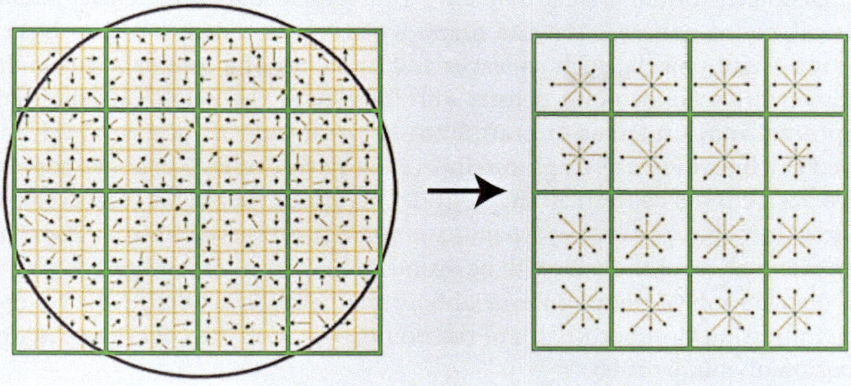

Figure 11.15 Sketch illustrating the transformation of the gradient orientations to the key point descriptor of the SIFT operator; the circle outlines the influence sphere of the Gaussian decaying weight function (source: Lowe, 2004)

are needed. When analysing time series of satellite images from multiple sources, the manual identification of ground control points is a tedious and cumbersome task. In an attempt to automatically georeference multi-source satellite images, Long et al. (2016) implemented a modification of the SIFT operator. The SIFT operator has been patented in 2004. Incorporating the operator in commercial software involves the payment of royalties to the University of British Columbia.

11.5.6 Concluding Remarks

The detection of interest points depends on the grey values in the image. Interest operators are defined as points of which grey values change when moving in any direction – this is in contrast to edges. To detect these locations, two orthogonal differential operators or the Laplace operator have to be exploited, which detect sudden jumps in grey values within an image patch. Differential operators are sensitive to noise and the Laplacian, which is based on two times differentiation of the image function is even very sensitive to noise. Therefore their running on an image should be accompanied by smoothing. A lot of research has demonstrated that Gaussian smoothing is best for a wide range of images. Although the first interest operators were developed in the mid-1970s, the improvement of existing interest operators and the design of novel ones is still an active research area. Adaptations are often necessary to improve performance, such as speed, to match images recorded under strongly differing lighting conditions and to match images captured over time spans of years.

11.6 Geometric Features from Point Clouds

The distribution of points of a 3D point cloud within a neighbourhood provides clues about the properties of the object present within that neighbourhood. In particular, the shape of the object provides a powerful geometric feature, which indicates the type of object. Geometrically, an object may have a variety of shapes, including linear, planar, cylindrical, cones, sphere, cube and fuzzy. By fuzzy, we mean that the object consists of multiple small objects, such as leaves and branches of a tree or a shrub. The shape depends on the scale or point density and the size of the neighbourhood. For example, captured from a manned aircraft flying at an altitude of 1 km, a power line wire is observed as a linear feature. Captured from a UAS, which flies 5 m above the power line, the wire has a curved cylindrical shape. The same is true for a cube. Observed from quite a distance the cube appears as perpendicular planes. Observed from a close-range position, only one plane of the cube will be visible.

The distribution of points in a neighbourhood can be modelled by the covariance matrix. Indicating the coordinates of the points by (x,y,z), the covariance matrix of the distribution of points reads:

$$\boldsymbol{C} = \begin{pmatrix} \sigma_x^2 & \sigma_{xy} & \sigma_{xz} \\ \sigma_{xy} & \sigma_y^2 & \sigma_{yz} \\ \sigma_{xz} & \sigma_{yz} & \sigma_z^2 \end{pmatrix} \tag{11.44}$$

The symmetric covariance matrix may be considered a structure tensor. As is the case for a 2 × 2 covariance matrix, for a 3 × 3 covariance, the eigenvalues can be computed. The three eigenvalues provide means to describe the geometry of the surface of the object part present in the neighbourhood. It is convenient to arrange eigenvalues in decreasing order, which means $\lambda_1 \geq \lambda_2 \geq \lambda_3$.

Eigenvalue computation of the covariance matrix of the X, Y, Z coordinates provides a means to distinguish linear features, planar features and volumetric features. These characteristics can be expressed using one measure: the Normalised Shape Indicator (NSI), which we define as follows:

$$\mathrm{NSI}_{3D} = \frac{\lambda_1 - \lambda_2 - \lambda_3}{\lambda_1 + \lambda_2 + \lambda_3} \tag{11.45}$$

The NSI can take values in the domain from minus one-third to plus one. If λ_1 will be large and λ_2 and λ_3 are both small, the NSI will be close to one and indicates the presence of a linear feature. If λ_1 and λ_2 are both large, while λ_3 is small the NSI will be close to zero and indicates the presence of a planar feature. If λ_1, λ_2 and λ_3 are all three large, the NSI will be close to minus one-third (1/3) and indicates a volumetric feature. These findings are summarised in Table 11.1. When applying the NSI, there will be intervals in which the shape is undefined. For example, if NSI = 0.5, the feature may be either classified as a linear or as a planar object. To avoid the wrong classification, the setting of four thresholds is feasible (Figure 11.16). These thresholds can be obtained from a learning stage in which a human operator identifies a number of prototypes.

Table 11.1 Relationship, indicated by V, between the size of the three eigenvalues of a 3D structure element spanned up by the 3D coordinates of a set of neighbouring points and the types of feature: no feature present, linear feature, planar feature and volumetric feature

	No feature	Linear	Planar	Volumetric
$\lambda_1 \approx 0$	V			
λ_1 large; $\lambda_2 \approx \lambda_3 \approx 0$		V		
λ_1 and λ_2 large; $\lambda_3 \approx 0$			V	
λ_1, λ_2 and λ_3 large				V

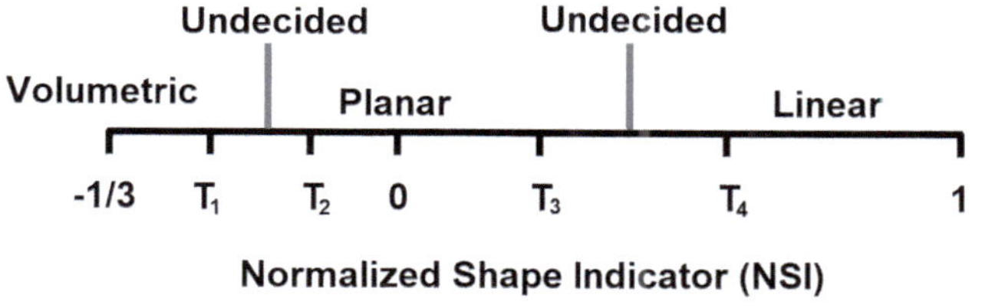

Figure 11.16 Normalised shape indicator (*NSI*) allows discrimination among linear, planar and volumetric shapes

The size of the eigenvalues uniquely determines whether the shape of an object is linear or planar. However, this is not the case for the diverse volumetric features. Volumetric features may consist of a variety of feature types, including cylindrical features, conical features, spherical features, cubical features and fuzzy features.

To distinguish among the variety of volumetric features, the normal vectors of the points within the neighbourhood have to be computed. The point density should be large enough to assume that the points around the point under consideration may be looked at as a plane. This means the object surface is considered to consist of piecewise planar shapes. The plane fitting can be done by a least-squares adjustment or by applying a RANSAC (random sample consensus) algorithm. If the normal vectors are parallel in one direction, while showing a regular pattern in the perpendicular direction, the object will have a cylindrical or conical shape. If the normal vectors are not parallel, but the deviations show a regular pattern in all directions, the object will have a spherical shape. The normal vectors of a cube pop up as a parallel for points on the same plane, while sudden jumps occur at the change of orientation of the planes. The pointing of the normal vectors of a fuzzy object will not show a few dominant directions but will be rather unstructured even chaotic.

In some scenes, particularly power line corridors through extensive, forested and uninhabited areas, the majority of objects will consist of three types which may be highly automatically recognised in airborne Lidar point clouds exploiting the height information and the above NSI (Figure 11.17). The objects are power lines (linear objects), trees, shrubs and other vegetation (fuzzy volumetric objects), and ground surface (planar objects).

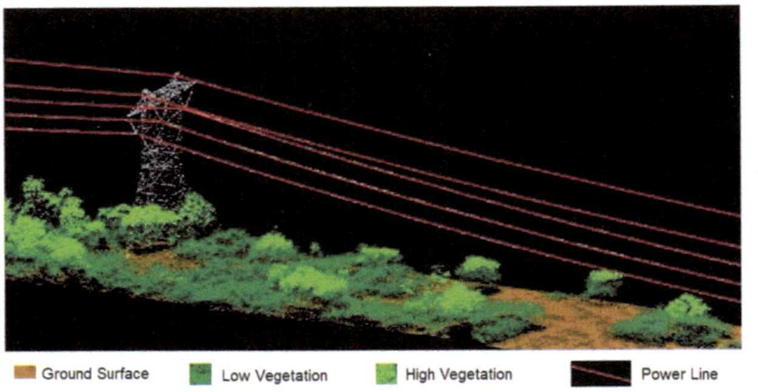

Figure 11.17 Power lines can be detected highly automatically in restricted scenes based on the normalised shape indicator (*NSI*) and height information

11.7 Similarity Measures and Decision Rules

An often used similarity measure for a broad diversity of applications, some lying far beyond pointcloudmetry, is the normalised cross correlation (Equation 11.9). This similarity measure requires the same dimension of the values of two data sets else one is comparing apples to oranges. In many cases, such as feature-based matching, the feature measures will consist of measurements of different dimensions. In multispectral classification, the feature measures do have the same dimension, which means grey values induced by the reflectance properties of objects. However, the number of features will be usually larger than two which complicates computation. An alternative for the cross correlation is to use a distance measure. The Euclidian or Mahalanobis distance measures are often used. In a classification problem, the feature vector of the prototypes may be yielded from a carefully chosen learning set, or from a set which has been prepared at forehand. In this section, we will first discuss two classification approaches applied to the individual pixels of multispectral images: vegetation index and multispectral classification. Next, we will have a look at feature-based matching as an extension to the matching issues discussed in Chapter 4.

11.7.1 Vegetation Index

In addition to operations carried out on the grey value of a single point, operations may be carried out on multiple measurements, or corrected measurements, of the same spot. An example of such an operation, which is often applied to multispectral images, is the computation of the Normalised Difference Vegetation Index (NDVI), which is defined as the ratio between the difference of the reflectance value of the near-infrared band (NIR: 0.7–1.5 μm) and the reflectance value of the red band (*R*: 0.62–7 μm) of the corresponding pixel, and the sum of the two values:

$$\mathrm{NDVI}(i,j)=\frac{\mathrm{NIR}(i,j)-\mathrm{red}(i,j)}{\mathrm{NIR}(i,j)+\mathrm{red}(i,j)} \tag{11.46}$$

The division by the sum of NIR and *R* aims at obtaining a normalised measure, which lies within the domain: $-1 \leq \mathrm{NDVI} \leq 1$. The NDVI is a simple but powerful and often

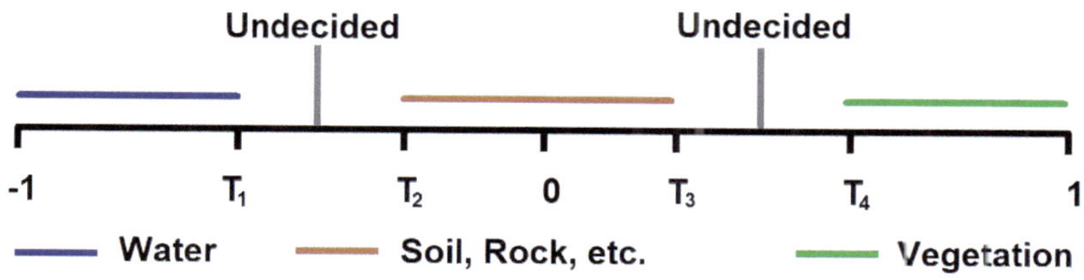

Figure 11.18 Normalised Difference Vegetation Index (*NDVI*) allows discrimination among three land cover types; four thresholds have to be set which may be derived from prototypes in a learning stage

used land cover classification method applied on multispectral images, which is able to distinguish among three land cover types: (1) vegetation, (2) rock and bare soil and (3) water. Given the reflectance curves of different object types (see Chapter 2), vegetation reflects a lot of NIR compared to red, resulting in an NDVI close to one. The reflectance values of NIR and red of bare soil and rock are similar resulting in an NDVI around zero. Since water bodies absorb NIR while reflecting a little bit in the red band, the NDVI for water will be close to minus one. As is the case for the NSI, discussed earlier, NDVI requires the setting of four thresholds (Figure 11.18), which can be determined by selecting a number of pixels, of which the land cover is known, in a training or learning stage. NDVI may play a role in the classification of laser point clouds, especially when each point is enriched with RGB values stemming from images, which are simultaneously recorded. NIR values are obtained from the reflectance values of the laser returns.

11.7.2 Machine Learning

The covariance matrix is a very powerful tool for many applications in the field of pointcloudmetry.

The inverse of the covariance matrix of two or more data sets can be used to assign weights to feature measures to obtain a similarity measure between the prototype features of a pixel or point and the feature measures of the unknown pixel or point. This similarity measure is used in the decision rule to determine to which class an unknown pixel belongs.

11.7.2.1 Multispectral Classification

To demonstrate the use of the covariance matrix in a classification problem, let us have a look at a simple machine learning problem in which two object types have to be classified based on two attributes. The distribution of the feature values of 16 prototypes for each of the two objects is shown in Figure 11.19. The feature values are obtained from the learning stage. They are distributed over two clearly distinguishable clusters, which is one of the main constraints put on the selection of suitable feature measures – feature measures should be discriminative.

Machine learning approaches, such as maximum likelihood classification, are often applied to the individual pixels of multispectral images. Machine learning classification approaches transfer values in a ratio scale into a nominal scale, particularly land cover

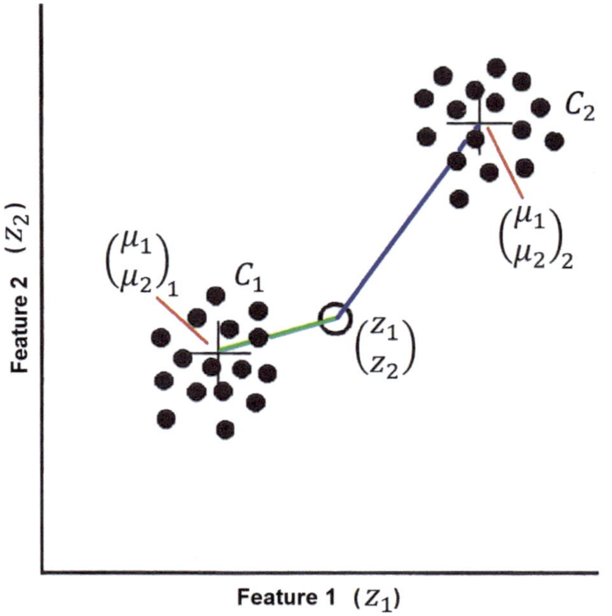

Figure 11.19 Two-dimensional feature space showing the distribution of the feature values of 16 prototypes of two objects obtained from a training (learning) stage; also shown is the location of the two feature values of an unknown object to be classified

classes. Machine learning has been extensively applied for land use and land cover classification of multispectral spaceborne and airborne images during the last four to five decades. The original intensity values, sometimes modified to correct for radiometric atmospheric distortions, are exploited as feature values. Next, they are compared with prototypes of the land cover classes and the pixel is assigned to the class with which its feature values show the best resemblance. In a maximum likelihood classification, this is done by computing the distance between the feature values of the pixel and the corresponding prototype feature values using the covariance matrix as a weighting instrument. This is done for each land cover class. The pixel is assigned to the class of which the feature values show the best similarity measure, provided that this similarity measure does not exceed a pre-described threshold. The latter means that the distance is so large that a class assignment becomes unreliable. The maximum likelihood similarity measure of a feature and its prototype k is computed as:

$$SM_k = \ln\left|\boldsymbol{C}_k\right| + \left\|d\right\|_k \tag{11.47}$$

where $|\boldsymbol{C}_k|$ is the determinant of the covariance matrix and $\|d\|_k$ a distance measure.

11.7.2.2 Mahalanobis Distance

Also in multispectral classification, the covariance matrix of the bands per land cover type plays a crucial role. The averages of the grey values per band of each object and their covariance matrix are found by taking training samples, which is the learning part of the classification. Next, the distances of grey values of an unknown point are computed to the average of each object, usually weighted according to their covariance matrix (Figure 11.19). The Mahalanobis distance and Euclidian distance are often used measures to compute the distance. The Mahalanobis distance (Mahalanobis, 1936) is a dissimilarity measure between two data sets, which are assumed to have the same statistical distribution expressed by the covariance matrix C_k:

$$\|d\|_k = \|M(i,j)\|_k = \sqrt{\left(z(i,j)-\mu_k\right)^T C_k^{-1}\left(z(i,j)-\mu_k\right)} \tag{11.48}$$

where

$z(i,j)$	Feature vector of point (i,j)
μ_k	Mean feature vector of class k
C_k^{-1}	The inverse of the covariance matrix of class k
$\|M(i,j)\|_k$	Mahalanobis distance of $z(i,j)$ to the mean of class k, expressing the similarity between the point under consideration and prototype values of class k

In a conventional multispectral classification task each and every pixel is classified separately. To save computation time, the computation of the determinant of the covariance in Equation 11.47 is often omitted. The effect on the quality of the classification is neglectable compared to other quality-degrading sources, discussed in the last part of this subsection. So, the pixel receives the class for which the Mahalanobis distance is minimum for all classes, and this distance is smaller than a given threshold. If the minimum is larger than this threshold, the pixel receives the class unknown. For classification purposes, the computationally expensive square root operation can be omitted since we are only interested in the range order of the outcomes. To demonstrate the use of the covariance matrix, let us have a look at the simple machine learning problem, discussed above, in which two object types have to be classified based on two feature measures. The distribution of the feature values of 16 prototypes of the two objects, obtained from a learning stage, is shown in Figure 11.19. The feature values are distributed over two clearly distinguishable clusters, which is one of the main constraints put on the selection of suitable feature measures – feature measures should be discriminative. The feature values of the sixteen prototypes of class one are listed in Table 11.2. These values have been obtained from a training stage and cover the domain 0–255, i.e. the grey values have a depth of 8 bits.

Table 11.2 Features values of 16 prototypes of the same class called training samples in multispectral classification

	1	2	3	4	5	6	7	8	9	10	11	12	13	14	15	16
z_1	160	145	141	152	148	155	157	151	147	149	154	146	144	151	159	155
z_2	92	94	102	105	110	100	103	97	104	98	104	107	93	99	105	108

A similar table can be made for class number two. The Mahalanobis classifier requires the mean feature values per class, the variances of the two feature values and the covariance.

$$\mu_1 = \begin{bmatrix} 150.9 \\ 101.3 \end{bmatrix} \quad \boldsymbol{C}_1 = \begin{bmatrix} 30.78 & 1.71 \\ 1.71 & 29.56 \end{bmatrix} \quad \boldsymbol{C}_1^{-1} = \begin{bmatrix} 0.0326 & -0.0019 \\ -0.0019 & 0.0339 \end{bmatrix}$$

From the variances and covariance also the normalised cross correlation can be computed (Equation 11.9): $R = 0.057$. This value shows that the feature measures are fairly well uncorrelated. This is also shown by the compactness and roundness of the two clusters in Figure 11.19. Feature independence is another important requirement for the selection of feature measures and this can be visually checked through scatter plots in feature space.

Similarly the mean, covariance matrix and inverse of the covariance can be computed for class two.

$$\mu_2 = \begin{bmatrix} 241.4 \\ 208.4 \end{bmatrix} \quad \boldsymbol{C}_2 = \begin{bmatrix} 44.3 & 2.6 \\ 2.6 & 18.4 \end{bmatrix} \quad \boldsymbol{C}_2^{-1} = \begin{bmatrix} 0.0228 & -0.0032 \\ -0.0032 & 0.0548 \end{bmatrix}$$

Given this information, the Mahalanobis distance can be computed for any unknown pixel and used as a decision rule to which class the pixel belongs. Let the feature vector of an unknown point read as $z_1 = 171$ and $z_2 = 122$, then:

$$\|M\|_1^2 = \left((171 - 150.9)(122 - 101.3)\right) \begin{pmatrix} 0.0228 & -0.0032 \\ -0.0032 & 0.0548 \end{pmatrix} \begin{pmatrix} (171 - 150.9) \\ (122 - 101.3) \end{pmatrix} = 27$$

The computation of $\|M\|_2^2$ is left to the reader. The outcome is $\|M\|_2^2 = 479$.

Since $\|M\|_1^2 < \|M\|_2^2$ and the value falls within a predefined threshold, the unknown pixel is assigned to class one, which would be expected based on the plot of the feature space. Figure 11.19 shows also the mean of the two features per object class as well as the feature vector of the unknown pixels, to which one out of the two classes has to be assigned (or the class unknown). The size of the two clusters is similar. When calculating the covariance matrix, this would mean that they would look very much the same. Since the covariance is close to zero, both matrices can be described by a single covariance measure:

$$\boldsymbol{C}_1 = \sigma_1^2 \boldsymbol{I} \quad \boldsymbol{C}_2 = \sigma_2^2 \boldsymbol{I} \tag{11.49}$$

where $\boldsymbol{I}$ is the 2 × 2 identity matrix, which means the two diagonal elements contain the values one and the two off-diagonal elements the values zero. Figure 11.19 also shows that the covariance matrices of the two clusters are similar, which means:

$$\sigma_1^2 = \sigma_2^2 = \sigma^2 \rightarrow \boldsymbol{C}_1 = \boldsymbol{C}_2 = \sigma^2 \boldsymbol{I} \tag{11.50}$$

So its use can be omitted, and Equation 11.48 becomes the minimum distance to the mean classifier. This classifier is also well known for many machine learning applications.

11.7.2.3 Quality Considerations

The classification of individual pixels has several drawbacks, including the following:

- Classification quality is sensitive to the presence of noise.
- Instrumental division of the electromagnetic (EM) spectrum into spectral bands doesn't necessarily correspond to the spectral signature of the desired land cover classes.
- The desired class or object type may consist of several subclasses with different spectral properties; in order to assign this class to a pixel, the individual subclasses have to be classified.
- Desired objects which have the same spectral properties may be indistinguishable.
- Pixels lying on the border between objects receive a mixed spectral signature, which makes them difficult to classify.
- Differing sun angles, caused by daytime or seasonal differences, cause differing illumination conditions. So images recorded at different times or capturing different parts of the earth require separate learning stages – the characteristics of the training samples are not invariant with respect to time and place. To partly eliminate the effects of illumination differences, band rationing can be applied.

The consequence is that per pixel classifiers often produce accuracies which are far from perfect; errors of 20% or more frequently occur. Several methods have been developed directed towards an improvement of the rather error-prone results of pixel-based classification. One of them is joining adjacent pixels together into regions based on some similarity measures prior to classification, a process known as region grown in the field of image processing. The regions may cover an entire object or parts of an object. The basic idea underlying this approach is that the division of a scene in pixels and considering these pixels as individual objects is artificial and arbitrary; the division depends on the accidental recording configuration. This so-called object-based image classification (OBIA) considerably improves classification results and has been implemented in several image classification software packages.

11.7.3 Feature-Based Matching

Within photogrammetry, interest points are extensively exploited for the automatic detection of tie-points in the overlaps of aerial images and the generation of digital elevation models. After detection of interest points, there are several ways to proceed to arrive at matches to find corresponding points in the overlaps of images. One of them is to carry out an area-based matching approach, such as normalised cross correlation or least-squares matching (Chapter 4). Another approach for finding matches is to assign attributes to the interest points. Next, the similarities between the attributes of the interest point in the target patch and candidate corresponding points in the search image patch are computed. This matching method is called feature-based image matching. The most elementary attribute of a pixel is its original grey value or colour. However, grey values and RGB values are not unique, and the grey values of corresponding points may differ because of illumination differences or because the overlapping images have been recorded by different camera sensors. Therefore, the attributes are derived from the grey values within a neighbourhood of the interest point.

Aerial surveys are carried out in a very structured way using a carefully designed flight plan. As a result, the scale of the images and the orientation of the images in space are quite similar. In addition, approximate values of the six exterior orientation parameters (Chapter 4) have been obtained from global navigation satellite system (GNSS) and inertial measurement unit (IMU) instruments onboard the aircraft. These measurements are pretty accurate although not at the level of accuracy one strives for in photogrammetry. Because of these specific characteristics of a photogrammetric aerial survey, the following attributes are suited to determine the similarity between a detected interest point in one image with candidate interest points in the other image:

1. Mean of the grey values of the image patch covered by the interest point
2. Variance of the grey values in the image patch
3. Number of pixels with a grey value smaller than the grey value of the interest point
4. Number of pixels with a grey value larger than the grey value of the interest point
5. Roundness measure (q)
6. Strength of the interest point (w)

Since the measures are computed over an image patch with a fixed size, e.g. a 7×7 window, the number of pixels with a grey value larger than the grey value of the interest point can be derived from the number of pixels with smaller values and its use doesn't add information. Therefore this measure is omitted resulting in five feature values. The similarity measure can be based on exploiting the Euclidian distance in feature space. The value domains and dimensions of the diverse measures usually differ. An image quantised with 8 bits has a value range from 0 to 255, while the roundness measures can have a value between zero and one. Therefore, normalisation has to be carried out so that all values fall within the range between zero and one. Some measures may contain more information or are more reliable than other measures. The level of importance can be quantified by introducing weights in the distance computation, resulting in a Mahalanobis type of distance.

Using the onboard measurements of the exterior orientation parameters and an indicative elevation of the interest point above a reference plane, the search area in the

Table 11.3 Similarity measures between the feature vector of an interest point in the target image and six candidate points in the search image (for explanation of meaning of symbols, see text)

	ξ_i	w_i	Γ_{1i}	Γ_{2i}	Γ_{3i}	Γ_{4i}	Γ_{5i}	Γ_{6i}
Mean	0.45	1	0.54	0.33	0,48	0.39	0,64	0.44
$\Delta N/N$	0,24	2	0.45	0.56	0.63	0.33	0.36	0.47
Variance	0.77	0.5	0.83	0.74	0.44	0.69	0.71	0.77
q	0.82	1	0.71	0.77	0.84	0.79	0.79	0.81
w	0.67	2	0.55	0.72	0.57	0,71	0.61	0.74
		6.5	**0.021**	**0.035**	**0.058**	**0.004**	**0.012**	**0.018**

other image can be computed from the interest point in the target image. In the example discussed in the sequel, this search patch contains six candidate interest points.

The five relevant feature measures of the interest point in the target image are collected in a vector containing the elements: ξ_i, $i = 1, \ldots, 5$. Similarly, the attributes of candidate interest points in the search patch are collected in six feature vectors: Γ_{ki}, $k = 1, \ldots, 6$; $i = 1, \ldots, 5$. The corresponding weights are collected in a weight vector: w_i, $i = 1, \ldots, 5$. In this case, a proper decision criterion is that the closest candidate point, which means the minimum distance point, in feature space, is assigned as the corresponding point. The weighted distance from the feature vector of the target point to each of the six candidate points can now be computed as follows:

$$\|d\|_k = \sqrt{\sum_{i=1}^{n} \frac{1}{w_i} \cdot \sum_{i=1}^{n} w_i \left(\xi_i - \Gamma_{ki}\right)^2} \tag{11.51}$$

For classification purposes, we are only interested in the range order of the outcomes. Therefore, the computationally expensive square root operation can be avoided. The bottom row of Table 11.3 contains the similarity measures computed as squared distances between the attributes of the interest point in the target patch and attributes of the six candidate interest points in the search patch.

The cell in the bottom row of column w_i contains the sum of the weights: 6.5. The row indicated with $\Delta N/N$ contains the ratio of the number of pixels with a grey value smaller than the grey value of the interest point and the total number of pixels within the image patch. The other indicators are explained in the text. According to the closest point decision criterion, candidate interest point number four should be labelled as the corresponding point.

11.8 Concluding Remarks

Images and raw laser point clouds are just the beginning of a processing chain, which aims at extracting information from these data sets which are useful for solving a particular geo-related real-world problem. To obtain the desired information intermediate steps are inevitable. The most important step is the detection of features. Many types of features may be derived from images and point clouds. The selection of features depends on the task at hand. Features are necessary for automated co-registration of images and laser scans, segmentation of images and point clouds, reconstruction of objects from point clouds, classification, object recognition, 3D mapping and other tasks. Feature detection also results in data reduction, which is beneficial for processing steps downstream of the pipeline.

One specific feature type is crucial in pointcloudmetry: interest points, which are required for co-registration and generation of digital elevation models from overlapping images. Finding corresponding points in the overlaps of two images or laser scans of the same scene is an essential but demanding pre-processing step. To obtain reliable matches, salient points, preferably located at blobs, T-junctions, corners or other junctions, have to be detected. These so-called interest points should be detected even when the viewing perspective changes. Next, descriptors have to be assigned to each interest point, which is used to determine the similarity of points in the overlaps. The descriptors are computed from the grey values of the image patch of the interest point and have to obey constraints, including distinctiveness, robustness to noise and invariance to geometric and radiometric transformations. In feature-based matching the descriptors of a point in the one image are compared with the descriptors of candidate corresponding points in the other image, a process which is similar to object recognition using a library of prototypes. The descriptors, or attributes, are stored in a feature vector, and the corresponding point is selected based on a distance measure in the feature space. The time needed for computing these distances depends on the size of the vector – the longer the vector the more computation time is required – and omitting computations which are irrelevant to the task at hand and would only increase the computation burden.

References

Bay, H., Ess, A., Tuytelaars, T., Van Gool, L. (2008) Speeded-up robust features (surf). *Computer Vision and Image Understanding*, 110(3), pp. 346–359.

Canny, J.F. (1986) A Computational Approach to Edge Detection. *IEEE Transactions on Pattern Analysis and Machine Intelligence*, 8, pp. 679–698.

Dreschler, L. (1981) Ermittlung markanter Punkte auf der Bildern bewegter Objekte und Berechnung einer 3-D Beschreibung auf dieser Grundlage. Ph.D. Thesis, University of Hamburg.

Dreschler, L., Nagel, H.-H. (1982) Volumetric Model and 3D Trajectory of a Moving Car Derived from Monocular TV Frame Sequences of a Street Scene. *Computer Graphics and Image Processing*, 20, pp. 199–228.

Dreschler, L.S., Nagel, H.-H. (1983) On the Selection of Critical Points and Local Curvature Extrema of Region Boundaries for Interframe Matching. In Huang, T.S. (Eds.), *Image Sequence Processing and Dynamic Scene Analysis. NATO ASI Series (Series F: Computer and System Sciences)*, Vol. 2. Springer, Berlin, Heidelberg.

Förstner, W. (1986) A Feature Based Correspondence Algorithm for Image Matching. In International Archives of the Photogrammetry, Vol. 26-III, Rovaniemi, pp. 1–17.

Förstner, W., Gulch, E. (1987) A Fast Operator for Detection and Precise Location of Distinct Points, Corners and Centres of Circular Features. In Proceedings of the ISPRS Intercommission Conference on Fast Processing of Photogrammetric Data, Interlaken, pp. 281–305.

Harris, C., Stephens, M. (1988) A Combined Corner and Edge Detector. In *Proceedings of the Alvey Vision Conference*, Manchester, pp. 147–151.

Koenderink, J.J. (1984) The Structure of Images. *Biological Cybernetics*, 50, pp. 363–396.

Lemmens, M.J.P.M. (1988) A Survey on Stereo Matching Techniques. *International Archives of the Photogrammetry, Remote Sensing*, 27(part B8), pp. V11–V23.

Lemmens, M.J.P.M. (1996a) Structure-based Edge Detection: Delineation of Boundaries in Aerial and Space Images. PhD Thesis, Delft University of Technology.

Lemmens, M.J.P.M. (1996b) A Survey of Boundary Delineation Methods. *International Archives of the Photogrammetry, Remote Sensing*, 31, pp. 435–441.

Lindeberg, T. (1994) Scale-space Theory: A Basic Tool for Analysing Structures at Different Scales. *Journal of Applied Statistics*, 21(2), pp. 224–270.

Long, T., Jiao, W., He, G., Zhang, Z. (2016) A Fast and Reliable Matching Method for Automated Georeferencing of Remotely-Sensed Imagery. *Remote Sensing*, 8(1), p. 56.

Lowe, D.G. (1999) Object Recognition from Local Scale-invariant Features. In *International Conference on Computer Vision*, Corfu, Greece, pp. 1150–1157.

Lowe, D.G. (2004) Distinctive Image Features from Scale-invariant Keypoints. *International Journal of Computer Vision*, 60(2), pp. 91–110.

Mahalanobis, P.C. (1936) On the Generalised Distance in Statistics. *Proceedings of the National Institute of Sciences of India*, 2(1), pp. 49–55.

Marr, D. (1982) *Vision: A Computational Investigation into the Human Representation and Processing of Visual Information*. Freeman and Company, New York, re-published by MIT press in 2010.

Marr, D., Hildreth, E. (1980) Theory of Edge Detection. *Proceedings of the Royal Society of London B*, 207, pp. 187–217.

Miksik, O., Mikolajczyk, K. (2012) Evaluation of Local Detectors and Descriptors for Fast Feature Matching. In *21st International Conference on Pattern Recognition*, Tsukuba, Japan, pp. 2681–2684.

Moravec, H.P. (1977) Towards Automatic Visual Obstacle Avoidance. In *Proceedings of the International Joint Conference on Artificial Intelligence*, Cambridge, Massachusetts, USA, p. 584.

Moravec, H.P. (1981) Rover Visual Obstacle Avoidance. In *International Joint Conference on Artificial Intelligence*, Vancouver, Canada, pp. 785–790.

Nixon, S.M., Aguado, A.S. (2012) *Feature Extraction & Image Processing for Computer Vision*. 3rd Edition, Elsevier, Academic Press, Amsterdam.

Ogundare, J.O. (2018) *Understanding Least Squares Estimation and Geomatics Data Analysis*. Wiley, Hoboken, NJ.

Pratt, W.K. (1991) *Digital Image Processing*. 2nd Edition, Wiley, New York.

Prewitt, J.M.S. (1970) Object Enhancement and Extraction. In Lipkin, B.S., Rosenfeld, A. (Eds.), *Picture Processing and Psychopictorics*. Academic Press, New York.

Point Cloud Processing Software

CHAPTER

12

The theoretical concepts of *pointcloudmetry*, as presented and discussed in the previous chapters, can only be deployed in practice by virtue of the existence of dedicated point cloud processing software. Indeed, pointcloudmetry wouldn't be possible without advanced and user-friendly commercial or open-source software. The software should be regularly upgraded to the ever-increasing point cloud sizes, point densities and user demands. This chapter, which is partly based on Lemmens (2014a–c, 2019), explores the characteristics and properties of point cloud processing software developed for the generation of digital elevation models and the creation of 3D models (3D maps) of outdoor and indoor scenes. Rather than describing the functionalities of individual software packages, it is more appropriate within the scope of this book to categorise the functionalities and look at the underlying design principles. This is because software is subject to continuous development efforts, and thus the abilities rapidly change over time. We start with the latter, continue with an indication of important functionalities and finish with a recap of cloud computing, a service which has gained increasing importance since 2015. To avoid confusion, note that cloud in cloud computing has nothing to do with the word cloud in point cloud – the words are identical, but they have completely different meanings within the two settings.

12.1 Design Principles

The origin of point cloud processing software may stem from a variety of backgrounds. For the sake of simplicity, we divide the software origins into four broad groups (Figure 12.1):

1. Manufacturers of point cloud-generating Lidar sensors
2. Manufacturers who aim at offering software for (semi-)automated information extraction from a broad variety of point clouds, including Lidar, photogrammetry, Radar and sonar
3. Manufacturers of photogrammetric software aimed at point cloud generation and information extraction from overlapping images
4. Manufacturers of application software

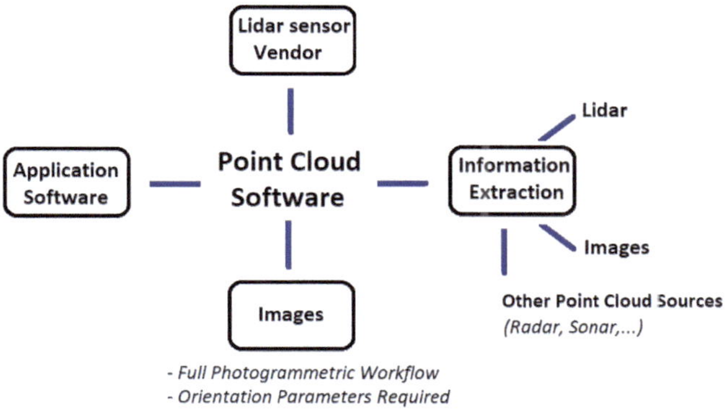

Figure 12.1 Schematic overview of the design backgrounds of point cloud processing software

In the formative years of laser scanning, point cloud processing software consisted of an assembly of software modules provided by the manufacturers of the laser scanner. Each software system was proprietary, and significant craftsmanship was required to extract from the point clouds meaningful information for the task at hand. Often the necessary craftsmanship was started during training sessions of the manufacturer. The offering of software by the manufacturers of terrestrial laser scanners (TLS), airborne Lidar sensors or mobile mapping systems was nourished by the early recognition that bringing laser scanners to the marketplace should be accompanied by software facilities to enhance the value of the outputs of their sensors or to make it usable for user applications. They complemented their hardware with proprietary software for managing, georeferencing, visualising, editing and exporting the outputs to dedicated software, such as computer-aided design (CAD) packages. One of the limitations at that time was the rather course point density compared to what can be achieved today.

On the other hand, software builders spotted commercial potential in offering tools for creating a broad pallet of end products from point clouds either generated by lasers or computed from overlapping images or other point cloud sources such as Radar or sonar. These manufacturers aim at developing software packages which are independent of sensor characteristics and thus can process point clouds acquired by a broad pallet of sensors. They focus on the information extraction part of the point cloud processing workflow. The software is not limited to targeted or specialised applications. Often, the point of beginning such software packages is that the point cloud has gone through the pre-processing stage of the workflow, including co-registration and georeferencing. The focus lies on (semi-)automated information extraction, visualisation, including fly-through animations, and editing, creation of triangular irregular networks (TINs) and products derived from these such as contour lines and profiles, volumes and 3D maps and 3D models. Inspection tools allow us to visually check and manually modify the data to enhance and extend the extracted information. Another important functionality is the creation of orthomosaics and 3D digital landscapes from images sometimes in combination with a Lidar DSM captured at the same time. Some software vendors have developed software modules on top of the core modules for a specific application such as the automated extraction of power lines from airborne Lidar point clouds. The development of specialised software is usually stirred up by demands from the user base.

The starting point of the design of photogrammetric software is the creation of point clouds from overlapping aerial images. Particularly the rapid rise of unmanned airborne systems equipped with a lightweight camera in combination with dense image matching technology based on semi-global matching has boosted the emergence of photogrammetric image matching software. Some manufacturers provide software for fully automatically establishing the entire photogrammetric workflow from interior and exterior orientation by bundle block adjustment, generation of DSMs, orthomosaics, 3D virtual landscapes and products derived from these. The core consists of image matching modules which compute, from the contrast in a patch around a pixel, features which are then matched with candidate points in the overlaps of the image. Other photogrammetric point cloud generating software packages require images which have already been gone through the bundle block adjustment process, i.e. the exterior orientation of each image has been computed by a separate photogrammetric software package.

Another point cloud processing software stems from the other end of the spectrum, which means the application domains. For example, civil engineers, architects and mining engineers who were used to a certain CAD system started to appreciate TLS or mobile laser scanning (MLS) point clouds, either collected by themselves or delivered by contractors. They asked vendors to add modules to their CAD system for processing these point clouds. Some manufacturers discovered new opportunities along the way and built dedicated modules on top of one or more base modules aimed at, for example, the mining industry or creating 3D models of crash sites. This development process is far from complete. New modules are being added all the time. Manufacturers of the more generic point cloud processing software may also enable linking of their software to widely used GIS packages as an add-on so that it can run on top of the GIS.

Often point cloud processing software originates in the form of a spin-off of academic research, resulting in start-ups under the wings of the university. Once the products and services offered have proven to be successful, often well-established geomatics firms may attempt to take over the start-ups in the next stage. Some start-ups sell their brainchild, while others withstand the allurement and continue to flourish flying on their own wings.

12.2 Functionalities

The functionalities of point cloud processing software differ widely. Gaining insight into the suitability of the software requires a good understanding of the offered functionalities. The functionality may start at the creation of the point cloud itself, as is the case for dense image matching software which creates a digital surface model (DSM) from overlapping imagery. The functionalities may be categorised in many ways. A first grouping is done by distinguishing three main groups: pre-processing tools, information extraction tools and inspection tools (Figure 12.2). We focus on a few important functionalities, in particular: storage formats, georeferencing, ground filtering, feature extraction, 3D mapping and visualisation and editing.

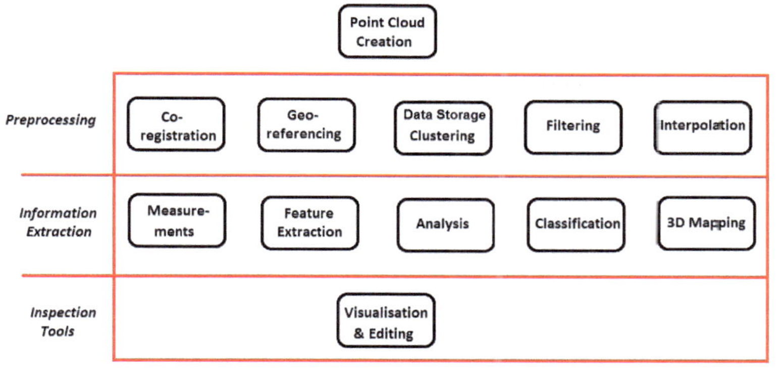

Figure 12.2 Functionalities of point cloud processing software categorised according to pre-processing, information extraction and inspection tools

12.2.1 Storage

A first challenge in the exploitation, analysis and mapping of point clouds is the huge file size: A 32-bit operating system can only handle files of up to a few gigabytes, and most computers are not designed for such large datasets. The amount of data produced by today's sensors is growing faster than the processing and storage capacity of database management systems. As a result, the storage of billions of 3D points is an issue in itself. The LAS storage format, initially developed for airborne Lidar, has become a de facto standard. The more generic E57 format, which is a vendor-neutral format like LAS and based on XML, supports metadata, 2D imagery and a broad pallet of point attributes. Nevertheless, proprietary formats are often preferred as they are optimised for system-specific outputs. For a number of information extraction tasks, such as classification and feature extraction/segmentation, nearby points should be stored close to each other in the database. This requires facilities for clustering of the points.

12.2.2 Georeferencing

A point cloud is a set of data points all of which should be represented in the same preferred coordinate system. Initially, the data is unorganised; software is designed to organise the unorganised and extract information from it. In most cases the initial point clouds consist of individual laser scans or are extracted from individual overlapping image pairs. These point clouds require co-registration and georeferencing. 3D patches of point clouds – i.e. individual scans – must be transformed into a single coordinate system to enable georeferencing or registration, which involves identifying common points in the overlaps. As a geometric transformation model, the 3D similarity (Helmert) transformation is usually applied (three translation parameters, three rotation parameters and one scale parameter). This process requires at least seven coordinates, i.e. two 3D ground control points (GCPs) and one height GCP. To obtain redundancy and thus the ability to detect outliers and to compute point precision through error propagation, more than seven GCPs, which are evenly distributed along the border of the site, should be measured, preferably with high precision land surveying equipment such as a differential global navigation satellite system (GNSS) receiver or a total station. The stitching of the single point clouds requires tie points in the overlaps. These are distinctive points in the scene or consist of targets placed, marked or painted in the overlaps. The latter requires more thorough planning of the survey to ensure that the marks are placed in the overlaps. Of course, the GCPs should be clearly identifiable. In addition to the indirect method of georeferencing described above, direct georeferencing can also be conducted in some cases. For example, a TLS can be centred over a known point and levelled as if it were a total station. The position can also be determined using a GNSS antenna mounted on top of the TLS.

12.2.3 Filtering

An important aspect is to clean the point cloud from unwanted points. The removal is called filtering or ground filtering if the point cloud concerns a digital surface model (DSM) of the terrain. Airborne Lidar, TLS as well as the DSM generated from the automatic matching of overlapping imagery capture all objects present on the site, including vegetation, cars, bikes, dogs and suchlike, all of which are irrelevant for the survey at hand and are thus "noise" from the surveyor's or user's point of view. Hence, the software should be able to remove such points, either through manual editing or preferably by automatic filtering. Most software contains tools to quickly and easily reduce the amount of this type of noise. Ground filtering is a specific method aimed at the removal of points reflected from vegetation when creating a 3D city model, or from vegetation and buildings when creating a bare-ground digital terrain model (DTM). In course of time, many methods have been developed, which is a token of the complexity of the ground filtering problem (see Chapter 10).

12.2.4 Feature Extraction

Key for specific industrial use, such as the 3D mapping of chemical plants, oil refineries and blast furnaces, is not only the manual measuring of length, height, distance, angle, area and volume, but also the extraction of geometric features such as lines and planes and the fitting of solid models through a set of points. The solid models may include planes, cubes, spheres, cylinders or user-specified objects such as valves and elbows. Vendors often offer plug-in modules, which provide a sequence of functionalities adapted to particular needs. Vendors provide "lite" versions aimed at users who do not need all the functionalities.

12.2.5 3D Mapping

For many applications, such as archaeology, architecture and 3D city modelling, 3D mapping from point clouds is an essential functionality. A realistic representation of buildings and their surroundings requires georeferenced aerial and/or street-level images to drape photo textures over facades, paved surfaces and so on (Figure 12.3). There are several ways to create a 3D representation of buildings, which visually may look quite the same. One of them is by creating TIN meshes. Next, each triangle is projected on the

Figure 12.3 Partially rendered 3D mesh model of St Paul's Cathedral, London, UK, using Leica HxMap 3D modeller obtained from a hybrid Lidar and camera system, which simultaneous captures Lidar point clouds and imagery (source: Lemmens, 2020; courtesy: Bluesky)

Figure 12.4 3D map of the commercial centre of Frankfurt, Germany, created from free and open-source Lidar data and satellite images; buildings, ground and streets consist of a textured mesh (source: Turbosquid.com)

corresponding aerial image and/or street view image for extraction of the RGB image content (Figure 12.4). Another way is by combining an airborne Lidar DSM with digital maps containing the footprints of buildings. By extruding the footprints depending on the height values provided by the Lidar DSM, an LoD1 3D map of buildings can be created (Figure 12.5).

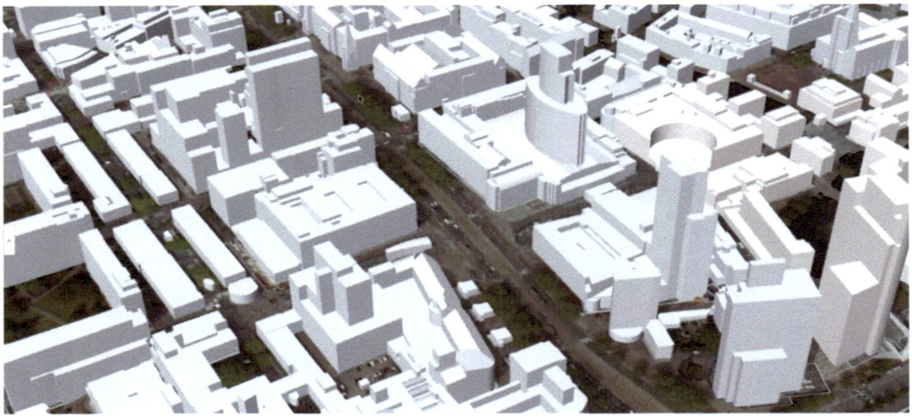

Figure 12.5 3D map at LoD1 of a part of Rotterdam automatically created by combining building footprints with airborne Lidar point clouds (courtesy: Municipality of Rotterdam)

12.2.6 Visualisation and Editing

Validation and analysis of point clouds require conversion of the x,y,z coordinates with attributes to black/white or colour (RGB) images. Removal of spurious points or adding new points requires editing facilities. Therefore, point cloud processing software should be able to visualise and edit point clouds with many billions of points at a reasonable rate; it should not take hours before a view appears on the screen. All the points may be depicted in a single colour and size, but a colour code may also portray the strength of the return signal, the elevation or the RGB values in case images are available. Inspection of the point cloud requires zooming, panning, rotating, fly-through and adjusting point size. Editing operations include the mouse clicking of individual points for the visual removal of outliers or a group of points by a human operator, addition of missing points or calculation of distances, angles, areas and volumes. The bounding box of a group of adjacent points may be identified by drawing a cube, a rubber surface or other volume shapes. Selection of a group of points allows geometric primitives such as planes, spheres, cylinders or non-uniform rational basis spline to be fit using an optimisation method – usually this will be based on least-squares adjustment or on the random sample consensus algorithm. The selection of a group of points also allows the manual removal of unwanted points such as those reflected on vegetation and buildings if one wants to create a bare-ground DTM.

12.3 Point Cloud Processing Software as a Service

The capabilities of today's geodata acquisition technologies – at the forefront with laser scanning and photogrammetry in the broadest sense – are tremendous. It has become astonishingly easy to capture urban areas, public buildings or historical sites through cameras or laser scanning or acquire open data from a diversity of sources. Even professionals without a surveying background can do the job, but such professionals require reliable, robust, easy-to-use and affordable processing software.

Storing data in the cloud has become increasingly popular among companies and the general public alike over recent years. The next logical development is to exploit software running on remote servers for data processing and deliver the end product via the internet to the user. This service has received the tag *cloud computing*. Here, the cloud refers to multiple computers which are connected through the internet. The computers may be distributed all over the world. The user doesn't need to define the computers he wants to use; he even doesn't know on which computers his task is processed or how many computers are used. All these aspects of deploying computer power are managed internally by the service provider. The user benefits from cloud computing facilities because he doesn't need to own and maintain the hardware and software. The user can adjust computer power and software abilities according to their needs. In other words, cloud computing offers scalable solutions.

The sustained rise of cloud computing is fuelled by the exponential growth in computational power, ever-faster access to storage devices and high-speed data transport over the internet resulting from broadband networks. These three technological advancements are allowing new business opportunities to flourish. The growing popularity of cloud services in recent years has been further boosted by the ease of internet access from almost any location across the globe.

Cloud services can be divided into three main models: Infrastructure as a Service (IaaS), Platform as a Service (PaaS) and Software as a Service (SaaS). IaaS creates a virtual desktop environment in which all data and all applications, including the operating system, are in the cloud. To process data and perform other IT tasks, the user can choose from a broad spectrum of software running on remote servers. In contrast, PaaS allows the user to deploy self-developed or self-purchased software using the hardware, programming languages, libraries and tools offered by the platform provider. The provider manages and controls the cloud infrastructure, but users have control over their applications. Meanwhile, SaaS goes a step further than PaaS: Users use the software supplied by the provider to process their own data and run it in a cloud infrastructure. Then, the data and the end product are safely returned to the user or shared with the user's customer via the internet. The provider hosts application software which can be accessed by users via the internet on either a subscription basis or a pay-per-use basis. Figure 12.6

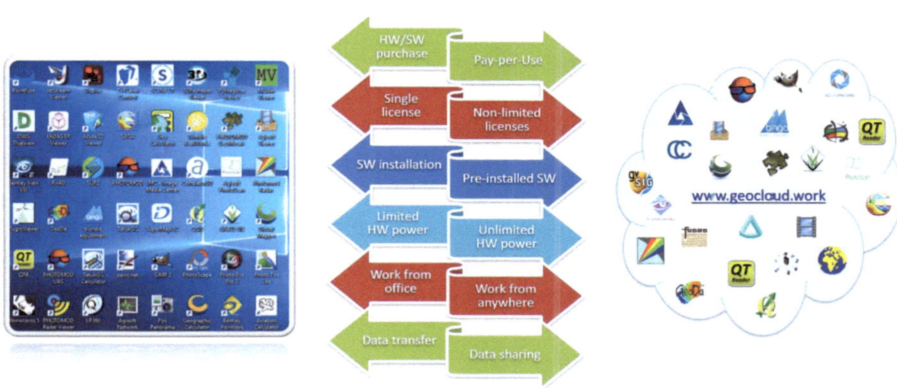

Figure 12.6 Software on a desktop versus cloud infrastructure (source: www.geocloud.work)

compares the main features of licensed software installed on a desktop against software offered as a service in a cloud infrastructure.

As a result of the advancements in smart software and miniaturisation of electronics, relatively small devices can capture a variety of sites, both indoors and outdoors, at high levels of detail and accuracy. This means that the focus of geomatics specialists is increasingly shifting from the exploration of geodata for certain applications to the acquisition and processing of geodata itself and the development of software to support automated processing. This shift in focus has also been reflected in the geomatics curriculum at many universities over the past decade. Besides this, professionals such as architects, constructors, cultural heritage conservators and urban green-space managers are becoming incidental users of geodata acquisition technologies. They lack the time and/or the inclination to get to grips with the nitty-gritty of these technologies. Instead, they merely want to be able to process the geodata – which they may have acquired using a handheld device or by pushing a trolley – into a file and format suitable for use in their BIM or CAD software. These casual users welcome services which make their jobs easier. They regard geomatics end products as raw materials for their own applications. It is simply a matter of uploading the data and indicating the desired processing steps in a digital form. Then, a little while later, the processed data becomes available to them in the desired format, and the user can focus on utilising it to create their actual product or service. Such users represent tremendous opportunities for providers who offer cloud-based geomatics software.

One example of a cloud computing service for geomatics software is GeoCloud. Launched in July 2018, the platform enables software vendors to host their geomatics software (Raizman, 2019). The service not only provides point cloud processing software but also covers all geomatics fields, including geodesy, photogrammetry, mapping, cartography, remote sensing, GIS, CAD, image processing, 3D city modelling and automatic object recognition. The platform has been built on Amazon Web Services (AWS), which is the world's largest cloud service infrastructure, and has inherited all its advantages. By installing their software on the platform, vendors benefit from doing business round the clock at lower marketing, pre-sale and support costs. After installation, vendors can test how well their software works. Small software producers, start-ups and other firms with limited financial resources can offer their software packages to a worldwide audience without high and risky investment. Established firms with a large consumer base also benefit from the platform since it enables them to extend their coverage globally and also to build awareness of their services and solutions among professionals outside of the geomatics field. After all, a single platform bringing together a variety of software is more likely to be found by smaller firms or casual users. Users do not need to invest in computers with powerful processors, but a fast internet connection suffices. A cloud infrastructure is very flexible for both users and software vendors since cloud computing provides a scalable and "always available" computer environment. Cloud computing enables users to work with software packages running on remote servers on either a pay-per-use or a subscription basis.

Cloud computing provides unified and easy access to a variety of geo-information software packages and data. While this is beneficial, the other side of the coin is that one has to rely on the service provider, which means relinquishing autonomy. One also depends on the security features with which the provider has surrounded his platform, and this is where trust comes in. Behind any digital service, there are not just robots, machines or computers but also an organisation run by people, who want to grow their business. This requires customers to keep coming back and recommend the organisation's services to others. This is a powerful driver for companies to do their utmost to satisfy customers. On the other hand, it is very conceivable that organisations that work with sensitive information will be reluctant to use cloud computing services, instead preferring to establish their own secure data processing environment.

The ongoing developments in information and communication technology (ICT) continue to ignite new business opportunities and ideas. One of them is offering software as a cloud computing service rather than on a licence basis. It is highly likely that these ICT advancements will further boost the use of geomatics tools and create greater public awareness of the benefits of geomatics in general and earth-related point clouds in particular.

12.4 Choosing Point Cloud Processing Software

To be useful for a particular task at hand, the millions or even billions of 3D points generated by a variety of active and passive sensors need to be stored, organised, combined, georeferenced, measured and analysed, and these tasks all require dedicated software running on fast computers supported by appropriate storage facilities. Many software tools, either proprietary or open source, are available to assist the human operator or analyst in the processing and analysis tasks of point clouds. To obtain insight into the possibilities of a particular software package and the suitability of the functionalities for the intended applications, conducting benchmarks is crucial. However, gaining insight into the capabilities of the software itself is just a first step in the selection. A generic package which can handle all types of sensor output and generate all types of the end product does not exist. Each package has its own peculiarities. Therefore, before purchase, not only the functionalities should be examined but also the ideas and philosophy underpinning the design of the software. Another important aspect includes the training facilities offered by the vendor. The user operating in a production environment needs also reliable and 24/7 support. It seems to be a rule of thumb that the capacity of sensors to generate data is continuously ahead of the capacity of computers to process the data. This fact of life induces the question: How does the vendor respond to the continuous technological advancements and expanding user requirements? For this, it is wise to look at how much the vendor invests in R&D. A large R&D team warrants a continuous stream of innovative software solutions. It is also worthwhile to query the vendor about any current or planned extensions, the ability to join software modules into one smooth workflow and interoperability with other software and services provided. On the other hand, researchers may prefer open-source software because the underlying algorithms can be scrutinised and adjusted if necessary.

References

Lemmens, M. (2014a) Background to Point Cloud Software. *GIM International*, 28(7), p. 11.

Lemmens, M. (2014b) Point Clouds (1): The Functionalities of Processing Software. *GIM International*, 28(6), pp. 16–19.

Lemmens, M. (2014c) Point Clouds (2): Survey of Processing Software. *GIM International*, 28(7), pp. 27–33.

Lemmens, M. (2019) Geoinformation Software as a Service: From Licensing to Subscription. *GIM International*, 33(4), pp. 35–37.

Lemmens, M. (2020) Simultaneous Capturing of Lidar and Imagery: How Major Cities May Benefit from a Hybrid Sensor System. *Geomatics World*, 28(2), pp. 20–22.

Raizman, Y. (2019) Personal Communication.

Pilot Studies and Applications

CHAPTER

13

As discussed in Chapter 1 of this book on *pointcloudmetry*, geomatics is a technology with a virtually infinite number of applications. As a special branch of informatics, geomatics focuses on the acquisition, processing and use of geodata for a plethora of applications. In many applications surveying techniques have been applied in a standardised way for decades or even centuries. When a new geodata acquisition technology is on the verge of emerging as promising surveying firms, researchers and managers of roads, railways, mines, power lines, forests and plants may read in their professional magazines that some colleagues already embraced the new, amazing technology. They want to know what is in for them. All conventional methods do have advantages and drawbacks. They were once chosen because they appeared to be the best of all that was available on the market at that time. However, technology is changing, and this provides opportunities to get the job done better, quicker and cheaper, while the data may have better quality for underpinning decisions. Revealing the opportunities is usually done through pilot or case studies, which often involve a comparison of two techniques, in particular laser scanning and photogrammetry in the case of *pointcloudmetry*. The type of platform may also be a topic of research.

This chapter reviews pilot studies conducted from 2009 onwards. The platforms carrying the sensors vary from tripod to unmanned airborne systems (UASs) and manned aircraft. The applications looked at include: corridor mapping, cultural heritage, mining, 3D mapping and forestry. First, we have a look at corridor mapping followed by cultural heritage documentation and mining. Next, the importance of point clouds for 3D mapping within the smart city and digital twin concept is considered. Following the section on forestry, we will have a brief look at the possibilities of serious gaming for conducting simulations in a virtual 3D environment. The chapter ends with a few concluding remarks.

TOTAL STATIONS

The reader may encounter the term total station in this chapter several times. For those who are not familiar with this workhorse of the land surveyor here, a brief explanation follows based on Lemmens (2016). A total station is an angle measuring device, called theodolite, integrated with an electronic distance measurement (EDM) unit. The integration enables the simultaneous measurement of horizontal and vertical angles, as well as slope distances using one device. The surveyor benefits in terms of portability, convenience and speed. Today, total stations have a wide variety of capabilities and are extensively exploited in cadastral surveying, in civil engineering and on construction sites. For many decades, and up until the 1960s, theodolites were the major instruments for collecting geodata referenced in a common coordinate system. The computation of coordinates also requires angles and distances. In the past, distances were measured with chains or measuring tapes. Without careful precautions, the chains and tapes produced inaccurate measurements. An important innovation that made distance measurements less cumbersome was EDM. Developed around 1940, EDM became commercially available in the 1960s. Microprocessors, solid-state memory, sensor technology, wireless communication and other innovations have been gradually incorporated into the devices over time. They further improved the efficiency of surveying and the quality of the measurements.

13.1 Corridor Mapping

A corridor is an elongated, narrow construction, which can extend over hundreds or even thousands of kilometres, while its width is in the order of tens or hundreds of metres. Corridors usually aim at transporting people, goods and services. This section considers the 3D mapping of roads, railways, railway tunnels and powerlines. The data acquisition technologies used are MLS, UAS photogrammetry and UAS Lidar.

13.1.1 Mobile Laser Scanning

This subsection focuses on mobile laser scanning (MLS) for road maintenance, road asset inventory and railway tunnel inspections.

13.1.1.1 Road Maintenance

Asphalt pavement is subject to deterioration due to warming and cooling, icing, car accidents, heavy traffic and overloading by trucks. Timely detection and mapping followed by rapid repair is essential to avoid worsening the road conditions. It is also necessary to avoid dangerous traffic situations leading to injuries. MLS is often used for road mapping since it is able to rapidly capture precise and dense point clouds of road pavement and road vicinity at normal traffic speed. Chen and Li (2016) used the RIEGL VMX-450 for capturing a road in Kingston, Ontario, Canada, in a pilot study to automatically detect pavement cracks. The feasibility study showed that the point density of the point cloud was dense enough to detect cracks at a centimetre level automatically (Figure 13.1). An earlier study conducted in south-east Finland demonstrated the suitability of MLS for road maintenance using commercial point cloud processing software (Schwarzbach, 2014). A section of 22 km of a two-lane road through the forest was captured forth and back from a van driven at 50–70 km/h. The van was equipped with a Trimble MX8 consisting of two Riegl VQ-250 scanners, four cameras and an

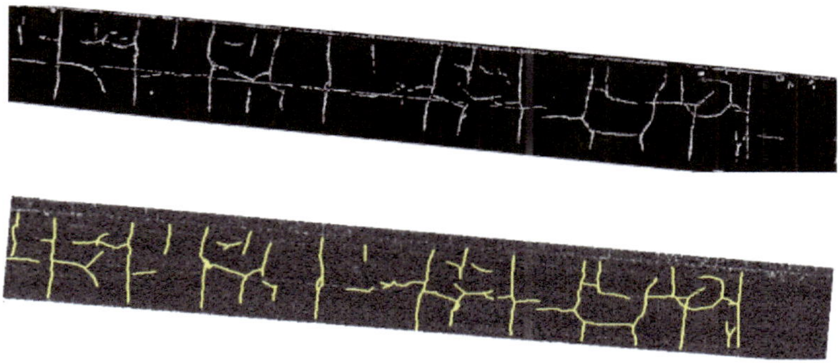

Figure 13.1 Automatically mapping road cracks in an MLS point cloud (top) and ground truth (bottom) (source: Chen and Li, 2016)

Applanix POS LV 520 positioning and orientation system. One camera recorded the road surface, and the other three pointed in the forward direction. Seventy-four GCPs were painted on both sides of three sections, each with a length of 500–600 m, at an interval of 50 m and measured using GNSS receivers and level instruments to provide accurate coordinates for georeferencing and accuracy assessment. Processing was done with Terrasolid's TerraScan, TerraMatch, TerraPhoto and TerraModeller. These modules run on top of Bentley's MicroStation.

GNSS signals were blocked by tunnels, buildings, trees, rocks or other obstructions near the road. This major error source could be tackled using GCPs and overlaps between the forward and backward scans. For high-accuracy work, the GCP spacing should not exceed 50 m for elevation and about 250 m for planar coordinates, provided that the elevations are measured with a levelling instrument to guarantee high accuracy. To detect road damages, a digital surface model (DSM) of the road was created on which an orthoimagery with a ground sample distance (GSD) of 1 cm was superimposed. This enabled the mapping of ruts and potholes with a depth of at least 2 cm. The mapping revealed the urgency for renewing the asphalt. The DSM allowed an accurate estimation of the surface renewal material. After thinning, the DSM was imported into a resurfacing machine. Thinning was necessary since the used road design software faced processing problems when confronted with massive point clouds.

13.1.1.2 Road Asset Inventory

MLS is suited for supporting highly accurate localisation and high-definition mapping purposes. This favourable property doesn't only support road maintenance as we have seen above but also road asset inventory (Ma et al., 2018). With respect to the mapping of roads and the signs painted on the pavement, pilot studies are conducted to automatically outline the paths that pass through a road intersection (transition lines) from MLS point clouds. Major automobile manufacturers all over the world together with digital map suppliers are eager to generate such driving lines from MLS data. Figure 13.2 shows that such road lines can be extracted semi-automatically with a high success rate.

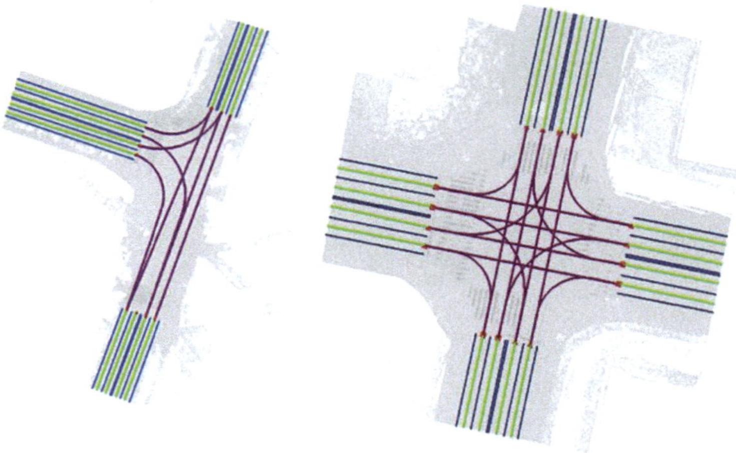

Figure 13.2 Road transition lines generated from high-density MLS point clouds (source: Ye et al., 2020)

13.1.1.3 Railway Tunnel Inspection

Tunnels undergo deformations due to the weight of the terrain above. To meet safety requirements, rail tunnels must be regularly inspected for deformations by comparing measurements captured over time. Usually, permanent benchmarks are measured with total stations by a survey crew – accurate but time consuming and expensive. Furthermore, a total station survey can interfere with the train timetables, causing delays and dissatisfied travellers. An alternative is the geometric modelling of tunnels from MLS point clouds, which can be done at sub-millimetre precision (Arastounia, 2016). Manual processing of the massive amount of points is costly, making automation key. A tunnel consists of a top part and a bottom part comprising the rail tracks and other railway infrastructure (Figure 13.3). Both parts suffer from terrain load, but the top part is the most affected and only this part needs to be modelled.

To test the feasibility of MLS for deformation monitoring, a 155-m trajectory of a rail tunnel was scanned with a Velodyne HDL 32E mounted on a rail car. Moving at 65 km/h, the trajectory was scanned within 10 seconds resulting in over 6 million points, which were registered in a uniform coordinate system. Using the X,Y,Z coordinates of each point, the tunnel's main axis was calculated and over 1,500 cross-sections were extracted at 10-cm intervals resulting in an initial model. The few parameters which describe an ellipse were computed from thousands of points using least-squares adjustment (LSA). Measurements not only contain outliers but also returns from objects mounted on the tunnel walls. These are visible as bulges in the data and have to be removed because they affect the modelling of the tunnel cross-sections as ellipses. LSA enables the removal of bulges and outliers automatically by computing the distances of the points to the ellipse computed from these points and removing the points with large distances to the ellipse using data snooping (Baarda, 1968). (For explanation of data snooping, see the text box on Non-Linear Least Squares Adjustment in Chapter 4; Section 10.3.1 on Weighted Linear Least Squares discusses some implementation issues.) This means the deviating

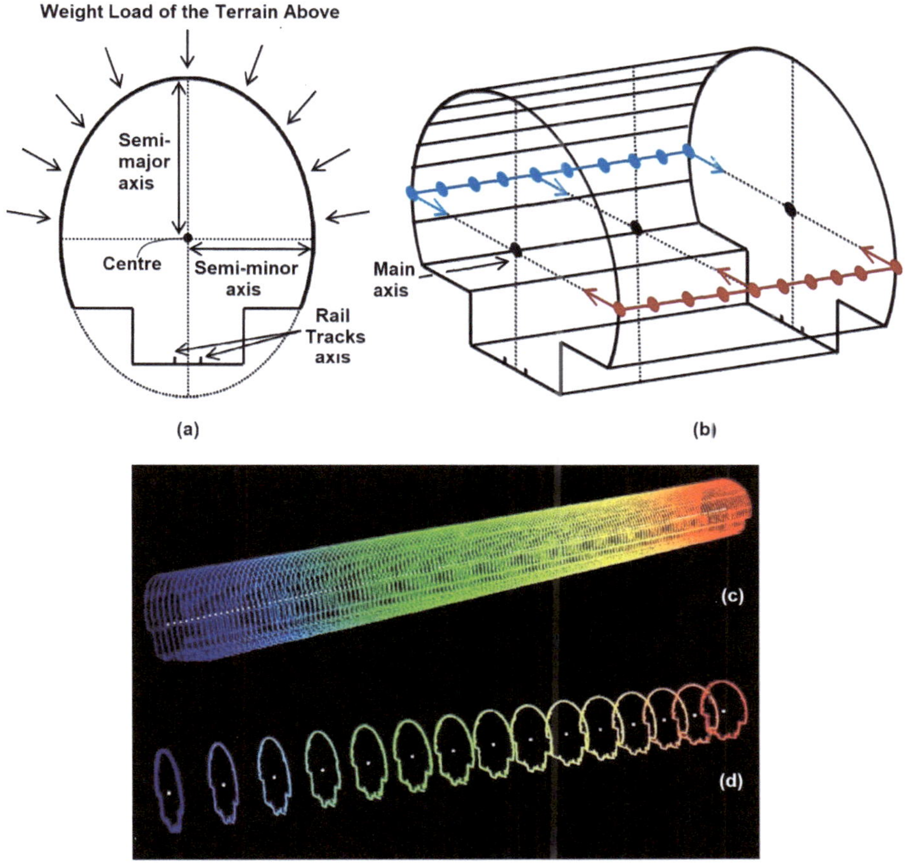

Figure 13.3 Model of a rail tunnel's cross-section (a), main axis computed as mid-point of two closest points (blue and red points) from opposite walls (b), the arrows are the normal vectors of the local best-fit planes; extracted cross-sections at 1-m (c) and 10-m intervals (d) (source: Arastounia, 2016; modified by ML)

points are treated as outliers. After the removal of deviating points, the parameters of the final ellipse are computed. LSA enables the computation of the standard deviations of the semi-major axis and semi-minor axis of the ellipse, which were 0.2 and 0.1 mm, respectively. The average deviation of the points which generated the ellipses was 1.2 cm. An MLS point cloud combined with automated processing allows more frequent inspections of deformation and improves safety.

13.1.2 UAS Photogrammetry for Railway Mapping

Monitoring and maintenance of rails, gravel beds, sleepers and extension works require accurate maps of railway trajectories and their surroundings. Ground surveys using total stations or GNSS are expensive and intervene in the regular train schedule. Aerial photogrammetry would be an alternative, but manned airborne surveys, carried out regularly, are costly. The continuous advancements in UAS photogrammetry triggered the performance of a pilot study (Karas and Safar, 2017). A 13.7-km trajectory of a non-electrified single-track railway in Drahany Highlands, Czech Republic, was chosen. The trajectory goes through various landscapes: flat agricultural fields, a narrow valley with steep slopes covered by tall conifers, and after reaching an altitude of 505 m, the track descends to meadows. A calibrated Canon EOS 700D was mounted on a modified Mikrokopter Hexa XL UAS. The camera has an image size of 22.3 × 14.9 mm and a resolution of 5,184 × 3,456 pixels, and the lens is a Canon EF 28 mm f/1.8 USM. To obtain a GSD of 1 cm, the average flight height was set at 65 m. Each image covered 52 × 35 m^2. The middle strip of three parallel strips recorded the axis of the railway, and the other two the surrounding area, including buildings. At some railway stations, where the single track disembogues into double or triple tracks, more flight lines were necessary (Figure 13.4). Taking two images per second allowed along-track overlaps of 80%. The across-track overlaps were 60%. Each flight captured a 350-m track, and to process the entire trajectory as a single block, the overlap between two adjacent tracks was set to 50 m resulting in 43 flights and 11,391 images.

For georeferencing, 173 GCPs were marked with fluorescent-green circles of 3–4 cm in diameter, which is 3.5 times the size of the required GSD. They were set up along the trajectory and measured using total stations. The angles and distances were transferred into 3D coordinates in the national Czech coordinate system (JTSK) system. Agisoft Photoscan Professional was employed for creating tie points, identification of GCPs and checkpoints, adjustment of the interior and exterior orientation parameters and the creation of the DSM. Trimble INPHO module DTMaster was used for DSM and digital terrain models (DTM) editing, and OrthoMaster and OrthoVista were used for orthorectification and seamless orthoimage creation.

The coordinates of points derived from the UAS photogrammetric end products were confronted with those measured with total stations. The planar coordinates of 1,375 checkpoints were measured in the digital landscape model created by superimposing the DSM with the orthoimage (1 cm GSD). This was done to improve the identification of the

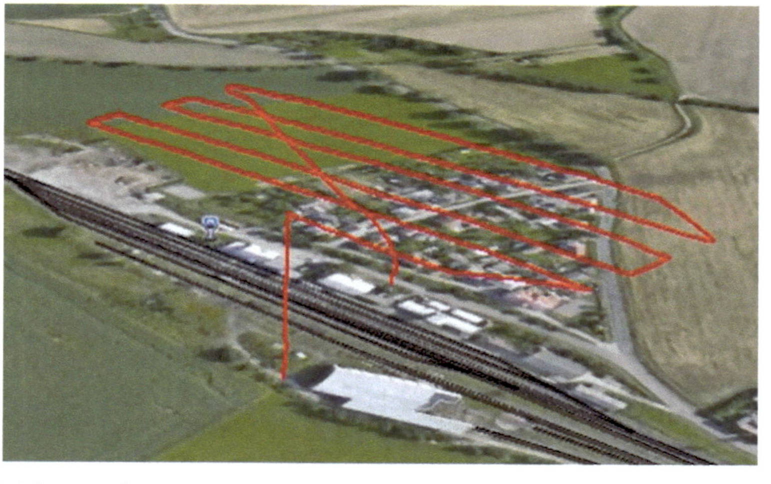

Figure 13.4 A large railway station required five flight lines instead of three for the single-track parts of the trajectory (courtesy: Upvision, Czech Republic)

points. For 80% of the points, the distances between orthoimage points and checkpoints were less than 20 mm; almost 90% of the points deviated less than 30 mm and 96.5% less than 40 mm. A GSD of 1 cm is high. Resampling the images to a GSD of 2 cm revealed that sleepers, platforms and other railway infrastructure can be mapped with a planar precision of 4 cm and a height precision of 6 cm. To guarantee high accuracy, 11,391 images were simultaneously processed in one single block. Processing on a single computer would take 2–3 months of non-stop computation. Since this is impractical, a computer network of 16 similar PCs (Intel Core i3 – 6100 CPU, 3.7 GHz, 8 GB RAM) was established. The computation of the DSM, DTM and orthoimage required 1,385 machine hours.

13.1.3 UAS Lidar for Power Line Mapping

Mapping of overhead power line corridors belongs to one of the first applications of airborne Lidar together with forestry. Especially when the corridors are long and go through rough, wooded areas, where ground measurements are cumbersome, labour intensive and costly, airborne Lidar provides an alternative. Trees underneath and alongside power lines may grow so tall that encroachment with wires and towers may cause a short circuit, forest fires and interruption of electricity provision. In vast and sparsely inhabited areas repair and mitigation require the deployment of expedition size. In addition, hurricanes, ice loads, landslides and other forces of nature may cause loss or dismantling of wires and towers. So, regular inspections have to be conducted all along the corridor.

Could UAS Lidar be a suitable alternative for airborne Lidar? GreenValley International, headquartered in Berkeley, California, USA, provided an answer by capturing a power line corridor with a length of 2,500 km in Guangdong Province, China (GreenValley International, 2018). The used LiAir UAS has been specially developed for power line mapping and consists of a DJI m600 Pro, which can stay in the air for 20 minutes, equipped with a Riegl mini-VUX1 – range accuracy 2 cm – GNSS devices and IMU (Figure 13.5). The maximum range of 250 m limits the flight height and requires careful flight planning because of the varying heights of the towers. Six teams conducted around five flights daily for 2 months in the spring of 2018. The point clouds were georeferenced

Figure 13.5 LiAir 250 equipped with Riegl miniVUX-1 Lidar sensor specifically developed for mapping the terrain beneath forest canopies (courtesy: GreenValley International)

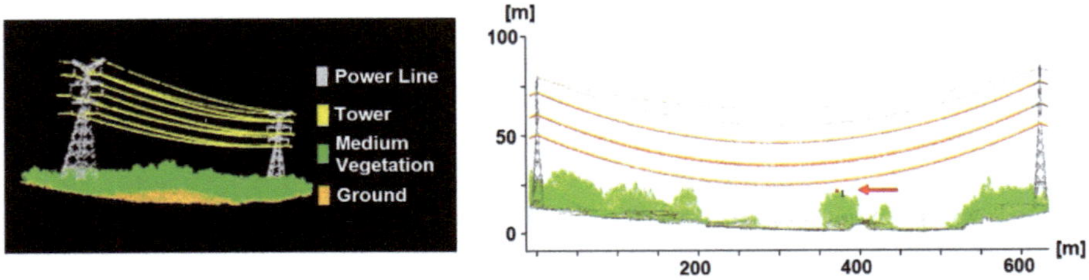

Figure 13.6 Result of automated classification of a power line corridor in China going through the forest (courtesy: GreenValley International)

and visually inspected. Machine learning algorithms, specially developed for power line classification, classified the points into power lines, transmission towers, ground, vegetation and buildings (Figure 13.6). The results of a machine learning exercise will never be perfect – manual editing was necessary for checking and correction. Hazardous locations were automatically detected. Future risks could be identified through simulations based on tree growth, damage caused by falling trees, wire sag under strong side-wind, ice coverage and high temperatures.

13.2 Cultural Heritage

Lidar and photogrammetric point clouds are well suited for documenting archaeological sites. Uncovering past constructions and relics of former civilisations – excavation – is invasive and thus destructive. Up until recently, documentation has been primarily done through total stations or GNSS receivers or through photogrammetry (aerial and close-range). Cultural heritage sites have usually a limited extent, and the ability of UAS photogrammetry to accurately capture a site of limited size within a short time span with high detail in a non-evasive way has made capturing the scene by cameras mounted on unmanned aircraft very popular among archaeologists and conservators. Another initiative to reduce costs is using smartphone cameras operated by photogrammetric laymen walking around the site.

13.2.1 Smartphone Cameras for Mapping Historical Buildings

Today, the quality of smartphone images is sufficient for 3D mapping. Combined with photogrammetric processing software running on a remote high-performance server, point clouds and 3D maps can be created highly automatically at low costs. This brings inexpensive 3D mapping in the reach of those tasked with the restoration of (heritage) buildings without the need for diving into the nitty-gritty of photogrammetry, as Nocerino et al. (2017) demonstrated in a large project – called Replicate. Three historical sites were captured by laymen, each site by three smartphones. After imagery uploading to the remote server, these laymen got feedback on sharpness, light conditions, gaps in the overlap of the images and other quality aspects. After completion of the image capture, the heavy computational burden of a bundle block adjustment,

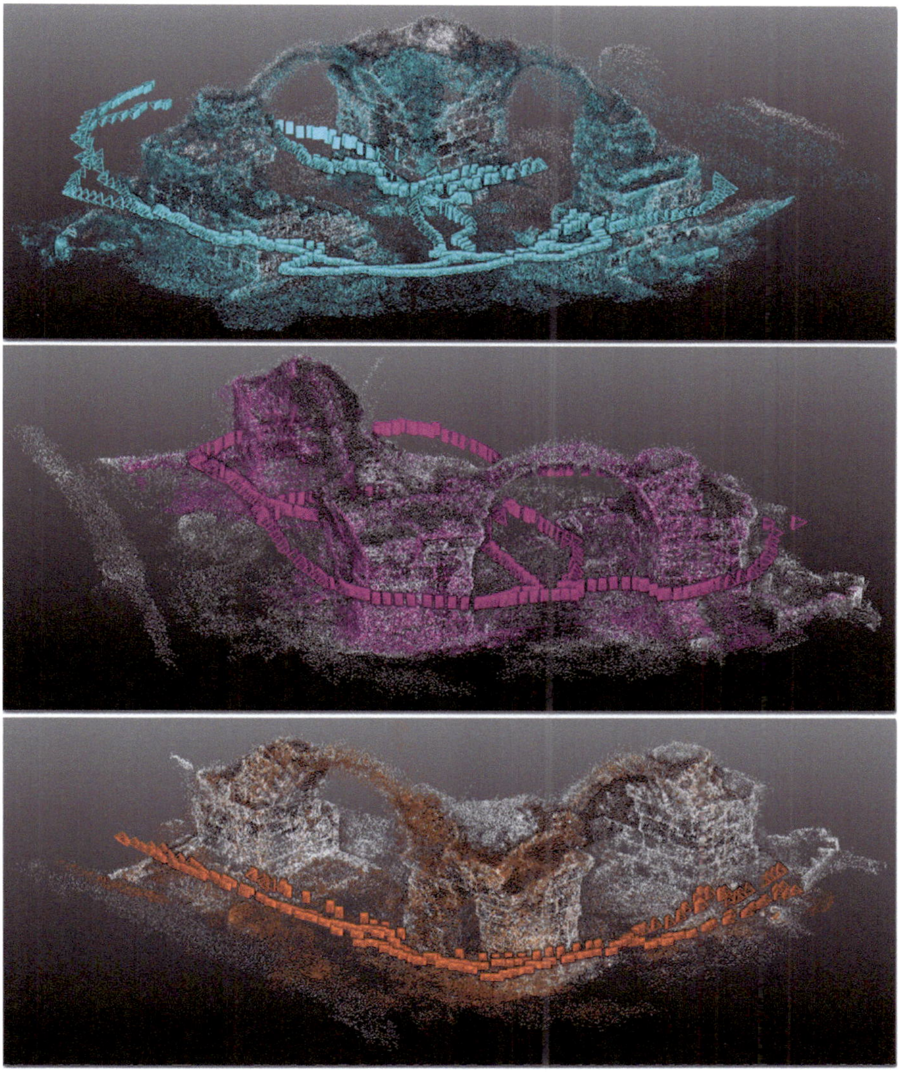

Figure 13.7 Three sparse point clouds of Saranta Kolones (coloured) generated from three smartphones; grey: the entire point cloud (source: Nocerino et al., 2017)

dense image matching and the creation of a TIN for generating 3D maps were executed on the remote server. Figure 13.7 shows the results for the Saranta Kolones monument at Pafos, Cyprus. Comparison with checkpoints showed that the geometric precision lies around 1 cm. The major challenge is not the positional precision but appeared to be the training and guidance of laymen through a proper image capture workflow especially when the objects are large and have complex shapes. Another challenge is the avoidance of occlusions caused by pedestrians, cyclists, cars, trucks and other moving objects.

13.2.2 UAS Photogrammetry for Cultural Heritage Mapping

UAS has attracted tremendous attention from surveyors and other geodata collectors, including archaeologists all around the world. Fixed-wing aircraft and multi-copters equipped with GNSS, IMU and RGB, near-infrared (NIR) or thermal infrared (TIR) cameras and increasingly Lidar have also evolved and matured into full-fledged surveying devices for creating height models, orthoimages, 3D maps and other geomatics products. Here we focus on UAS photogrammetry conducted in Rapa Nui, also called Easter Island, and Tintagel Island, Cornwall, UK.

13.2.2.1 Rapa Nui

Rapa Nui, the most isolated inhabited place on the planet, was first documented in European annals by Roggeveen. The Dutch admiral set foot on the island on Easter Sunday in 1722, and so it was baptised Easter Island. It has the size of Texel, the largest Wadden island in the Netherlands. Its gigantic statues, called Moai, inscrutable history and location in the middle of the Pacific Ocean made it famous. In the early 2000s the statues, which are UNESCO cultural heritage since 1995, were touched by laser light for the first time. Up to that time, photogrammetry and tapes were used. To lift a tip of its historical veil, the University of Hawaii wanted detailed and accurate 3D maps of the statues. The surveyors of D. Wellman Surveying, USA, brought a Cyrax 2500 TLS to the island to capture the statues and total stations to measure the 3D coordinates of GCPs for georeferencing the points (Wellman, 2003). The capturing of the Moai by TLS was a success, and HafenCity University Hamburg, Germany, continued the documentation of the Moai in 2007 using a Trimble GX TLS in a collaborative effort with the German Archaeological Institute (Kersten et al., 2010). Today photogrammetry is back on Rapa Nui, thanks to unmanned airborne systems.

UASs can be easily transported, and this is ideal for creating point clouds, orthoimages and digital landscapes of remote spaces. A UAS survey of the entire island, conducted in September 2015, took two operators using two UASs simultaneously for 10 days (Norambuena and Sainz, 2016). During 52 flights – path lengths varying from 25 to 28 km – 21,840 images were recorded with an along/across-track overlap of 90%/70% and 12 cm GSD. Sixty-seven GCPs were measured with double frequency GNSS in static mode, resulting in a planar precision of better than 5 cm and a height precision of 10 cm. After aerotriangulation, RMSE appeared to be 3.6 cm and the GSD error was 0.77. Aerotriangulation and creation of a DSM, orthoimagery and a digital map at a scale of 1:500 were carried out with Agisoft PhotoScan Pro. The DSM contains 11 points/m^2. Figure 13.8 shows a small portion of the orthomosaic of the entire island. Note that the shadows of the statues do not point in the same direction. This indicates that this is an orthomosaic created from images captured at different times of the day. From the point cloud, two raster DSMs were generated: one DSM with a grid spacing of 56 cm and the other a high-resolution DSM with a grid spacing of 30 cm. Figure 13.9 shows two parts of the digital landscape model created by superimposing the orthomosaic over the DSM. Also, elevation contour lines were computed. Figure 13.10 shows a part of the digital map consisting of the orthomosaic superimposed with contour lines. One operator spent 50 working days processing the imagery to end products.

Figure 13.8 Part of the orthomosaic of Rapa Nui created by UAS photogrammetry (source: Norambuena and Sainz, 2016)

Figure 13.9 Two parts of the digital landscape model of Rapa Nui created by UAS photogrammetry (source: Norambuena and Sainz, 2016)

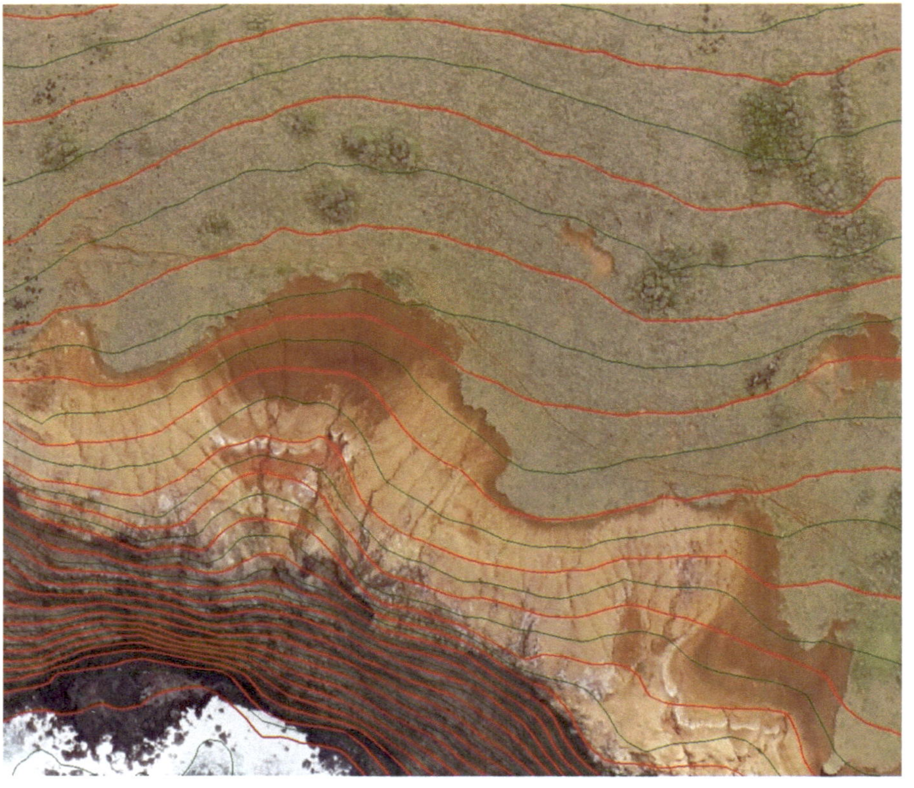

Figure 13.10 Part of the digital map with contour lines derived from the DSM (source: Norambuena and Sainz, 2016)

13.2.2.2 Tintagel Island

Point clouds are also the source for printing accurate and detailed 3D models, which may be exhibited in the visitor centre of a cultural heritage site. The ruins of the Tintagel Castle area in Cornwall, UK, have an association with the legends of King Arthur and are a major tourist attraction. In the 19th century archaeological investigations started. A new exhibition brings King Arthur to life and shows the island changing over 1,500 years of history – from a thriving Dark Age settlement to a medieval fortress, through to romantic ruin. The centrepiece is a printed 3D model of the island and adjacent mainland created from UAS photogrammetry (Blogg, 2016). Due to project deadlines, the survey had to be executed during winter, a windy and gloomy period in this part of the world. Poor light conditions require long exposure times which cause blur. The shooting of sharp images with a short exposure time requires capturing a wide dynamic radiometric range, which means storage in raw image file format using a high-quality camera. The relatively flat parts could be captured by vertical images taken from a fixed-wing UAS. To obtain uniform details, the cliffs and the ruins had to be captured with a multicopter shooting oblique images. Processing with Agisoft Photoscan resulted in a point cloud with an RMSE of less than 5 cm for the *X*, *Y* and *Z* coordinates. The 42 million points were converted to a TIN. Before inputting the TIN into the 3D printing machine, the meshes had to be cleaned from small spikes, tunnels, holes and non-manifold faces and intersecting polygons. The TIN created from all 42 million points was impractical to work with. Thinning to 15 million points produced the best 3D model.

13.2.3 Airborne Lidar

The ancient Maya site of Copan, Honduras, was captured by a Lidar system mounted on a manned aircraft. Copan was prosperous and mighty for four centuries emerging in the fifth century when the Roman Empire started to decline. Excavation of temples and pyramids started in 1850. From 2006 to 2008, the maps of 150 years of documentation archaeological research were digitised. To improve and extend the area covered, an airborne Lidar survey was conducted for 4 days in May 2013 using a Leica ALS50 Phase II as part of the interdisciplinary project MayaArch3D (von Schwerin et al., 2016). MayaArch3D aims at joining distributed archaeological data together through WebGIS tools to stimulate collaborative archaeological research. The ALS50 Phase II system captures four returns for each emitted laser pulse. To obtain the intended 15 points/m^2 plus, the area was captured twice through an across-track overlap of over 50%. The average first return density was 21.57 points/m^2. The density of last returns, which means returns reflected from the ground surface, depends on topography and permeability of the vegetation canopy and was 2.91 points/m^2 on average. Where vegetation density is high, most pulses don't reach the ground compromising map quality.

The points were classified into bare earth, archaeological ruins and modern buildings using proprietarily automated and manual ground filtering software. The classified points were the basis for creating a bare earth model (DTM). A DTM plus archaeological structures were also created. The filters separated bare earth from vegetation but were unable to separate bare earth from archaeological structures. The semi-automatic filters separated mounds, which are collapsed structures, from topographic features such as natural hills, but were not able to distinguish mounds, which have a small vertical extension – 14% of the features classified as bare earth were actually mounds. The filters also removed important archaeological objects, a characteristic of Lidar filters that is customised to separate vegetation from manmade structures. Nevertheless, airborne Lidar point clouds can improve archaeological maps and identify hidden structures as laser pulses partly penetrate the jungle canopy. The MayaArch3D efforts enable archaeologists to employ 2D and 3D WebGIS tools to collaborate online and bring distributed archaeological data together (Figure 13.11).

13.2.4 TLS for Mapping Technological Heritage

Cultural heritage not only concerns millennia-old relics of disappeared civilisations but also industrial and technological monuments built in the 19th and early 20th centuries. One such monument is the Shukhov Tower in Moscow, Russia, located a few kilometres south of the Kremlin. The radio tower, built between 1919 and 1922 and called after its designer V.G. Shukhov (1853–1939), is an icon of Soviet broadcasting and a world-renowned monument of Russian avant-garde architecture. Radio broadcasting was conducted from 1922 until 2001 and TV broadcasting from 1937 until 1967. The slender

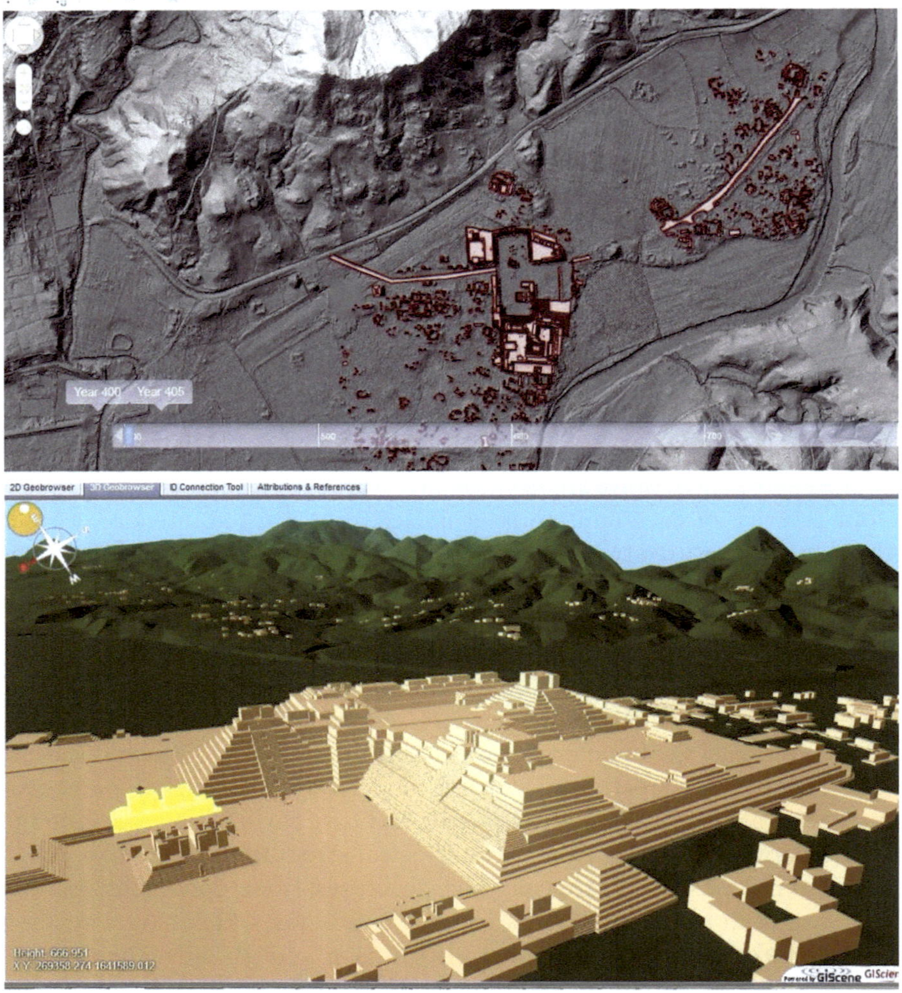

Figure 13.11 2D Geobrowser (top) showing archaeological structures of the ancient Maya site of Copan, Honduras, overlaid on an airborne Lidar DTM; scene viewer (bottom) showing 3D city model overlaid on DTM (source: von Schwerin et al., 2016)

construction consists of six hyperboloids constructed of steel rods riveted together with horizontal rigidity hoops, which form a hyperboloid of revolution (Figure 13.12). Antennas are mounted on top of the hyperboloids. The original design has been pretty well preserved, but its state of severe corrosion testifies to the centennial age of the 150-m-high steel construction. The icon has been long under the threat of demolition.

To support restoration and improve documentation, a TLS survey was conducted in December 2011 using a Leica ScanStation2 and a FARO Focus3D; see Figure 13.13 (Leonov et al., 2015). The capture of the exterior of the tower all the way around was done by scan positions on the ground outside the tower, while the lower sections from the inside were scanned from a ground position in the centre of the tower. The

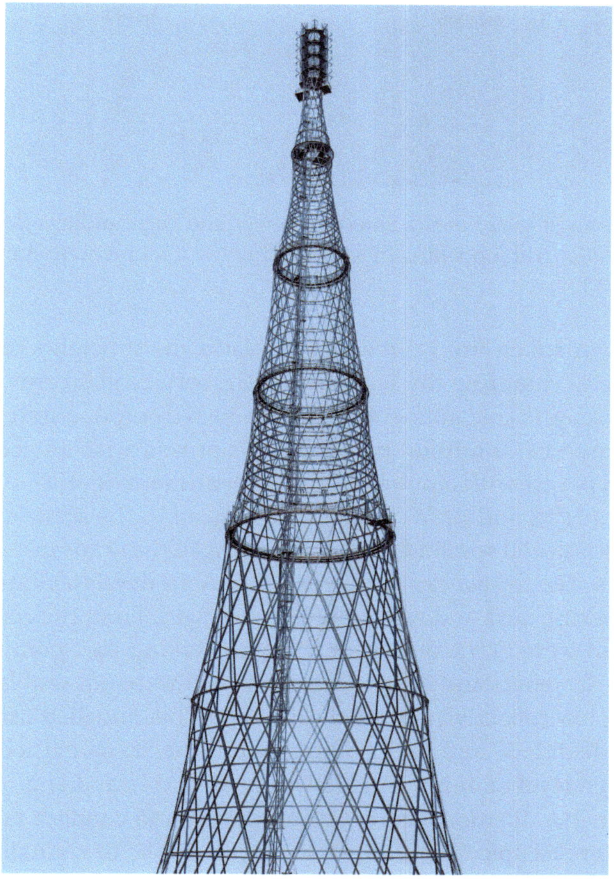

Figure 13.12 3D model of the Shukhov Tower, Moscow, Russia, obtained from TLS point cloud (courtesy: Mikhail Anikushkin, Trimetari Consulting, Saint Petersburg, Russia)

Figure 13.13 Laser scanners used in the TLS point cloud acquisition project of Shukhov Tower (courtesy: Mikhail Anikushkin, Trimetari Consulting, Saint Petersburg, Russia)

Figure 13.14 TLS point cloud of the Shukhov Tower and surroundings (left), view on a part of the 3D model (middle) and modelling of a joint using the documentation made in 1947 (source: Leonov et al., 2015)

upper sections were taken from three service platforms at heights of 100 m, 125 m and 141 m, respectively. Mounting the scanners on the service platforms was challenging as December in Moscow is cold and windy, and the heavily eroded state of the steel did not add to safety. Around 100 million points were captured with an accuracy of 7 mm, of which 65 million points with a point spacing at centimetre level.

Standard Leica HDS and B&W targets were placed in the scene as artificial markers and measured with a total station for georeferencing of the scans in the Moscow geodetic reference systems. The fieldwork took three operators 2 days. In contrast, the 3D modelling took 300 working days and was done in two stages. First, the rods and rings visible in the point cloud were extracted. Next, the painstaking part commenced: thousands of rivets, gussets, fasteners and other elements which were not visible in the point cloud due to occlusion, low reflectivity or small size had to be modelled using drawings of the construction made in 1937 and 1947 (Figure 13.14). The 3D model accurately documents the tower at a high level of detail and allows constructors and engineers to renovate or rebuild the tower. The 3D model allows us to inspect and conduct measurements easily in the office. A stereoscopic tour allows the general public to virtually climb the tower.

13.2.5 Site Characteristics and Survey Method

In this section, airborne Lidar, close-range photogrammetry, UAS photogrammetry and terrestrial laser scanning have been considered. Which survey method is best suited for which site? Historic England (2018) provides criteria to decide whether an airborne Lidar point cloud is suited for capturing a particular cultural heritage site. The main criteria include the following:

- The archaeological survey should aim at producing a detailed (3D) map.
- Therefore, the archaeological objects should be sufficiently large to be visible in the point cloud.
- Since airborne Lidar data are expensive, the size of the area should extend 50 km^2.
- Archaeological sites covered by wood particularly rely on airborne Lidar since this is the only remote sensing technique able to capture the ground surface in detail.
- Is an airborne Lidar point cloud of the area available off-the-shelf?
- Is processing hardware/software and the skills to use it available?

Research conducted by Rüther et al. (2009), who surveyed the Wonderwerk Cave, South Africa, and Banse et al. (2015), who surveyed the necropolis of Al-Ayn in Oman, sets light on the strengths and shortcomings of various ground-based surveying methods for cultural heritage documentation. The studies reveal that the usability of total stations, GNSS receivers, terrestrial laser scanning and ground-based photogrammetry for archaeological applications depend on various site characteristics, including the following:

- *Occlusion*, i.e. one object or part of an object becomes invisible due to the presence of other objects between the instrument and the occluded object. Minimising occlusion requires careful sensor set-up. Capturing the site with 80% overlap between adjacent images reduces data gaps due to occlusion.
- *Light conditions.* A cave, for example, is dark and often deep, which makes imaging unfeasible. Photogrammetry would result in time-consuming fieldwork and processing of the images. When captured in daylight, shadow obscures the joints of the quarry blocks, which impedes dense image matching. Shadows are absent in laser scans.
- *Sky view.* Measuring with a GNSS receiver requires a line of sight between the points and the satellites. Therefore, GNSS receivers are unsuited for jungles, subsurface sites and urban canyons.
- *Number of points.* Regularly shaped manmade structures consisting of facades and roofs can be modelled as planar shapes, which require the measurement of relatively few numbers of characteristic points. In these cases the option of surveying with total stations or GNSS receivers is feasible. Irregularly shaped, natural surfaces, such as the walls of a cave, require tens of thousands of points, pushing away the conventional land surveying from practical use.
- *Geometric accuracy* of the coordinates of the points. Similar distances between the sensors (camera or laser scanning) and objects result in similar accuracies.
- *Workload* to capture and process the collected data. The time of fieldwork and workload of data processing of TLS and ground-based photogrammetry are similar. The most time-consuming part for both is the measurement of GCPs.
- *Volume and weight of the equipment.* Although TLS scanners are becoming gradually more portable and compact, the older TLS systems still appear less than perfectly suited for surveying remote sites or to quickly capture scenes such as construction sites because the equipment is bulky, heavy and difficult to move around. In particular, this is annoying under harsh conditions such as those present in deserts. In the words of Neubauer (2007), who carried a Riegl LMS Z420i scanner combined with a calibrated Nikon D100 digital camera to the Cheops Pyramid and the Sphinx of Egypt in the framework of the Scanning of the Pyramids Project 2004 (Neubauer et al., 2005): 'You might wish for a lighter scanner, but when the equipment faultlessly works for a full ten hours of data collection, even during a sand-storm, you appreciate its robustness.' Today TLS devices are handier without compromising robustness.
- *Costs of equipment and processing.* Archaeologists welcomed TLS, but the equipment is relatively expensive. They increasingly use lightweight cameras which are suited for photogrammetric data capturing. They are cheap and can be easily transported to remote areas. Software, which allows self-calibration, produces

accurate 3D maps and other documentation. Dense image matching creates point clouds in which small objects show great detail. Open-source software and cloud computing services enable the cheap production of point clouds.

13.3 Subsurface Mining

Underground mines consist of a complex of long, narrow tunnels, corridors, excavation chambers and shafts. The spaces are often cramped, while temperature and humidity conditions may be harsh. Managers require information on production volumes, the progress of planned extension of tunnels, the volume of material moved and places of potential collapse due to deformation or subsidence. They may also want to compare the as-built tunnels with the design, for example, to check whether the diameter is sufficient for transporting mining machines. Total stations, inertial navigation systems and levelling have limited applicability beneath the surface. Such equipment is appropriate for mapping the layout of tunnels, monitoring deformations and measuring points for creating simple 3D maps. Detailed 3D maps of tunnels would take too much time. TLS may be a cost-effective alternative (Wang et al., 2013). With a Trimble FX 3D phase-shift scanner, 20 km of tunnels of a gold mine in China located 650 m below the surface were captured. Ranges were measured with millimetre-level precision up to a distance between sensor and object of 100 m depending on reflectivity. Recording at the full field of view of 360° × 270°, a single scan produced 65 million points in 5 minutes. Target spheres were placed as tie points in the overlaps of the scans for co-registration purposes. For transforming the scans into the coordinate system of the mine, marks were attached to the tunnel walls and measured with reflectorless total stations. One GCP per four scans provided the required precision. From the georeferenced point cloud, pipes and conduits, ducts, drainage channels and other objects were manually extracted and modelled as planar and curved surfaces, tubes and other regular geometric shapes using software tools. The data format should allow us to import the data into software that mine-surveyors are used to and feel comfortable with. The survey and mapping of the gold mine took 1 year.

13.4 Gravel Quarry Mining

Dense point clouds of gravel quarries taken within regular time intervals enable the calculation of volume extraction, monitoring of production progress and supporting daily logistics such as determining routes of dump trucks and optimal location of the diggers. Managers of a gravel quarry require information on the amount of gravel dug within a certain time span and the volume of gravel on the stock. The surveys should disrupt the digging work as least as possible as this would waste time and thus reduce revenues. Also for this application, UAS photogrammetry comes into view as an alternative to total stations, GNSS and TLS. Some mining activities create artificial lakes, which are often used for leisure after the exhaustion of the resources.

13.4.1 Consumer-Grade UAS Photogrammetry

UAS photogrammetry is exploited all over the world for capturing quarries and open-pit mines. Usually, the UAS has been specially designed for 3D mapping purposes, which means the UAS is a professional device and thus relatively expensive. However, in some parts of the world, for example, Ukraine, the budgets are tight, while professional ground surveys are expensive. Sufficient points on steep slopes and ledges have to be accurately measured to calculate extracted volumes. This may be cumbersome and tedious because of the extent of the area or its inaccessibility. Ongoing earthworks often destroy GCPs, requiring monthly checks and repairs, while safety is also an issue. Is a consumer-grade copter equipped with a camera an alternative? Gorkovchuk et al. (2017) investigated its suitability using the Rafalivka basalt quarry, located in the Rivne province, as a pilot study. The quarry with a depth of 50 m was flown with the low-cost DJI Phantom 3 Pro quadcopter equipped with a 12-MP camera with a fixed focus. The copter can stay in the air for half an hour allowing it to capture the 70-ha quarry in two flights. Capturing images in automatic mode can be done by designing a flight plan with Pix4D or DroneDeploy (Figure 13.15). Around the borders, 11 GCPs were distributed and 7 temporary GCPs were placed within the quarry as checkpoints. The coordinates of the 18 GCPs were measured with a total station. The batteries required a warming-up because of the temperature drop below 15°C. To diminish the effect of the strong wind, the initial planned flight height of 120 m was lowered to 60 m. Although less strong at the altitude, the wind was strong enough to increase power consumption; after 15 minutes, the battery was exhausted and had to be changed, thus increasing the number of planned flights from two to six. The survey could be completed within one

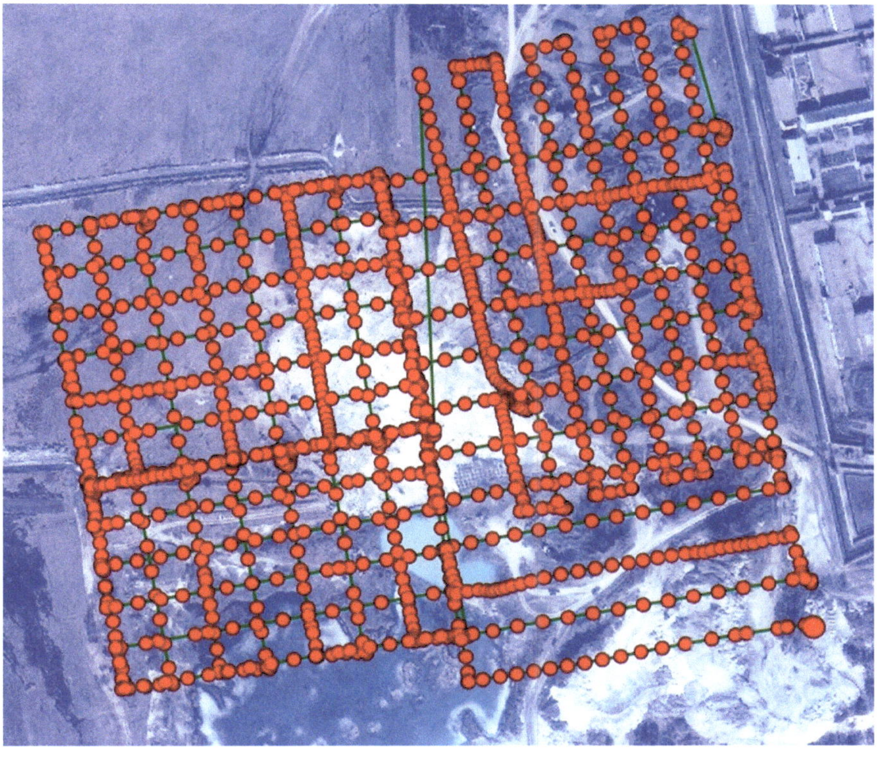

Figure 13.15 Flight paths and exposure locations of the UAS survey over a basalt quarry in Ukraine (source: Gorkovchuk et al., 2017)

workday, after 12 km of flight resulting in 1,324 images with 90% overlap and a GSD of 3 cm. After collection of the imagery, focal length, principal point and other camera parameters were computed simultaneously in the aerotriangulation procedure. The interior orientation parameters were stable over time. Next, a DSM point cloud was created using image matching from which a DEM was derived by removing vegetation, buildings, vehicles, quarry machinery and other non-ground points. Finally, orthomosaics, contour maps and volumes were generated from the DEM and original images (Figure 13.16). The seven checkpoints revealed a 3D RMSE of 7 cm, which is well within the national standards and allows accurate volume calculations. Because of the number of images, the computation on a desktop with a Core i7 processor, 64 GB RAM and a GTX980Ti graphic card took 7 hours. The computation on a laptop with a Core i7 Mobile processor, 8 GB RAM and a GTX 960 M graphic card took 26 hours. The 64 GB RAM of the desktop enabled the exploitation of full-resolution images, whereas the laptop could only process the images at half-resolution. The lower resolution slightly reduced the positional accuracy by 3 cm compared to the full-resolution solution but significantly reduced the level of detail. The high processing times of highly overlapping images limit the use of consumer-grade UAS in small areas. Permanent GCPs may

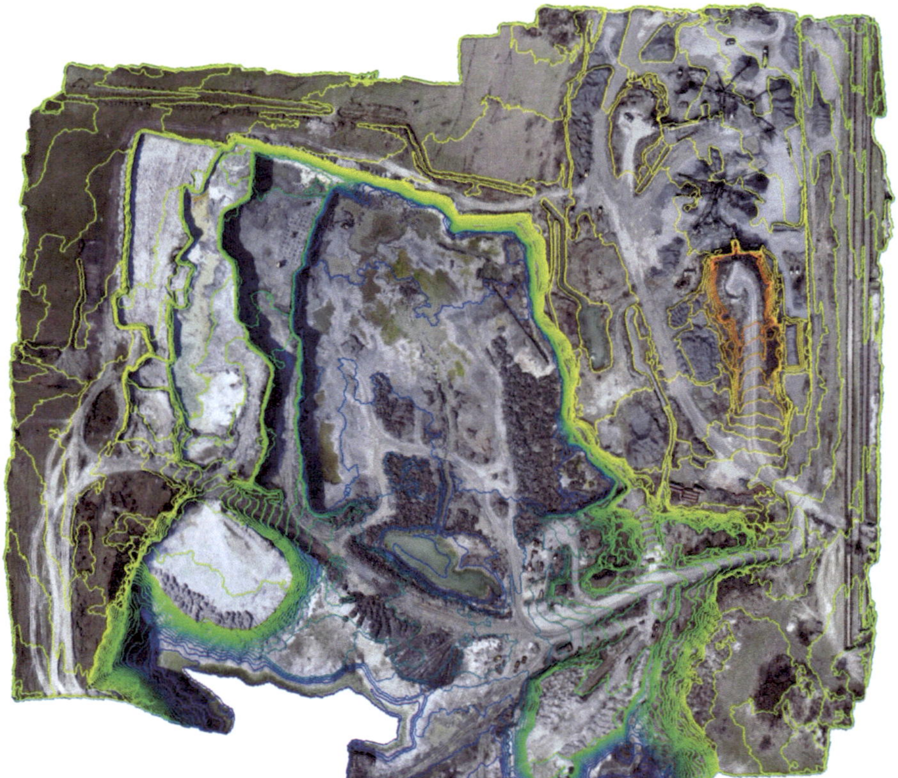

Figure 13.16 Contour lines superimposed on the orthomosaic of a basalt quarry in Ukraine (source: Gorkovchuk et al., 2017)

increase accuracy and reduce survey time but requires regular checks on damage or destruction.

13.4.2 TLS Versus UAS Photogrammetry

Mining managers can choose from a variety of geodata acquisition techniques to determine volume extraction and create DTMs and orthomosaics, all products which help proper management of mining sites. The choice of a particular technology depends on many factors, including:

- Accuracy
- Costs
- Weather conditions through the year
- Impact on the existing workflow
- Safety
- Size, shape and layout of the site

The accuracy of the point clouds determines the accuracy of the volume of extracted material, productivity rates and projections of future earnings. Accurate estimation of excavated volumes is also important for determining whether the mine can meet the orders from clients. So accuracy is the basic factor which determines the suitability of a survey method and is usually a major performance criterion. There is a direct relationship between costs, revenues and profit. The costs include the purchase, maintenance and lifespan of the equipment, operational expenses, impact on the workflow and additional training costs. Mining sites are often located in parts of the world that have harsh environmental conditions. A UAS is sensitive to harsh weather conditions and dust. When a survey team has to wait days to conduct a survey, this adds to inefficiency and costs. The acquisition of point clouds is always followed by a processing/analysis stage in which the information, which is essential for the task at hand, is generated, preferably in a highly automated way. The costs of the workflow and the costs of changing existing workflows are factors to be considered. Costs are also associated with interruption of the production activities. The safety of the surveyors is a crucial factor. The layout of a mine determines accessibility and visibility from surveying positions.

Sometimes the accuracy and level of detail of one technology are better than the other. Nevertheless one may choose the "inferior" technology because of the beneficial properties of the other factors. To test the suitability of TLS compared to UAS photogrammetry, a four-hectare gravel pit in the Olten region of north-western Switzerland was captured with both technologies (Weber and Lerch, 2018). Factors investigated included accuracy and density of the point clouds, in-office preparation time, on-site data acquisition time, processing time and costs. Preparation for the survey involved determining the optimal distribution of GCPs and TLS scanner positions and designing the flight plan of the UAS. The site was flown with a senseFly eBee Plus RTK/PPK, which is a professional fixed-wing UAS aircraft with proprietary flight planning and management software (Figure 13.17). Take-off and landing place was a meadow next to the pit. With Agisoft PhotoScan, a DSM and an orthomosaic were generated. The site was also captured with a Trimble SX10 scanning total station positioned at three TLS

Figure 13.17 The SenseFly eBee Plus RTK/PPK equipment (left) and Trimble SX10 scanning total station (right) used in a comparison pilot study (source: Weber and Lerch, 2018)

stations outside the pit and two at the bottom of the pit. On average, the TLS survey took 45 minutes per station, adding up to an on-site survey of a half-day. To georeference the TLS stations and to assess the accuracy of the UAS flights, nine GCPs were laid out and measured with a Trimble R10 GNSS receiver.

The georeferenced TLS and UAS point clouds were analysed in CloudCompare and Autodesk CAD Civil 3D. With accuracy at the millimetre level, TLS points are in the order of ten times better than UAS points. TLS also produces higher point densities. RTK-enabled UAS photogrammetry provides sufficient levels of detail and accuracy even when no GCPs are used. For larger sites, UAS photogrammetry provides greater efficiency and improved surveyor safety compared to TLS. For deep pits – the pilot pit had a depth of 40 m – occlusions in the UAS images require interpolations causing a decrease in the accuracy of occluded areas.

13.4.3 Integration of TLS and UAS Photogrammetry

In the above, researchers investigated the pros and cons of TLS versus UAS photogrammetry for mapping an open gravel pit mine. On the other hand, some researchers advocate the integrated use of TLS and UAS photogrammetry particularly for capturing open-pit mines in remote and difficult-to-access areas. For example, Tong et al. (2015) demonstrate its suitability by conducting pilots at three open-pit phosphate mines in Yunnan province, China. The integration of both surveying methods is stirred by the observation, also found by Weber and Lerch (2018), that the precision of TLS points is an order of magnitude better than for UAS photogrammetry. As a result, the approach is not just based on merging the point clouds from TLS and UAS photogrammetry but to exploit the high precision performance of TLS. This is done by using the TLS points as GCPs for georeferencing the UAS point clouds. Since there are many TLS points distributed over the entire site – in contrast to the usually sparse GNSS networks of which the points are distributed along the pit border – the precision of the UAS points can be considerably improved. Tong et al. (2015) achieved a decimetre-level accuracy for the photogrammetric DSM and the orthomosaic. This approach is suited for application in remote mountainous areas where accessibility is low and the safety of the survey crew may be at risk.

13.5 3D City Maps

The advantages of 3D city maps were recognised many centuries ago. Armed with pencil and paper, surveyors and cartographers in the 16th and 17th centuries exploited their craftsmanship to produce bird's-eye views which combined a 2D map with perspective views of buildings (Figure 13.18). Hundreds of years later, aerial images processed with digital photogrammetric workstations enabled the extraction of 3D coordinates of corners of buildings and other key points, albeit in a labour-intensive manual process. The various types of imagery (satellite, aerial and street level), airborne, mobile and terrestrial Lidar and topographic and 2D maps containing building footprints allow 3D city models to be created highly automatically. As airborne Lidar – and, later on, mobile mapping systems – became more mainstream, this boosted the research into automated 3D mapping.

3D maps are essential for the realisation of the digital twin which is key to the popular smart city concept. Capturing and representing the 3D space of a city is challenging because of the level of detail required and the vast amount of data. Automated processing may help. The experiences gained in the city of Istanbul show that 3D city mapping has many issues, challenges and limitations, including excessive data storage requirements, the need for manual editing, incompleteness and other data quality issues (Baskaraca et al., 2019). The core data sets for creating the 3D map of Istanbul are an airborne Lidar

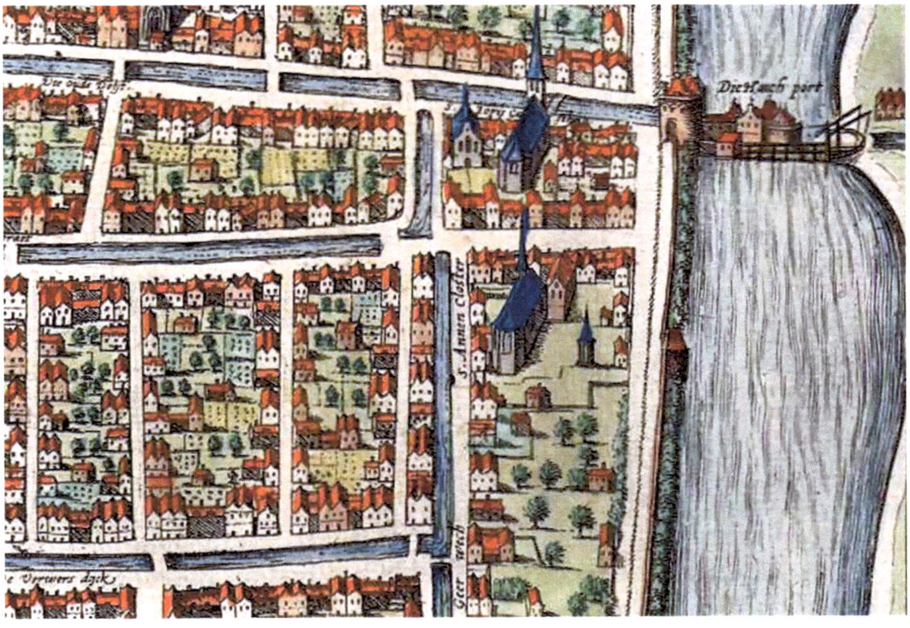

Figure 13.18 This bird's-eye view of a part of Delft, the Netherlands, drawn in 1580, is an example of the early need for 3D maps

DSM and DTM, aerial images recorded simultaneously with the Lidar surveys and the building footprints extracted from the 1:1,000 base map created from aerial images. The Lidar point cloud and images were captured during helicopter surveys – flight height 500 m – conducted in 2012 and 2014. The helicopter was equipped with a Q680i Lidar system from Riegl, a DigiCam 60 Mpixel camera, an AeroControl GNSS/IMU navigation system and an IGI CCNS-5 flight management system. The average point density of the Lidar point cloud is 16 points/m^2. The images were recorded with a GSD of 5 cm and with 60% along-track and 30% across-track overlap. To ensure high geometric accuracy, eight GNSS base stations were used. Of course, the survey of urban areas with an extent of 5,400 km^2 requires thorough flight planning.

The Lidar data was stored in 17,000 LAS files each covering an area of 500 m by 700 m and geometrically corrected using proprietary software of the manufacturer of the laser scanner. Next, a DSM, a bare ground DTM and a normalised DSM (nDSM), all three with a GSD of 25 cm, were created. An nDSM is a DSM in which the heights of buildings and other objects refer to the ground surface instead of a local or national reference system. Orthoimages were created from the DSM and simultaneously recorded aerial images. Next, classes were assigned to the points: ground, building, and low, medium and high vegetation using MicroStation V8i Connect, TerraSolid and TerraScan software. The automatic classification resulted in an accuracy of 90%, which induced the need for tedious manual repair efforts. A major source of error was confusion between nearby objects such as buildings and trees (Figure 13.19). Next, the classified points were combined with building footprints and orthoimages using TerraModeller to automatically generate around 1.5 million building outlines at the level of detail (LoD) 2. Next, the building blocks were augmented by automatically adding roofs using TerraModeller. Around 3,800 landmarks such as mosques and churches were mapped with higher geometric details of facades and roofs (LoD3) than other buildings based on architectural 3D CAD files. Also, bridges and other complex structures had to be manually mapped. Since no georeferenced photos taken at street level were available, the buildings couldn't be photo-textured (Figure 13.20). To increase the level of detail and to cover facades with an image texture, mobile mapping surveys will be conducted using a car equipped with laser scanners and panoramic cameras. Narrow streets will be captured by backpack MMS. The capturing of the 32,000 km of roads and streets of Istanbul will produce 2.73 petabytes of panoramic image data.

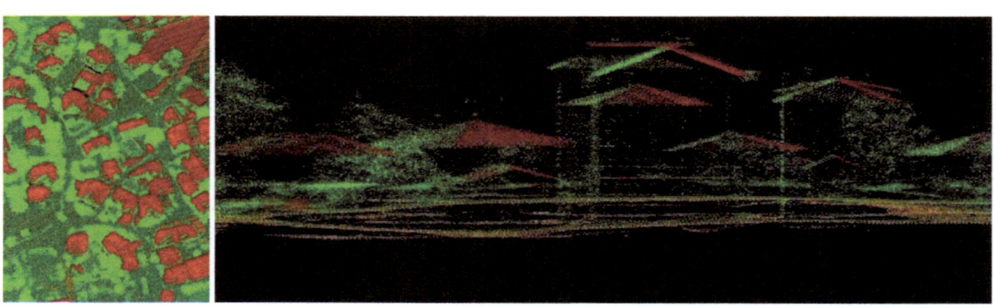

Figure 13.19 Top view on the classification of buildings (red) and vegetation (green) in a part of the city of Istanbul (left) and profile showing the misclassification due to cluttering of buildings and vegetation (right) (source: Baskaraca et al., 2019)

Figure 13.20 Untextured LoD 2 and LoD 3 buildings of the 3D city model of downtown Istanbul (source: Baskaraca et al., 2019)

Despite all the photogrammetric methods and laser scanning research, automated 3D mapping has not yet materialised; today's detailed 3D city maps require cumbersome manual efforts. 3D mapping requires the right data, the right software tools and highly skilled geomatics professionals. Regular updating is a must. Public authorities often become enthusiastic about 3D maps in the wake of the smart city and digital twin concept and the release of financial resources. However, the costs of annual or biannual updating to preserve the value of the 3D maps may be underestimated. Maintenance is a prerequisite – the very nature of geo-information means that damage originating from substandard products may not be immediately visible. Mistakes made today may – a decade or two later – cause society significant losses, not only moneywise but also in terms of injuries, accidents, demolished houses and human suffering.

13.6 Forestry

Although the general public may look at a forest as a natural habitat in which wildlife and biodiversity should be handled with care, the assembly of trees is an asset for the owners of the land in the first place. Each and every tree has a market value. The major difference between productive forest land and agricultural land lies in the time span between planting and harvesting. To grow up agricultural crops need a time span of weeks or months and seldom more than 1 year, while the time span of forest growth covers decennia. The management of wood production forests is as important as the management of farmland, mines, quarries or palm oil plantation. Wood is a major resource used for hammering roof frames, chiselling and graving furniture, collecting wood for fuel and many other applications. For centuries, wood production relied on time-consuming field inspections. Throughout the centuries, foresters have searched for suitable technologies to relieve the cumbersome fieldwork. Since film cameras could be mounted on aircraft, remote sensing techniques have been applied to support forest inventories (Colwell, 1965).

The grand success of airborne Lidar has been primarily ignited by foresters who recognised its value as a valuable information source for managing wood production forests. Today airborne Lidar forms the basis of many forest management systems. Wood production forests in Canada, Finland, Estonia and Russia cover large areas of boreal forest, and this means that good governance of these areas requires capture from an aerial viewpoint. For example, Canada's boreal forests extend from the east coast to the west coast over a distance of almost 5,000 km and cover over 3 million km^2 (Brandt, 2009). To provide an impression of the sheer size: This is the area covered by the entire subcontinent of India. The European Union covers an area of 4.5 million km^2, which is just one and a half times bigger than the forest area in Canada. Monitoring remote and extensive forest resources across such a large area is challenging, and airborne Lidar, aside from multispectral satellite imagery, is a very helpful technology (Hopkinson et al., 2011).

13.6.1 Tree Parameters

Forest means trees, and trees mean timber and timber means construction materials, doors, windows, furniture and chipboard. Forestry is an economic activity like any other, and the value of a forest stand planted for production purposes depends on tree type, the volume of the trees, the tree spacing or density and growth rate. Foresters regard the following parameters as key features describing the characteristics of trees (Jupp and Lovell, 2007; Hyyppä et al., 2018):

- Tree height
- Crown base height, which is the distance from the foot to the lowest green branches or group of branches
- Diameter at breast height
- Diameter at a prespecified height, for example 6 m
- Volume
- Basal area
- Growth rate
- Age
- Biomass
- Leaf area index

Some of these parameters, such as diameter and height, can be measured or computed from direct measurements (Figure 13.21). For example, volume can be calculated from the height and diameters measured at several positions between the foot and top assuming a conical tree shape or from one diameter using a geometric cone-tapering model. Other parameters can only be estimated through physical modelling or statistical approaches.

Measuring the diameter at several heights in the field is time consuming and thus costly. A reliable alternative is to quantify the diameter of a tree at the breast height of the forester by measuring the circumference. Breast height, which is internationally

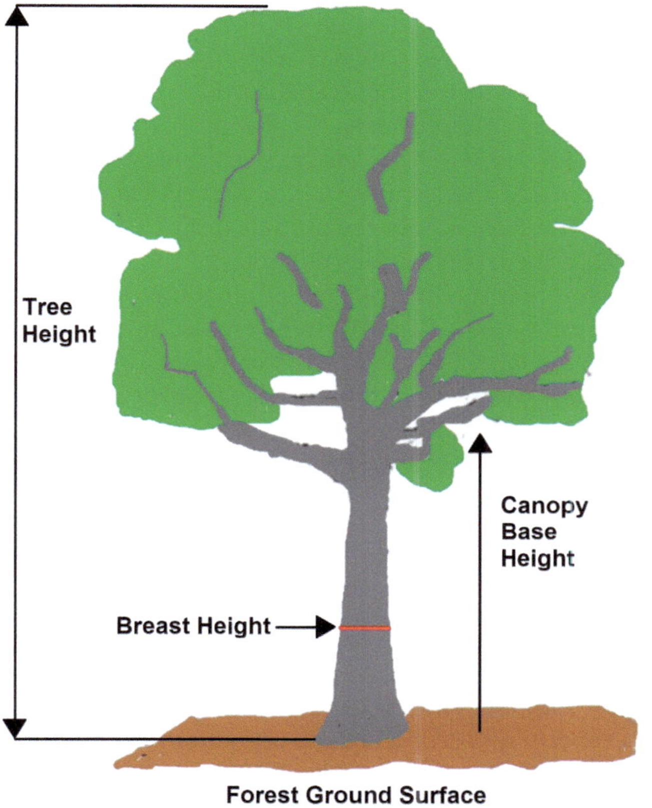

Figure 13.21 Measurable parameters of a tree

standardised to 1.3 m above ground level, relates to individual trees and can only be determined from airborne Lidar if the point density is in the order of 50 points/m^2 or higher and the trees are free standing (Bucksch et al., 2014). Foresters usually quantify tree density by *basal area*. Basically, this is the average planar area covered by the trees in a plot divided by the area of that plot. The basic parameters for its computation are the number of trees and the areas of their cross-sections in a plot. Commonly the basal area is expressed in m^2/ha or ft^2/acre. Dividing a unit of area by a unit of area provides a dimensionless number. So, the basal area isn't actually an area but a ratio between two areas, expressed in different units of area. A hectare (ha) is equivalent to 10,000 m^2. Dividing the basal area expressed in m^2/ha by 100 results in a percentage of the plot being covered with trees. In addition, foresters are interested in volume, vertical structure, canopy crown height, canopy crown base, age and species. These are all parameters which provide insight into the economic value, the growth stage and optimal harvest time. They can be derived, directly or indirectly, from airborne Lidar point clouds. This is why foresters are so enthusiastic about this active sensor – it enables them to obtain the parameters, which are so important for their forest management tasks and so cumbersome to acquire through fieldwork, by conducting analysis in the office.

13.6.2 DTM and CHM

The accuracy of many parameters, which characterise a forest, depends on DTM quality (Simpson et al., 2017). Airborne Lidar enables the creation of reliable, accurate and dense DTMs of the forest ground surface. Laser pulses are able to reach the ground and return to the Lidar receiver. Discrete return airborne Lidar captures at least two peaks: the first return and the last return. The first return may stem from the top of the canopy and the last return from the ground. Together with data on precipitation and soil quality, DTMs enable the determination of the suitability of land for forest plantation. Assuming that the first returns originate in reflections on the top of the canopy, a canopy height model (CHM) can be created by subtracting the DTM from the canopy surface model (CSM). A CSM is basically a forest DSM. Since the first return may stem from lower vegetation, the tree heights in the CHM are systematically underestimated – a limitation already known from the beginning (Hyyppä et al., 2018). By annually capturing the forest around the same time, insight is gained into tree growth.

13.6.3 Fire Monitoring and Biodiversity

Discrete multiple return Lidar systems record the peaks of the return. In contrast, full-waveform Lidar samples the return of each emitted pulse at fixed time intervals. A typical time interval of 1 nanosecond (ns) corresponds to a sampling distance of 15 cm. This means a tree with a height of around 20 m is sampled at 150 intervals. This provides a high-resolution vertical cross-section of the vegetation, which enables wildlife habitat mapping and analysis of vegetation density, vertical structure and ladder fuels (Kelly and Di Tommaso, 2015). Full-waveform point clouds may require a hundredfold of computer storage compared to discrete point clouds, and their processing time will increase accordingly if the commercial software package is able to process full-waveform point clouds at all.

The great foresters' interest in airborne Lidar is based on the ability of laser pulses to partly penetrate the forest canopy and so to obtain estimates of the vertical structure of the forest. Images don't have this ability. Because airborne Lidar reveals forest stratification the point clouds are extensively used in forest fire modelling and biomass estimation (Maguya, 2015). A fire starting on the ground will propagate through shrubs and small trees and may reach the tree canopy. The lower vegetation is ladder fuel for the fire to climb to the crowns. This may destroy timber yield and endanger wildlife and the lives of people living in the neighbourhood. Combining airborne Lidar point clouds with multispectral images enables the production of accurate fuel type maps. After a fire, regular airborne Lidar surveys enable the estimation of the damage and vegetation recovery (Gajardo et al., 2014). Although forests are commercially exploited, they are also home to bird species, insects and other animals, and valued florae. From forest stratification, information on biodiversity can be deduced. The determination of the relationship between airborne Lidar point clouds and patterns of biodiversity is still a research subject (Müller and Vierling, 2014).

Important airborne Lidar point cloud parameters include (Laes et al., 2008):

- Point density per square metre
- Number of returns per pulse, which should be at least two (first and last) to enable the generation of DTMs and CHMs
- Laser beam divergence; a narrow beam provides a higher percentage of points returned from the ground and allows higher flight heights
- Scan angle determines the swath width; the smaller the scan angle, the more ground points will be captured, but this adds to costs since more flight lines are required
- Swath overlap; the higher the overlap, the higher the point density, resulting in an increase in the accuracy of the DTM and the CHM
- Season of survey determines the leaf quantity of deciduous trees; the fewer the leaves, the better the ground will be captured
- Precision of the resulting 3D coordinates
- Costs of the Lidar survey up to the delivery of the end products

Does the grand appreciation of airborne Lidar mean that airborne or satellite images have no value anymore? Sunbeams which reach the forest ground will be only rarely captured by the camera. Assuming that the DTM, which described the forest ground surface, does not change much over time, a CSM can be derived from overlapping images using dense image matching and from subtracting the DTM from the CSM, a CHM. Compared to Lidar, images have the advantage that they can capture not only NIR but also RGB. The vegetation index provides information on the health condition of trees. Compared to a Lidar survey, the purchasing of satellite images is relatively cheap.

From the start of the use of airborne Lidar, foresters were enthusiastic about the wealth of information of the generated point cloud, especially because a part of the laser pulses was able to penetrate the forest canopy and reach the ground. Primarily airborne Lidar point clouds provide foresters with information about the canopy height of a stand or individual tree, the structure of the trees and the earth's surface.

13.7 Serious Gaming

Facility managers of industrial sites may no longer be satisfied with static 3D models but demand animations to improve maintenance and repair planning and to guarantee workforce safety. A pilot project demonstrated the suitability of serious gaming tools for managing industrial sites (Leonov et al., 2016). In June 2014 two hectares of a 220 kV power substation built in 1964 were scanned from 37 positions using a Leica P20 (Figure 13.22). The power station, located near Abakan and Chernogorsk, Russia, covers six hectares. The scanning took two workdays. The strong electric power fields required strict safety rules to avoid damage to the scanner. The point spacing of a few millimetres produced a very dense point cloud which was thinned to obtain a workable data set. After pre-processing with Leica Cyclone software, the point cloud was transferred to a 3D textured model using Blender 3D. This free and open-source software enables the

Figure 13.22 Terrestrial laser scanning of power substation (courtesy: Mikhail Anikushkin, Trimetari Consulting, Saint Petersburg, Russia)

Figure 13.23 Virtual activities in the 3D model of a power station created from a TLS point cloud (courtesy: Mikhail Anikushkin, Trimetari Consulting, Saint Petersburg, Russia)

creation of 3D models and the creation of games, movies and artworks. 3D models of vehicles and workers were generated. Built upon the Unity3D, an engine primarily used for developing computer games, users can move around these 3D models as in a game (Figure 13.23). Users can observe the site in mono and stereo; 3D views enhance the real-world experience of the virtual scene. Three use cases convinced the facility managers of the value of serious gaming.

13.8 Concluding Remarks

The review of applications in this chapter demonstrates that the mother of all point clouds, called geomatics, is a technology with a nearly infinite number of applications. That is the nice thing about geomatics; it is a collection of a great variety of technologies from which all geosciences and geo-management tasks may benefit. The more benefits, the more craftsmanship is required. And craftsmanship is what geomatics studies, all over the world, provide.

References

Arastounia, M. (2016) Automated As-Built Model Generation of Subway Tunnels from Mobile LiDAR Data. *Sensors*, 16(9), p. 1486.

Baarda W. (1968) *A Testing Procedure for Use in Geodetic Networks.* Publications on Geodesy 9, 2(5), Netherlands Geodetic Commission, Delft, The Netherlands.

Banse, P., Kluge, P., Runne, H. (2015) Comparing DIM and TLS: Capturing an Archaeological Site in Oman. *GIM International*, 29(7), pp. 27–29.

Baskaraca, P., Buyuksalih, G., Rahman, A.A. (2019) 3D City Modelling of Istanbul: Issues, Challenges and Limitations. *GIM International*, 33(1), pp. 16–19.

Blogg, A. (2016) Reconstructing a Mythical Past in 3D: Creating a Printed 3D Model of Tintagel Island from UAS Photogrammetry. *GIM International*, 30(3), pp. 33–35.

Brandt, J. (2009) The Extent of the North American Boreal Zone. *Environmental Reviews*, 17, pp. 101–161.

Bucksch, A.K., Lindenbergh, R.C., Rahman, M.Z.A., Menenti, M. (2014) Breast Height Diameter Estimation from High-density Airborne Lidar Data. *IEEE Geoscience and Remote Sensing Letters*, 11(6), pp. 1056–1060.

Chen, X., Li, J. (2016) A Feasibility Study on Use of Generic Mobile Laser Scanning System for Detecting Asphalt Pavement Cracks. In *ISPRS Archives. 41 International Archives of the Photogrammetry Remote Sensing and Spatial Information Sciences*, XLI-B1, pp. 545–549.

Colwell, R.N. (1965) The Extraction of Data from Aerial Photographs by Human and Mechanical Means. *Photogrammetria*, 20, 211–228.

Gajardo, J., García, M., Riaño, D. (2014) Applications of Airborne Laser Scanning in Forest Fuel Assessment and Fire Prevention. In Maltamo, M., Næsset, E., Vauhkonen, J. (Eds.), *Forestry Applications of Airborne Laser Scanning*, Springer, Dordrecht, pp. 439–462.

Gorkovchuk, D., Gorkovchuk, J., Hutnyk, B. (2017) Low-cost UAS Photogrammetry for Mining: Exploring Consumer-grade Copters for Quarry Mapping. *GIM International*, 31(11), pp. 20–23.

GreenValley International (2018) Power Line Corridor Mapping: Methodologies to Acquisition and Data Processing. https://greenvalleyintl.com/applications/power-line-inspection-using-airborne-lidar/

Historic England (2018) *Using Airborne Lidar in Archaeological Survey: The Light Fantastic.* Swindon, Historic England. HistoricEngland.org.uk/research/methods/airborne-remote-sensing/lidar/

Hopkinson, C., Michael A., Wulder, M.A., Coops, N.C., Milne, T., Fox, A., Bater, C.W. (2011) *Airborne Lidar Sampling of the Canadian Boreal Forest: Planning, Execution & Initial Processing.* SilviLaser, Hobart, Tasmania.

Hyyppä, J., Yu, X., Kaartinen, H., Kukko, A., Jaakkola, A., Liang, X., Wang, Y., Holopainen, M., Vastaranta, M., Hyyppä, H. (2018) Forest Inventory Using Laser Scanning. In Shan, J., Toth, C.K. (Eds), *Topographic Laser Ranging and Scanning: Principles and Processing.* 2nd Edition, CRC Press, Boca Raton, FL.

Jupp, D.L.B., Lovell, J.L. (2007) *Airborne and Ground-based Lidar Systems for Forest Measurement: Background and Principles.* CSIRO Marine and Atmospheric Research, Canberra, Australia.

Karas, J., Safar, V. (2017) UAS Photogrammetry and Railway Mapping: Pilot in Czech Republic Attains High-precision. *GIM International*, 31(4), pp. 27–29.

Kelly M, Di Tommaso S. (2015) Mapping Forests with Lidar Provides Flexible, Accurate Data with Many Uses. *California Agriculture*, 69(1), pp. 14–20.

Kersten, T., Lindstaedt, M., Mechelke, K., Vogt, B. (2010) Terrestrial Laser Scanning for the Documentation of Archaeological Objects and Sites on Easter Island. Computer Applications and Quantitative Methods. In Melero. J., Cano, P. (Eds.), *Computer Applications and Quantitative Methods in Archeology*, Granada, Spain, BAR International Series 2494, Archaeopress, Oxford, UK.

Laes, D., Reutebuch, S., McGaughey, B., Maus, P., Mellin, T., Wilcox, C., Anhold. J., Finco, M., Brewer, K. (2008) *Practical Lidar Acquisition Considerations for Forestry Applications*. RSAC-0111-BRIEF1. U.S. Department of Agriculture, Forest Service, Remote Sensing Applications Center, Salt Lake City, UT.

Lemmens, M. (2016) Total Stations: The Surveyor's Workhorse. *GIM International*, 30(10), pp. 20–25.

Leonov, A., Anikushkin, M., Buynov, A. (2016) Serious Gaming for Facility Management: Bringing Dynamics into the 3D Model of a Power Substation in Russia. *GIM International*, 30(9), pp. 20–23.

Leonov, A. V., Anikushkin, M.N., Ivanovb, A.V., Ovcharov, S.V., Bobkov, A.E., Baturin, Y.M. (2015) Laser Scanning and 3D Modeling of the Shukhov Hyperboloid Tower in Moscow. *Journal of Cultural Heritage*, 16(5), pp. 551–559.

Ma, L., Li, Y., Li, J., Wang, C., Wang, R., Chapman, M.A. (2018) Mobile Laser Scanned Point-Clouds for Road Object Detection and Extraction: A Review. *Remote Sensing*, 10, p. 1531.

Maguya, A.S. (2015) Use of Airborne Laser Scanner Data in Demanding Forest Conditions. Ph.D. thesis, Lappeenranta University of Technology, Finland.

Müller, J., Vierling, K. (2014) Assessing Biodiversity by Airborne Laser Scanning. In Maltamo, M., Næsset, E., Vauhkonen, J. (Eds.), *Forestry Applications of Airborne Laser Scanning*. Springer, Dordrecht, pp. 357–374.

Neubauer, W. (2007) Laser Scanning and Archaeology: Standard Tool for 3D Documentation of Excavations. *GIM International*, 21(10), pp. 14–17.

Neubauer, W., Doneus, M., Studnicka, N., Riegl, J. (2005). Combined High Resolution Laser Scanning and Photogrammetrical Documentation of the Pyramids at Giza. *The International Archives of Photogrammetry, Remote Sensing and Spatial Information Sciences*, 36–5/C34/1, pp. 470–475.

Nocerino, E., Poiesi, F., Locher, A., Tefera, Y.T., Remondino, F., Chippendale, P., Van Gool, L. (2017) 3D Reconstruction with a Collaborative Approach Based on Smartphones and a Cloud-Based Server. *International Archives of the Photogrammetry, Remote Sensing and Spatial Information Sciences*, XLII-2(W8), pp. 187–194.

Norambuena, P., Sainz, J. (2016) UAS Survey of Rapa Nui: Capturing the World's Most Remote Inhabited Island. *GIM International*, 30(6), pp. 17–19.

Rüther, H., Chazan, M., Schroeder, R., Neeser, R., Held, C., Walker, S.J., Matmon, A., Horwitz, L.K. (2009) Laser Scanning for Conservation and Research of African Cultural Heritage Sites: The Case Study of Wonderwerk Cave, South Africa. *Journal of Archaeological Science*, 36, pp. 1847–1856.

Schwarzbach, F. (2014) Road Maintenance with MMS: Accurate and Detailed 3D Models Using Mobile Laser Scanning. *GIM International*, 28(7), pp. 34–37.

Simpson, J.E., Smith, T.E.L., Wooster, M.J. (2017) Assessment of Errors Caused by Forest Vegetation Structure in Airborne LiDAR-Derived DTMs. *Remote Sensing*, 9(1101), doi:10.3390/rs9111101.

Tong, X., Liu, X., Chen, P., Liu, S., Luan, K., Li, L., Liu, S., Liu, X., Xie, H., Jin, Y., Hon, Z. (2015) Integration of UAV-Based Photogrammetry and Terrestrial Laser Scanning for the Three-Dimensional Mapping and Monitoring of Open-Pit Mine Areas. *Remote Sensing*, 7(6), pp. 6635–6662.

von Schwerin, J., Richards-Rissetto, H., Remondino, F., Grazia Spera, M., Auer, M., Billen, N., Loos, L., Stelson, L., Reindel, M. (2016) Airborne LiDAR Acquisition, Post-processing and Accuracy-checking for a 3D WebGIS of Copan, Honduras. *Journal of Archaeological Science: Reports*, 5, pp. 85–104.

Wang, J., Li, L., Yan, J. (2013) The Digital Mine: 3D Modelling of a Gold Mine Using Laser Scanning. *GIM International*, 27(11), pp. 29–31.

Weber, A., Lerch, Th. (2018) Comparing Workflow and Point Cloud Outputs of the Trimble SX10 TLS and senseFly eBee Plus Drone. In *FIG Congress*, Istanbul, Turkey.

Wellman, D. (2003) Rapa Nui: Easter Island: Archaeological 3D Laser-scanning in the South Pacific. *GIM International*, 17(8), pp. 40–43.

Ye, C., Li, J. Jiang, H., Zhao, H., Ma, L., Chapman, M. (2020) Semi-Automated Generation of Road Transition Lines Using Mobile Laser Scanning Data. *IEEE Transactions on Intelligent Transportation Systems*, 21(5), pp. 1877–1890.

Index of Applications

Index of Formulas

Index